26th
International Meeting on Fundamental Physics

26th
International Meeting on Fundamental Physics

Selected Topics on Electroweak Interactions, Neutrinos and QCD

A Review of High Energy Colliders

Isla de La Toja, Galicia, Spain *1–5 June 1998*

Editor

B. Adeva
Universidade de Santiago de Compostela

World Scientific
Singapore • New Jersey • London • Hong Kong

Published by

World Scientific Publishing Co. Pte. Ltd.

P O Box 128, Farrer Road, Singapore 912805

USA office: Suite 1B, 1060 Main Street, River Edge, NJ 07661

UK office: 57 Shelton Street, Covent Garden, London WC2H 9HE

British Library Cataloguing-in-Publication Data
A catalogue record for this book is available from the British Library.

ELECTROWEAK INTERACTIONS, NEUTRINOS AND QCD
A Review of High Energy Colliders

ISBN 981-02-3940-8

Printed in Singapore by Uto-Print

LOCAL ORGANIZING COMMITTEE

B. Adeva (Chairman)	University of Santiago de Compostela
M. Aguilar-Benítez	CIEMAT, Madrid
A. Ferrer	IFIC, Univ. Valencia/CSIC
J. Fuster	IFIC, Univ. Valencia/CSIC
A. López-Agüera	University of Santiago de Compostela
C. Mañá	CIEMAT, Madrid
M. Martínez	IFAE, Univ. Autónoma Barcelona
S. Peris	University of Barcelona
A. Pich	University of Valencia

INTERNATIONAL ADVISORY COMMITTEE

E. Fernández	IFAE, Univ. Autónoma Barcelona
A. Morales	University of Zaragoza
C. Pajares	University of Santiago de Compostela
A. Ruiz	University of Cantabria
J. Sánchez-Guillén	University of Santiago de Compostela
R. Tarrach	University of Barcelona
F.J. Yndurain	Univ. Autónoma Madrid

TECHNICAL SECRETARY

M. Fatás University of Zaragoza

SPONSORS

Comisión Interministerial de Ciencia y Tecnología (CICYT)
Universidad de Santiago de Compostela

FOREWORD

The XXVI International Meeting on Fundamental Physics was held in Isla de La Toja, a seaside resort in the Galician Rias Bajas, Spain, from June 1st to 5th, 1998.

A number of topics was covered, attempting to review main recent experimental results in electroweak interactions, QCD, and neutrino physics, from the high energy accelerators LEP II, Tevatron Collider, HERA, and SPS. In addition, some contributions were included in astro-particle physics, dark matter searches, as well as an outlook of the physics program of future particle colliders.

The sponsorship of the Meeting was shared by the CICYT (Comision Interministerial de Ciencia y Tecnologia), the University of Santiago de Compostela, and the Instituto de Física de Altas Energías which is part of it.

I warmly thank all the contributors to the Proceedings, the participants, and the above institutions, for making possible the success of the Meeting. As in previous ocasions (this is the XXVI edition) the meeting served to reassemble the spanish community in High Energy Physics, and to provide a real measure of its activity, ambitions, and predictible future. Indeed, the spanish institutions contributed this time to nearly 90of maturity in this field.

Special thanks should be given to Joan Fuster, Carlos Maña, Santiago Peris, Toni Pich, and Manel Martínez, members of the organizing committee, who contributed with excellent ideas. Also to Angeles López-Agüera for her help in organizational matters, and to Máximo Pló, for the same reason.

I am particularly grateful to Manuel Aguilar-Benítez, one of the founders of this meeting, who provided the real driving force this time. I thank Antonio Ferrer, head of the H.E.P. Specialized Group of the R.S.E.F. for his support to the meeting, and help stablishing its future scope.

Finally, I would like to express my gratitude to Mercedes Fatás, who really carried out the organization of the conference in a very efficient way, as well as the preparation of these proceedings.

CONTENTS

Low x QCD tests at HERA

F. BARREIRO
Universidad Autónoma de Madrid, Spain

I summarize the most important results obtained to date by the H1 and ZEUS Collaborations on Deep Inelastic Scattering (DIS) at low x. The main emphasis is on a discussion of QCD tests for the proton structure function.

1 The proton structure function

One of the most interesting results obtained at HERA so far, has to do with the dramatic increase of the proton structure function F_2 at low Bjorken x values [1,2]. These measurements allow the extraction of the gluon density in the proton [3,4] down to $x \sim 10^{-5}$.

Although the general framework to discuss deep inelastic scattering is established and well known since two decades, the interpretation of these data

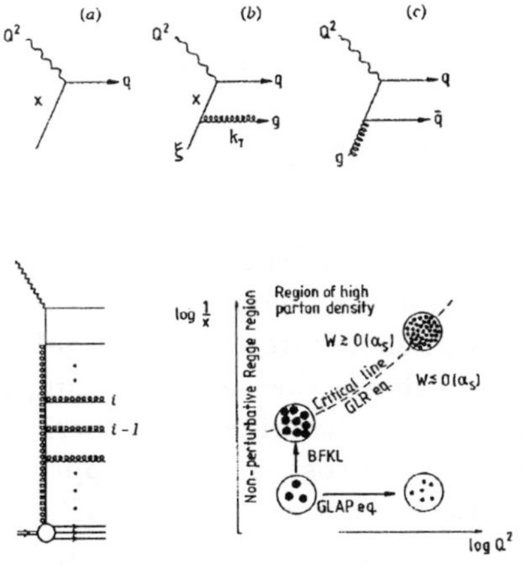

Figure 1: Feynman diagrams for DIS and the $x - Q^2$ plane

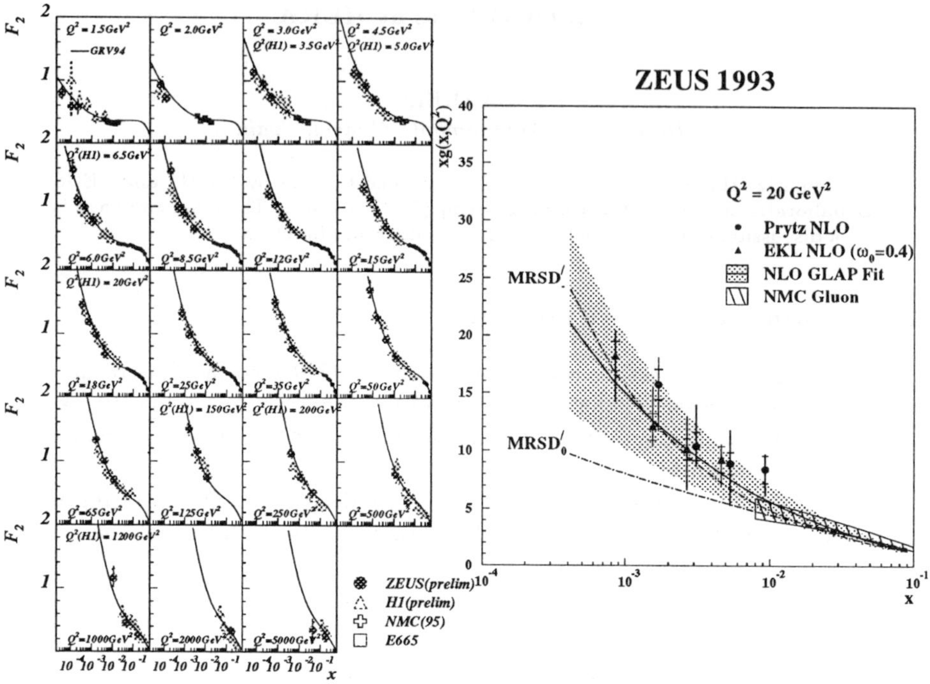

Figure 2: Summary of F_2 measurements and the gluon density determination

is subject to some controversy. Two lines of thought are generally followed to describe the data.

- On the one hand, those who advocate that this dramatic increase of the proton structure function can be obtained from singular [5,6] (non-singular [33]) parton densities at moderate $Q_0^2 \sim 4\ GeV^2$ (resp. at very small $Q_0^2 \sim 0.5\ GeV^2$), which are then evolved using the well known GLAP equations [8]. This procedure describes the experimental data very well, as illustrated in Fig. 2, at the cost of having approximately twenty parameters which enter into the parametrization of valence and sea quark, as well as gluon densities at an input Q_0^2 value. In addition to assumptions about their functional forms.

- On the other, those who argue that since at very low x values, the boson gluon fusion mechanism is the dominant source of LO corrections to the Born level cross-sections and the kernel $P_g \to gg$ is singular, one would expect that when including higher order QCD corrections, one would have to sum multigluon exchange ladders. Depending on the approximation used to perform this sum, see Fig.1, one encounters deviations from the GLAP linear evolution equations. In fact, at fixed Q^2 the x-dependence is most generally given by the BFKL equations [9], which predict the parton densities to behave like $x^{-\lambda}$ with

$$\lambda = \frac{3\alpha_s(Q^2)}{\pi} \cdot 4log\ 2 \sim 0.5 \tag{1}$$

However, theory does not tell us where in the (Q^2, x) plane lies the transition region beyond which the expansion in terms of $ln\frac{1}{x}$ is important. In addition, these evolution equations are not yet numerically implemented in the parametrizations discussed above.

We think that in order to clarify the issue one should try

- ♣ to search for specific fingerprints of the BKFL equations, as discussed by Mueller [10] and others [11];
- ♣ to make detailed comparisons between analytic NLO QCD predictions and experimental data to see if one could isolate regions of phase space where discrepancies might appear.

In this context, we would like to remind the reader that since the behaviour at small x of the proton structure function is connected to the singularities of the operator product expansion matrix elements, one has two specific predictions, depending on whether these singularities lie to the left or to the right of those of the anomalous dimension matrix [12], such that either

- the proton structure function should grow faster than a log but slower than a power [13] in x, i.e.

$$\begin{aligned} F_2(x, Q^2) &= C_0\{\frac{9log\{(logQ^2/\Lambda^2)/(logQ_0^2/\Lambda^2)\}}{(33-2n_f)\pi|logx|}\}^{\frac{1}{2}} \cdot \\ &\cdot exp\sqrt{\frac{144|logx|}{(33-2n_f)}\{log\frac{logQ^2}{logQ_0^2}\}} \end{aligned} \tag{2}$$

  leading to a double asymptotic behaviour as discussed by Ball and Forte [14]

- or else the proton structure function should behave as a power in x i.e. $x^{-\lambda_S}$ with λ_S a Q^2 independent constant except for possible variations due to crossing flavour thresholds [15,17,16] which should depend on Q^2. This behaviour is the same as that calculated by Witten for $\gamma^*\gamma$ scattering [18].

In the next sections I would like :

* to present NLO parametrizations for F_2 and make extensive comparisons with the most recent data [23,24].

* to extract the gluon density in the HERA kinematic regime and to present predictions for $R(x, Q^2)$ which will be measured soon at HERA.

2 LO and NLO predictions

Perturbative QCD provides evolution equations for the structure functions, in such a way that if we know them at a given Q_0^2, we can predict them at any other Q^2 value, assuming that both Q_0^2 and Q^2 lie in a range where perturbation theory applies.

The GLAP equations represent one of the forms in which the evolution in Q^2 can be expressed, whereas the most direct result of the operator product expansion (OPE) approach is expressed in terms of moments of structure functions. Both approaches are equivalent.

Once we know the evolution equations, the goal would be to find the functional form $F(x, Q^2)$ for the structure functions such that used at an input Q_0^2, the resulting "evolved" function would continue to be the same $F(x, Q^2)$ simply calculated at the new Q^2 value. No such functional form has been found so far, thus any simple analytical form chosen as input is not invariant as a function of Q^2, in contradiction with the fact that any other value Q_1^2 could have been chosen as the starting point for the evolution.

On the other hand, some properties of the structure functions are known to be compatible with the GLAP evolution equations, in particular the behaviour at the end points $x = 0, 1$ and certain sum rules. So, in a certain sense, we know exact solutions to the GLAP evolution equations, albeit only locally and not for the whole x range. These results were derived in [20] and [21] at the leading (LO) and next to leading (NLO) order.

Let us consider $F_2(x, Q^2)$ for DIS $e-p$ scattering. It can be written as

$$F_2(x, Q^2) = F_S(x, Q^2) + F_{NS}(x, Q^2) \tag{3}$$

where $F_S(x, Q^2)$ ($F_{NS}(x, Q^2)$) refer to the singlet (resp. non-singlet) pieces, whose evolution equations are different. Indeed the singlet part and the gluon momentum density $F_G(x, Q^2)$ evolve together, whereas the non-singlet part is decoupled from gluons and evolves independently.

2.1 LO predictions

To LO, the moments evolution have a very simple form

$$\mu_{NS}(n,Q^2)=\left(\frac{\alpha_s(Q_0^2)}{\alpha_s(Q^2)}\right)^{\mathbf{d(n)}}\cdot\mu_{NS}(n,Q_0^2) \tag{4}$$

for the non-singlet piece and

$$\vec{\mu}(n,Q^2)=\left(\frac{\alpha_s(Q_0^2}{\alpha_s(Q^2}\right)^{\mathbf{D(n)}}\cdot\vec{\mu}(n,Q_0^2) \tag{5}$$

where $\vec{\mu}(n,Q^2)$ is the two-component vector

$$\vec{\mu}(n,Q^2)=\left(\begin{array}{c}\mu_S(n,Q^2)\\ \\ \mu_G(n,Q^2)\end{array}\right) \tag{6}$$

defined as usual

$$\mu(n,Q^2)=\int_0^1 dx x^{n-2}F(x,Q^2) \tag{7}$$

and $\mathbf{d(n)}$ and $\mathbf{D(n)}$ are proportional to the anomalous and anomalous dimension matrix, whose exact expressions are

$$\mathbf{d(n)}=\frac{16}{33-2n_f}\left(\frac{1}{2n(n+1)}+\frac{3}{4}-S_1(n)\right) \tag{8}$$

and

$$\mathbf{D(n)}=\frac{16}{33-2n_f}\left(\begin{array}{cc}\frac{1}{2n(n+1)}+\frac{3}{4}-S_1(n) & \frac{3n_f}{8}\frac{n^2+n+2}{n(n+1)(n+2)}\\ \frac{n^2+n+2}{2n(n^2-1)} & \frac{9}{4n(n-1)}+\frac{9}{4(n+1)(n+2)}+\frac{33-2n_f}{16}-\frac{9S_1(n)}{4}\end{array}\right) \tag{9}$$

where

$$S_1(n)=n\sum_k\frac{1}{k(k+n)} \tag{10}$$

and n_f the number of flavours.

From the fact that Q_0^2 and Q^2 are arbitrary values in the range of applicability of perturbation theory, it follows that the moments have to be of the form

$$\mu_{NS}(n,Q^2)=\left[\alpha_s(Q^2)\right]^{-\mathbf{d(n)}}\cdot b_{NS}(n) \tag{11}$$

and

$$\vec{\mu}(n,Q^2) = \left[\alpha_s(Q^2)\right]^{-\mathbf{D(n)}} \cdot \vec{b}(n) \tag{12}$$

with $b_{NS}(n)$ and $\vec{b}(n)$ being independent of the squared momentum transfer.

If we try a Regge inspired functional form

$$F_{NS}(x,Q^2) \underset{x\to 0}{=} B_{NS}(Q^2)\cdot x^{\lambda_{NS}(Q^2)} \tag{13}$$

and

$$F_S(x,Q^2) \underset{x\to 0}{=} B_S(Q^2)\cdot x^{-\lambda_S(Q^2)} \tag{14}$$

for the structure functions near the end point $x=0$, the dependence of $B(Q^2)$, $\lambda(Q^2)$ upon Q^2 is related to the moments in Eqs. 11 and and 12 for the value of n at which they diverge.

The results which can be found in [21] and [20] are such that λ_{NS} and λ should be Q^2 independent, with furthermore $\lambda_{NS} < 1$ and $\lambda_S > 0$, and

$$B_{NS}(Q^2) = B_{NS}\cdot\left[\alpha_s(Q^2)\right]^{-d(1-\lambda_{NS})} \tag{15}$$

and

$$B_S(Q^2) = B_S\cdot\left[\alpha_s(Q^2)\right]^{-d_+(1+\lambda_S)} \tag{16}$$

In addition, the gluon structure function ought to be proportional to the singlet piece of the structure function i.e.

$$F_G(x,Q^2) = B_G(Q^2)\cdot x^{-\lambda_S} \tag{17}$$

with

$$B_G(Q^2) = \frac{d_+(1+\lambda_S) - D_{11}(1+\lambda_S)}{D_{12}(1+\lambda_S)}\cdot B_S(Q^2) \tag{18}$$

where by $d_+(n)$ we denote the largest eigenvalue of the anomalous dimension matrix $\mathbf{D(n)}$.

2.2 NLO predictions

The extension of these results to the NLO is tedious, the result being of the form

$$B_{NS}(Q^2) = B_{NS}\cdot\left[\alpha_s(Q^2)\right]^{-d(1-\lambda_{NS})}\cdot\{1+\frac{\alpha_s(Q^2)}{4\pi}\cdot v_{NS}(1-\lambda_{NS})\} \tag{19}$$

$$B_S(Q^2) = B_S \cdot \left[\alpha_s(Q^2)\right]^{-d_+(1+\lambda_S)} \cdot \{1 + \frac{\alpha_s(Q^2)}{4\pi} \cdot v_S(1+\lambda_S)\} \tag{20}$$

and

$$B_G(Q^2) = B_S \cdot \left[\alpha_s(Q^2)\right]^{-d_+(1+\lambda_S)} \cdot \{B_{GS} + \frac{\alpha_s(Q^2)}{4\pi} \cdot v_G(1+\lambda_S)\} \tag{21}$$

with $v_{NS}(1-\lambda_{NS})$, $v_S(1+\lambda_S)$ and $v_G(1+\lambda_S)$ being known functions of the exponents λ_{NS} and λ_S given in [21].

The last two of these functions have been fitted to simple rational expressions and the results for $n_f = 3, 4$ are presented in Fig. 2 together with similar fits to B_{GS} which is defined in Eq. 18 as the proportionality factor to LO between $B_G(Q^2)$ and $B_S(Q^2)$.
Note that the gluon structure function, at the end point $x = 0$ is completely determined by the quark singlet structure function which in turn is the dominant piece of $F_2(x, Q^2)$. In fact,

$$F_G(x, Q^2) \underset{x \to 0}{=} \frac{B_{GS} + \frac{\alpha_s}{4\pi} \cdot v_G(1+\lambda_S)}{1 + \frac{\alpha_s}{4\pi} \cdot v_S(1+\lambda_S)} \cdot F_S(x, Q^2) \tag{22}$$

Since $F_S(x, Q^2)$ is defined in terms of quark densities in the proton as

$$F_S(x, Q^2) = < e_i^2 > \sum_i x \cdot q_i(x, Q^2) \tag{23}$$

where the sum runs over quark flavours with charge given by e_i. The gluon density is the given by

$$xG(x, Q^2) = \frac{1}{< e_i^2 >} \cdot F_G(x, Q^2) \tag{24}$$

Let us turn now to a discussion of the limit $x \to 1$. We have considered

$$F(x, Q^2) \underset{x \to 1}{=} A(Q^2) \cdot (1-x)^{\nu(Q^2)} \tag{25}$$

The dependence of the functions $A(Q^2)$ and $\nu(Q^2)$ is now related to the evolution of the moments in the limit $n \to \infty$, yielding

$$A_{NS}(Q^2) = A_{NS} \cdot \left[\alpha_s(Q^2)\right]^{-d_0} \cdot \frac{\Gamma(1+\nu_0)}{\Gamma(1+\nu(Q^2))} \tag{26}$$

$$A_S(Q^2) = A_S \cdot \left[\alpha_s(Q^2)\right]^{-d_0} \cdot \frac{\Gamma(1+\nu_0)}{\Gamma(1+\nu(Q^2))} \tag{27}$$

with

$$\nu(Q^2) = \nu_0 - \frac{16}{33 - 2n_f} \cdot log\ \alpha_s(Q^2) \tag{28}$$

and

$$d_0 = \frac{16}{33 - 2n_f} \cdot \{\frac{3}{4} - \gamma_E\} \tag{29}$$

with γ_E Euler's constant.
The gluon momentum density is determined by the singlet structure function, i.e.

$$F_G(x, Q^2) = \frac{2}{5} A_S(Q^2) \cdot \frac{(1-x)^{\nu(Q^2)+1}}{(\nu(Q^2)+1) \cdot log \frac{1}{1-x}} \tag{30}$$

Finally from the fact that $d(1) = 0$ and $d_+(2) = 0$, the following sum rules can be derived

$$\int_0^1 x^{-1} \cdot F_{NS}(x, Q^2) dx = \left(1 - 3 < e_i^2 >\right) \{1 + k \cdot \frac{\alpha_s}{4\pi}\} \tag{31}$$

with k a known and very small constant [21] and

$$\int_0^1 \{F_S(x, Q^2)\ +\ F_G(x, Q^2)\} dx = < e_i^2 > \cdot \{1 - \alpha_s \cdot \frac{5}{9\pi}\} \tag{32}$$

3 Parametrizations for the proton structure function

In this section, we write down approximate parametrizations compatible with the exact conditions discussed above.

3.1 LO parametrizations

Let us consider the singlet piece in the form

$$F_S(x, Q^2) = \{B_S(Q^2) \cdot x^{-\lambda_S}\ +\ C_S(Q^2)\}(1-x)^{\nu(Q^2)} \tag{33}$$

where

$$B_S(Q^2) = B_S \cdot \left[\alpha_s(Q^2)\right]^{-d_+(1+\lambda_S)} \tag{34}$$

and $\nu(Q^2)$ given by Eq. 28.

The singlet part is the most important contribution at small x, so that the predicted behavious near $x = 0$ has to be implemented through the term $B_S(Q^2) \cdot x^{-\lambda}$. Note from Eq. 33 that it is divergent when $x \to 0$ and that the coefficient $B_S(Q^2)$ grows rapidly with Q^2, becoming asymptotically dominant.

This part is responsible for the increase of the structure function at small x as a function of Q^2 as experimentally observed.

On the other hand, in the limit $Q^2 \to 0$, it becomes negligible, in fact it vanishes in the on-shell limit. This is the region where we expect Pomeron-like contributions to remain. This term we parametrise with the help of the second piece in Eq. 33 proportional to $C_S(Q^2)$, whose relative weight will decrease as a function of increasing Q^2 but remain important at low and intermediate Q^2 values.

We would like to remark that the dependence of both functions $B_S(Q^2)$ and $C_S(Q^2)$ can be determined by implementing the limits at $x = 0, 1$. Thus,

$$C_S(Q^2) = \; -B_S \cdot \left[\alpha_s(Q^2)\right]^{-d_+(1+\lambda_S)} + A_S \cdot \left[\alpha_s(Q^2)\right]^{-d_0} \cdot \frac{\Gamma(1+\nu_0)}{\Gamma(1+\nu(Q^2))} \tag{35}$$

For the non-singlet part we could try a parametrization of the form

$$F_{NS}(x, Q^2) = \{B_{NS}(Q^2) \cdot x^{-\lambda_{NS}} \; + \; C_{NS}(Q^2) \cdot x\} \cdot (1-x)^{\nu(Q^2)} \tag{36}$$

but since its contributions turns out to be small, we will use simply

$$F_{NS}(x, Q^2) = \; B_{NS}(Q^2) \cdot x^{-\lambda_{NS}} \cdot (1-x)^{\nu(Q^2)} \tag{37}$$

with

$$B_{NS}(Q^2) = B_{NS} \left[\alpha_s(Q^2)\right]^{-d_0} \frac{\Gamma(1+\nu_0)}{\Gamma(1+\nu(Q^2))} \tag{38}$$

In Eq. 36, λ_{NS} is related to the intercept of the leading Regge trajectory contribution to this piece, $\lambda_{NS} = 0.45$.

In one wishes to restrict the number of free parameters, one can use Eq. 31 to yield

$$B_{NS}(Q^2) = \left(1 - 3 \cdot < e_i^2 >\right) \cdot \frac{\Gamma(1 + \lambda_{NS} + \nu(Q^2))}{\Gamma(\lambda_{NS})\Gamma(1+\nu(Q^2))} \tag{39}$$

For the gluon momentum density, the prediction will be

$$F_G(x, Q^2) = \; \{B_G(Q^2) \cdot x^{-\lambda_S} \; + \; C_G(Q^2)\} \cdot (1-x)^{1+\nu(Q^2)} \tag{40}$$

3.2 Parametrizations at the NLO

At the next to leading order, the only significant modification to the expressions given above refer to an extra term of the form $1 + \frac{2\alpha_s}{3\pi} log^2(1-x)$ which will

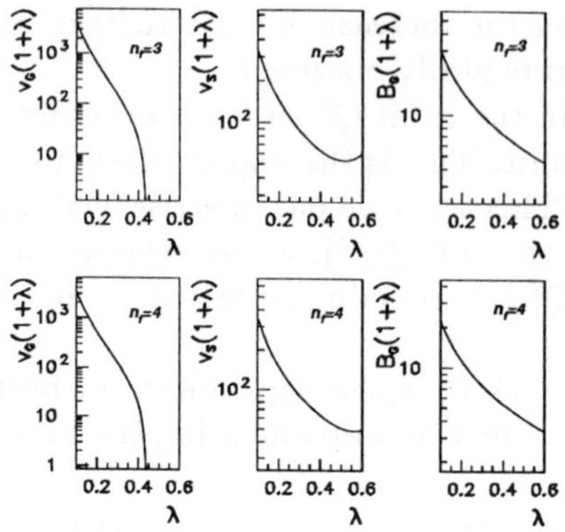

Figure 3: The dependence on λ of B_{GS}, v_G and v_S

be only noticeable at large x. Therefore we have for the dominant singlet part

$$F_S(x,Q^2) = \{B_S(Q^2)\cdot x^{-\lambda_S} + C_S(Q^2)\}(1-x)^{\nu_1(Q^2)}\cdot\{1+\frac{2\alpha_s}{3\pi}log^2(1-x)\} \quad (41)$$

with $\nu_1(Q^2)$ being given by the following expression

$$\nu_1(Q^2) = \nu(Q^2) - \alpha_s\{\frac{4}{3\pi}\Psi(\nu(Q^2)+1) + a_1\} \quad (42)$$

The coefficient $B_S(Q^2)$ is given by Eq. 19, and $C_S(Q^2)$ is such that Eq. 35 is satisfied i.e. including NLO terms

$$C_S(Q^2) = \; -B_S\cdot\left[\alpha_s(Q^2)\right]^{-d_+(1+\lambda_S)} \; + \; A_S\cdot\left[\alpha_s(Q^2)\right]^{-d_0}\cdot\frac{e^{g(\alpha_s)\alpha_s}\cdot\Gamma(1+\nu_0)}{\Gamma(1+\nu(Q^2))} \quad (43)$$

with

$$g(\alpha_s) = a_0 + a_1\cdot\Psi(1+\nu(Q^2)) + \frac{2}{3\pi}\{\Psi^2(1+\nu(Q^2) - \Psi'(1+\nu(Q^2))\} \quad (44)$$

Here Ψ denote the logarithmic derivative of the Γ function, a_0 and a_1 are constants essentially independent of the number of flavours. Numerically $a_0 = -1.18$ and $a_1 = 0.66$. For the non-singlet part we take

$$F_{NS}(x,Q^2) = \; B_{NS}(Q^2)\cdot x^{-\lambda_{NS}}\cdot(1-x)^{\nu(Q^2)}\cdot\{1+\frac{2\alpha_s}{3\pi}log^2(1-x)\} \quad (45)$$

with $B_{NS}(Q^2)$ fixed in such a way as to satisfy the sum rule given in Eq. 31, whose NLO corrections are negligible. These parametrizations are similar to those used by the authors of [27] to fit fixed targed lepton nucleon scattering data.

At this point we would like to remark that these predictions are valid for a range in Q^2 where the number of flavours is fixed.

4 Comparison with experimental data

We can think of two possibilities to compare experimental data to theoretical predictions. The simplest is exemplified in our fits to the '94 ZEUS shifted vertex data [22] to the expression given by Eqs. 14 and 20 in NLO. This has also been tried in [12]. In order to be in the kinematical region where this term in the singlet structure function is dominant, we restricted ourselves to Q^2 values above 3 GeV^2 and Bjorken x below 0.01. The QCD scale parameter $\Lambda_{\overline{MS}}$ was fixed as it will be discussed below. The reason for this has to do with the fact that the normalization factor B_S and the strong coupling constant are tightly correlated, so we prefer to fix the coupling constant to values measured elsewhere. The number of excited flavours was assumed to be constant and the quality of the fits are similar irrespective of whether we take $n_f = 3$ or $n_f = 4$. Disregarding the overall normalization factor, B_S, the relevant fitted parameter λ_S turns out to be 0.32 ± 0.01 for four excited flavours. The χ^2 is 25 for 33 experimental points. The agreement between data and NLO QCD predictions is quite good as illustrated in Fig. 4, even down to small Q^2 values where the applicability of perturbation theory could be questioned. The solid lines in Fig. 4 represent the results of the fits.

The gluon density extracted from these fits also agrees well with that obtained by ZEUS using the Prytz[35] and/or Ellis-Kunszt-Levin methods[36], which relate the gluon density to the logarithmic derivative of the proton structure function. This is shown in Fig. 3. Our smaller error band has to do with the fact that the gluon density determination within our formalism, is subject essentially to the same statistical errors as the proton structure function itself.

The criticism to the procedure developed so far is clear from Fig. 4. A fast rise of the cross section for virtual γ-p scattering is observed at large W^2 values, but this rise is sitting on top of a non-negligible plateau which moreover exhibits clear leading twist behaviour. Questions like, what is the dependence of λ upon variations in the limits used to define the kinematical region over which the fits are done, or whether one should subtract from the low x region a contribution smoothly coming from the large x (i.e. small W^2) domain, have to be answered in a quantitative way.

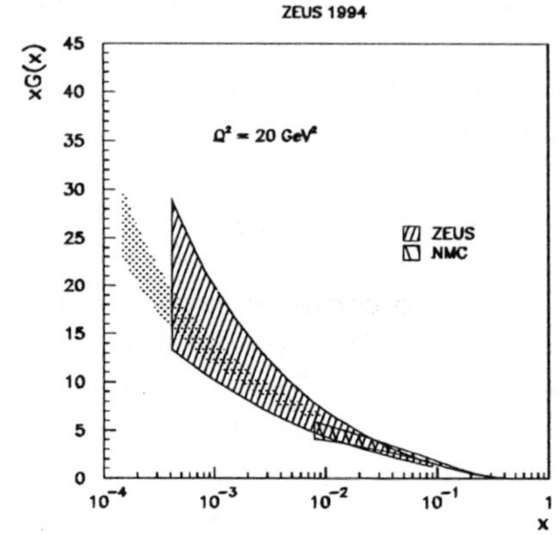

Figure 4: The gluon density determined from shifted vertex ZEUS data compared with previous determinations

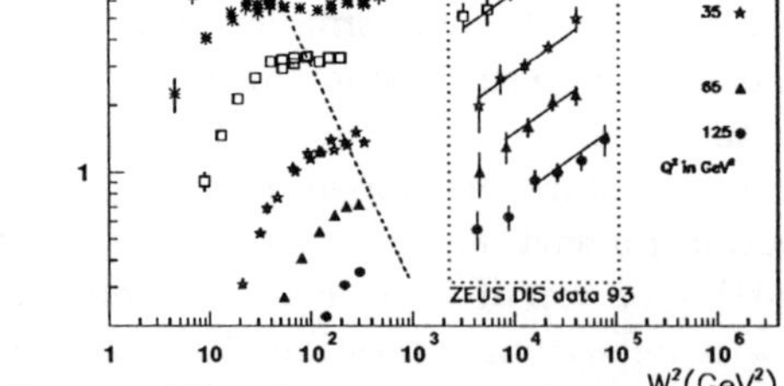

Figure 5: The $\gamma^* - p$ cross section as a function of W

In order to do this, we consider a different approach in which we have performed fits to the data in the whole (x, Q^2) region covered by the experiments. We would like to recall that four parameters are involved in our fits, if we consider a region in Q^2 with a fixed number of flavours. A coefficient B_S to give the normalization of the singlet piece to F_2, a coefficient A_S which serves to define the subleading contribution to the singlet piece, λ_S which defines the growth rate for small x and ν_0 which fixes the behaviour of the structure function at large x values. In principle λ_S could be different for different numbers of excited flavours. The strong coupling constant is fixed through $\Lambda_{\overline{MS}} = 263\ MeV$ for four flavours as given in [25]. The dependence of the QCD scale parameter on the number of flavours is taken as in [26]. Although reasonable fits can be obtained with $n_f = 4$ independent of Q^2, we find that the quality of the fits improve by considering $n_f = 3$ below 10 GeV^2 and $n_f = 4$ above.

We are aware that this is an approximation, since calculations using the GRV parametrization show that the charm quark contribution rises smoothly with increasing Q^2. Once F_2^{charm} is measured, this could be subtracted from the experimental F_2 values and we could fix $n_f = 3$ in the complete Q^2 range. The effect of the bottom quark excitation, due to its charge, is smaller. In fact, the data does not seem to require a fifth flavour at high Q^2 values. These are the main sources of uncertainty in our determination of the gluon density and of $R(x, Q^2)$ which we cannot take into account in a model independent way right now. The H1 data [23] has been fitted over the entire x region and for Q^2 larger than 3.5 GeV^2. In Fig. 5 we show the results of the fits. We also show the extrapolation from our fits to Q^2 down to 1.5 GeV^2 where the validity of the perturbative expansion is questionable. The χ^2 obtained is less than 1 per degree of freedom. One should note that in the first two Q^2 bins, the predictions tend to be above the data systematically. There could be several reasons for this: next to NLO corrections, only two flavours in the proton are excited, or more likely non-perturbative effects, such as higher twist, begin to emerge.

We would like to point out that at very high Q^2 values, i.e. $Q^2 \geq 800\ GeV^2$, the term $C_S(Q^2)$ which we adscribe to the soft pomeron contribution and decreases with increasing Q^2, tends to get into small but negative values, probably more due to the oversimplified parametrization in the whole (x, Q^2) range than to a genuine physical effect. Since our parametrizations should not be valid for very small and very high Q^2 values, we fitted the ZEUS data [24] over the whole x range in the region $6.5 \leq Q^2 \leq 800.0\ GeV^2$. The results of the fits are presented in Fig. 6 and the corresponding gluon densities in Fig. 7.

The values of the parameters obtained from fitting the H1 [23] and ZEUS

Table 1: Summary of the fitted parameters

	B_S	λ_S	A_S	ν_0
$H1$	$(5.4 \pm 0.1) \cdot 10^{-4}$	0.357 ± 0.005	0.46 ± 0.02	1.25 ± 0.10
$ZEUS$	$(4.3 \pm 0.1) \cdot 10^{-4}$	0.344 ± 0.005	0.48 ± 0.01	1.12 ± 0.04

data [24] are summarized in Table 1. We have tried some variations in the fitting procedure like considering different λ_S values for different regions in Q^2 namely $Q^{\leq}11\ GeV^2$ where we take $n_f = 3$, $11 \leq Q^2 \leq 100\ GeV^2$ where we take $n_f = 4$ and above $100\ GeV^2$ with $n_f = 5$. The values for λ_S turn out to be 0.32, 0.34 and 0.36 respectively. We have also considered the term $C_S(Q^2)$ to be proportional to x^ϵ, with ϵ a free parameter. The resulting fitted value for it turned out ao be compatible with 0.

We do not expect the parameter ν_0 to be well determined by the fits, since it is linked to the behaviour of the structure functions at large x values, while the HERA data fitted in this analysis concentrate in the low x domain. A more reliable determination would require to incorporate to the fits data from fixed target experiments.

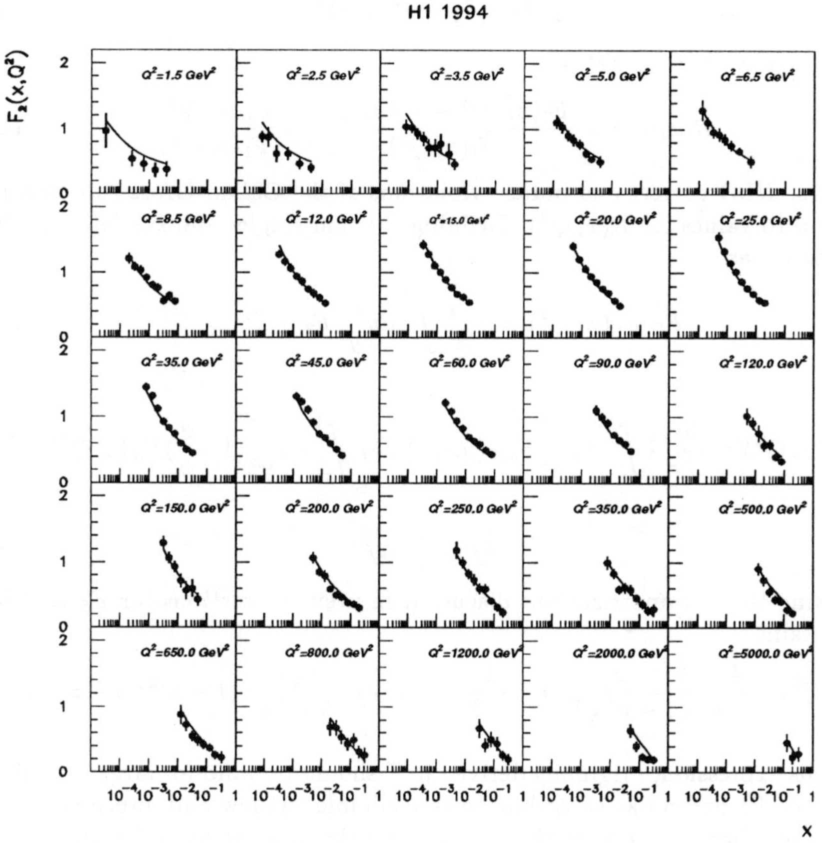

Figure 6: The H1 1994 data along with the results of the fits described in the text

5 Predictions for $R(x, Q^2)$

As it is well known, to LO the Callan-Gross relation holds, namely

$$F_2(x, Q^2) = F_1(x, Q^2) \tag{46}$$

in such a way that the ratio

$$R(x, Q^2) = \frac{F_2(x, Q^2) - F_1(x, Q^2)}{F_1(x, Q^2)} = \frac{F_L(x, Q^2)}{F_1(x, Q^2)} \tag{47}$$

vanishes. NLO corrections induce violations of the Callan-Gross rule, leading to non-zero values for $R(x, Q^2)$. Defining the non-singlet and singlet contributions to F_L as

$$F_{L,NS}(x, Q^2) = \frac{4\alpha_s}{3\pi} \int_x^1 dy \frac{x^2}{y^3} F_{2,NS}(y, Q^2) \tag{48}$$

and

$$F_{L,S}(x, Q^2) = \frac{4\alpha_s}{3\pi} \{ \int_x^1 dy \frac{x^2}{y^3} F_{2,S}(y, Q^2) + \delta_f \int_x^1 dy \frac{x^2}{y^3} (1 - \frac{x}{y}) F_G(y, Q^2) \} \tag{49}$$

with

$$\delta_f = \frac{3}{2} \cdot n_f \tag{50}$$

and using the parametrizations discussed in previous sections for F_S and F_G, one obtains

$$R(x, Q^2) = \frac{4\alpha_s}{3\pi} \left(\frac{x}{1 + \nu(Q^2)} + \frac{1}{2 + \lambda_S} \{1 \; + \; \delta_f \cdot \frac{B_{GS}}{3 + \lambda_S} \cdot (1 - x)^2 \} (1 - x) \right) (1 - x) \tag{51}$$

This expression, originally derived in[28] should be valid for large Q^2 values, where the Pomeron like contribution is negligible. At low and intermediate Q^2 values, one has to consider the full parametrization for both F_S and F_G, so that one obtains

$$R(x, Q^2) = \frac{4\alpha_s}{3\pi} \left(\frac{x}{1 + \nu(Q^2)} + \frac{1}{2 + \lambda_S} \left\{ \frac{B_S(Q^2) \cdot x^{-\lambda_S} + \frac{2+\lambda_S}{2} \cdot C_S(Q^2)}{B_S(Q^2) \cdot x^{-\lambda_S} + C_S(Q^2)} + \right.\right. \tag{52}$$

$$\left.\left. + \delta_f \cdot \frac{1}{3 + \lambda_S} \cdot \frac{B_G(Q^2) \cdot x^{-\lambda_S} + \frac{(2+\lambda_S)(3+\lambda_S)}{6} \cdot C_G(Q^2)}{B_S(Q^2) \cdot x^{-\lambda_S} + C_S(Q^2)} (1 - x)^2 \right\} (1 - x) \right) (1 - x)$$

Notice that both expressions are proportional to α_s. In Fig. 8 we present the predictions from Eq. 52, and also as a measure of the importance of the Pomeron-like term, those derived from Eq. 51.

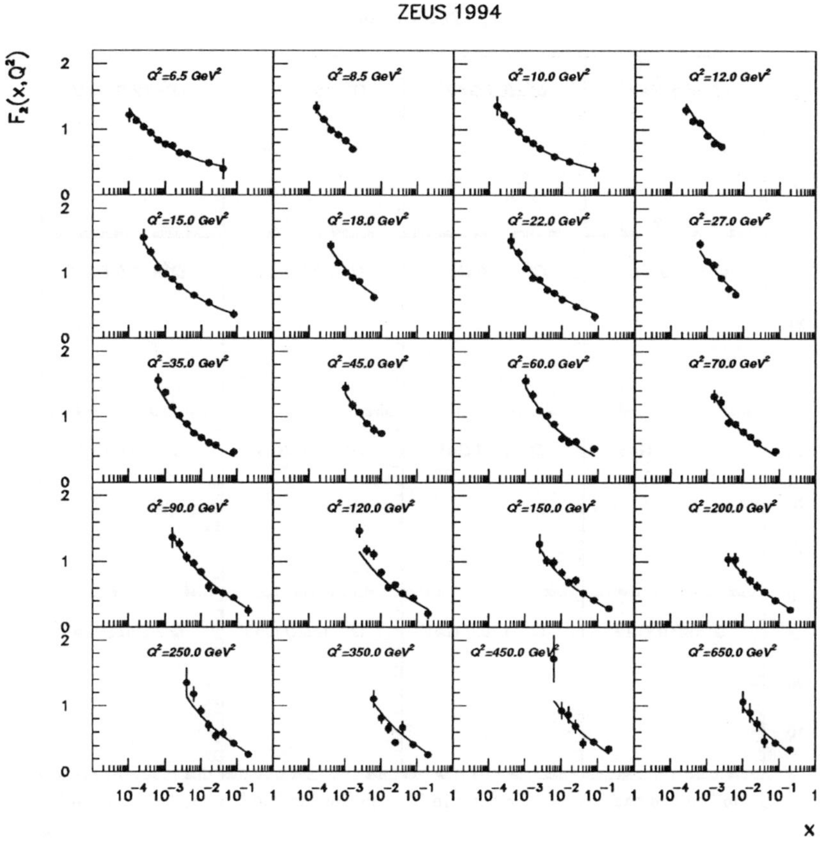

Figure 7: The ZEUS 1994 data along with the results of the fits described in the text

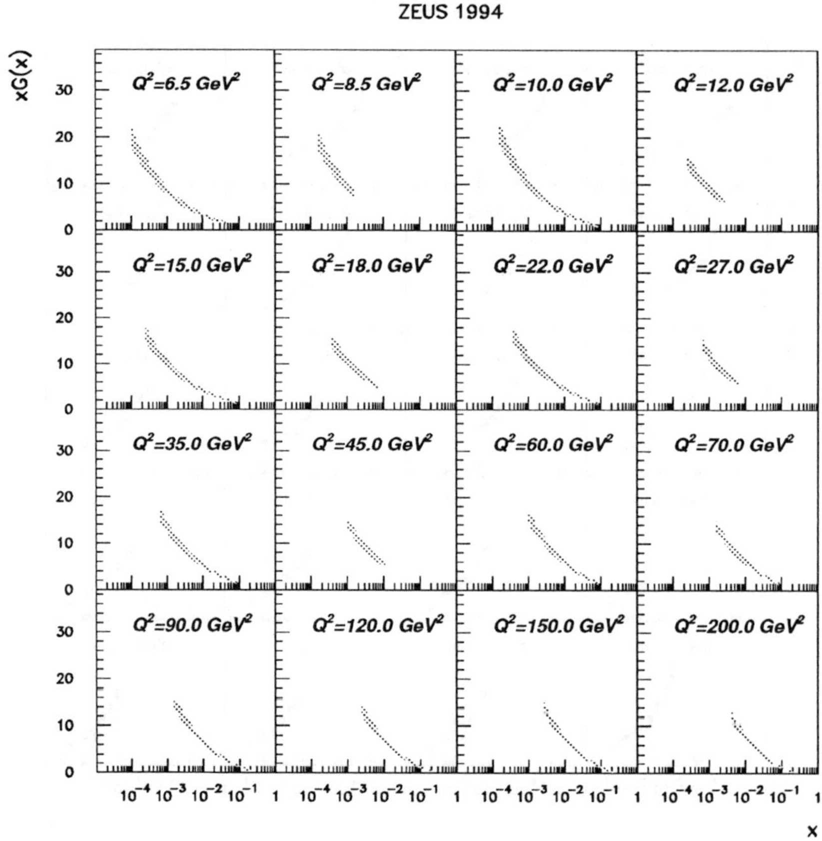

Figure 8: The gluon density extracted from the ZEUS 1994 data

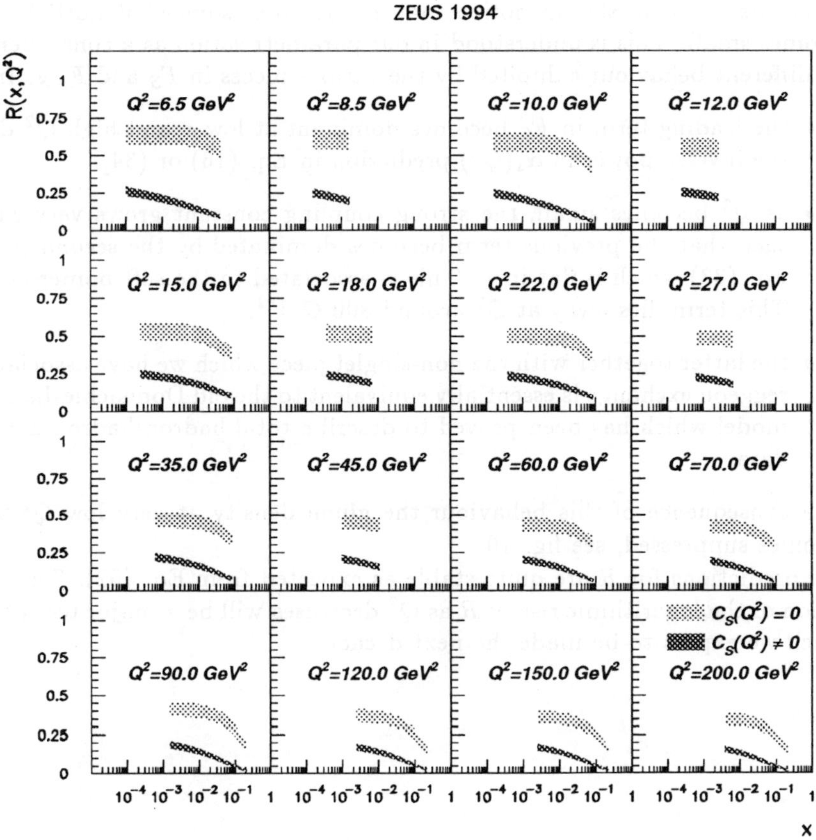

Figure 9: Predictions for $R(x,Q^2)$ from fits described in the text

6 The limit $Q^2 \to 0$

I would like to start this section by making the observation that surprisingly, the fits I have described above extrapolate reasonably well[39] to the very low Q^2 region where perturbative QCD should not apply. This is illustrated in figs. 9, 10 and 11. Data from the E665 Collaboration have also been included in the fits[38]. One should notice that F_2 becomes somewhat flattish as Q^2 becomes small. This is understood in our parametrisation as a consequence of the different behaviour exhibited by the various pieces in F_S and F_{NS}. In fact

- the leading term in F_S becomes dominant at low x and high Q^2 due to the inverse power in $\alpha_s(Q^2)$ prediction in Eq. (16) or (34).
- as Q^2 becomes small, the strong coupling constant grows very rapidly such that the previous term becomes dominated by the second piece in Eq. (33) which is flat in x. This is associated to the soft pomeron term. This term dies away at Q^2 around 800 GeV^2.
- the latter together with the non-singlet piece which we have associated to reggeon exchange is essentially equivalent to the old Donnachie-Landshoff model which has been proved to describe total hadron-hadron x-section data.

As a consequence of this behaviour the gluon density at very low Q^2 values becomes suppressed, see fig. 10.
The predictions for R are quite stable as expected from Eq. (52). The observation of the logarithmic rise of R as Q^2 decreases will be a major test of QCD which we expect to be made the next decade.

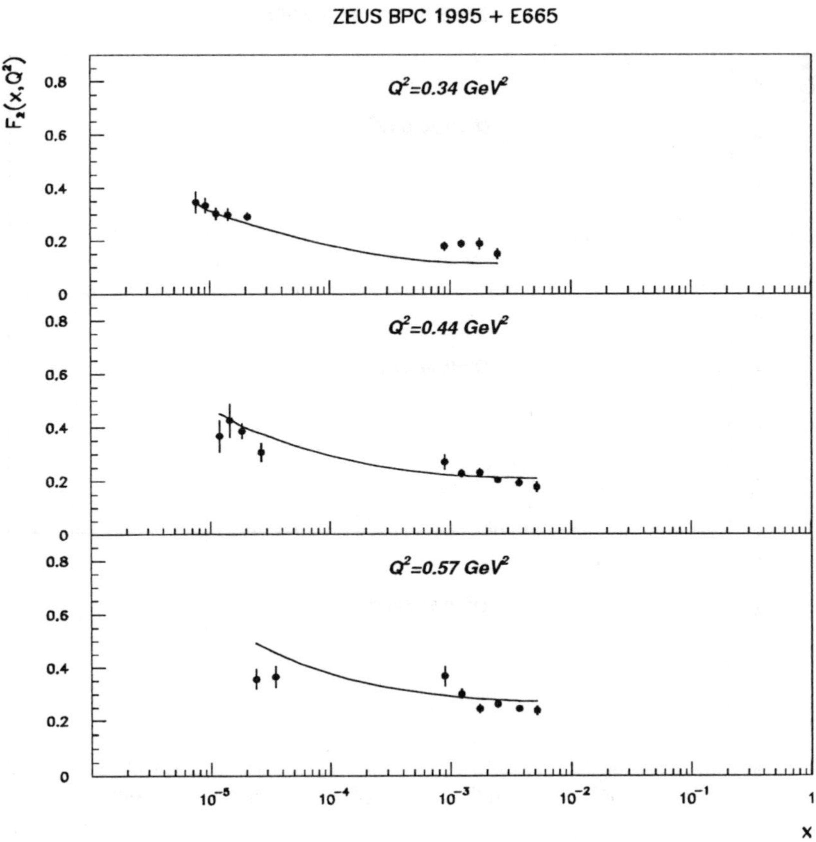

Figure 10: Extrapolation of the NLO described above to $Q^2 \leq 1\ GeV^2$

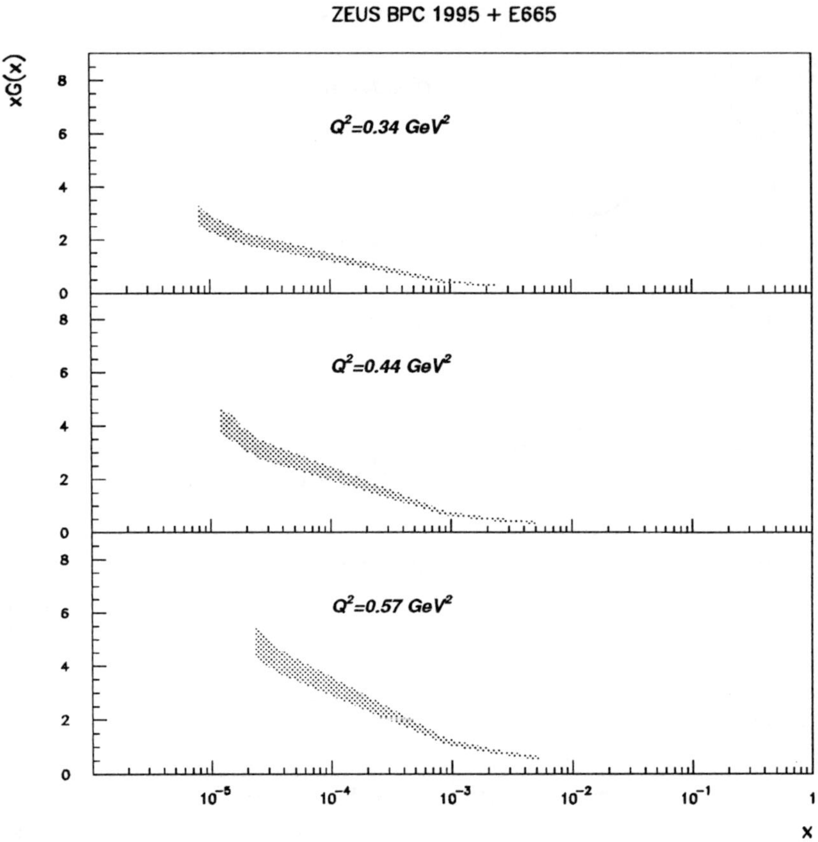

Figure 11: The gluon density for $Q^2 \leq 1\ GeV^2$ values

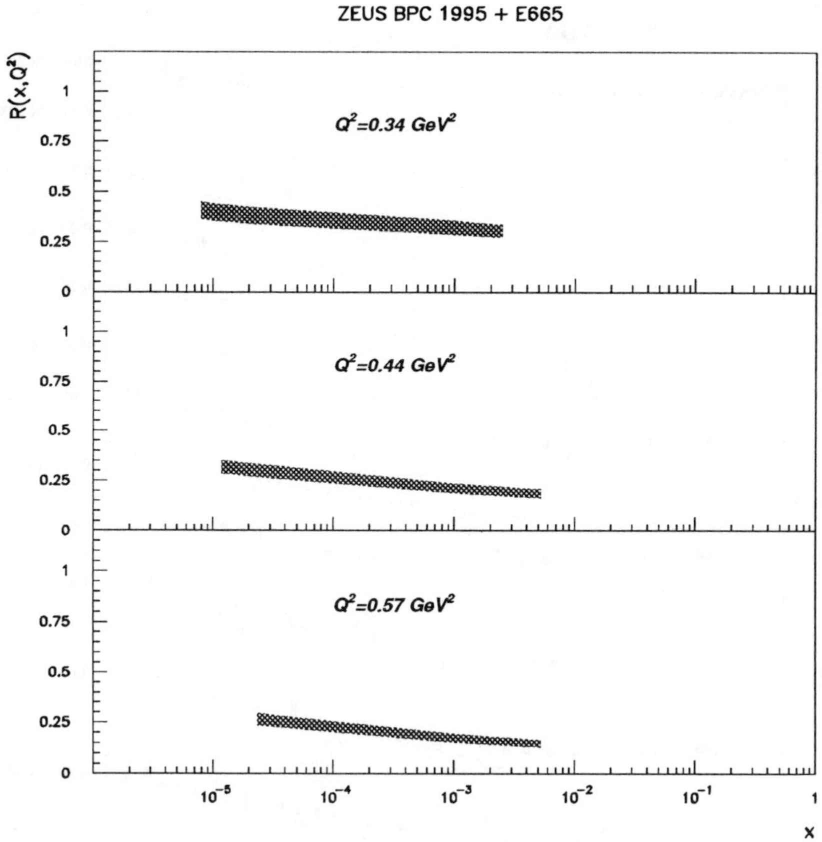

Figure 12: Predictions for R for $Q^2 \leq 1\ GeV^2$ values

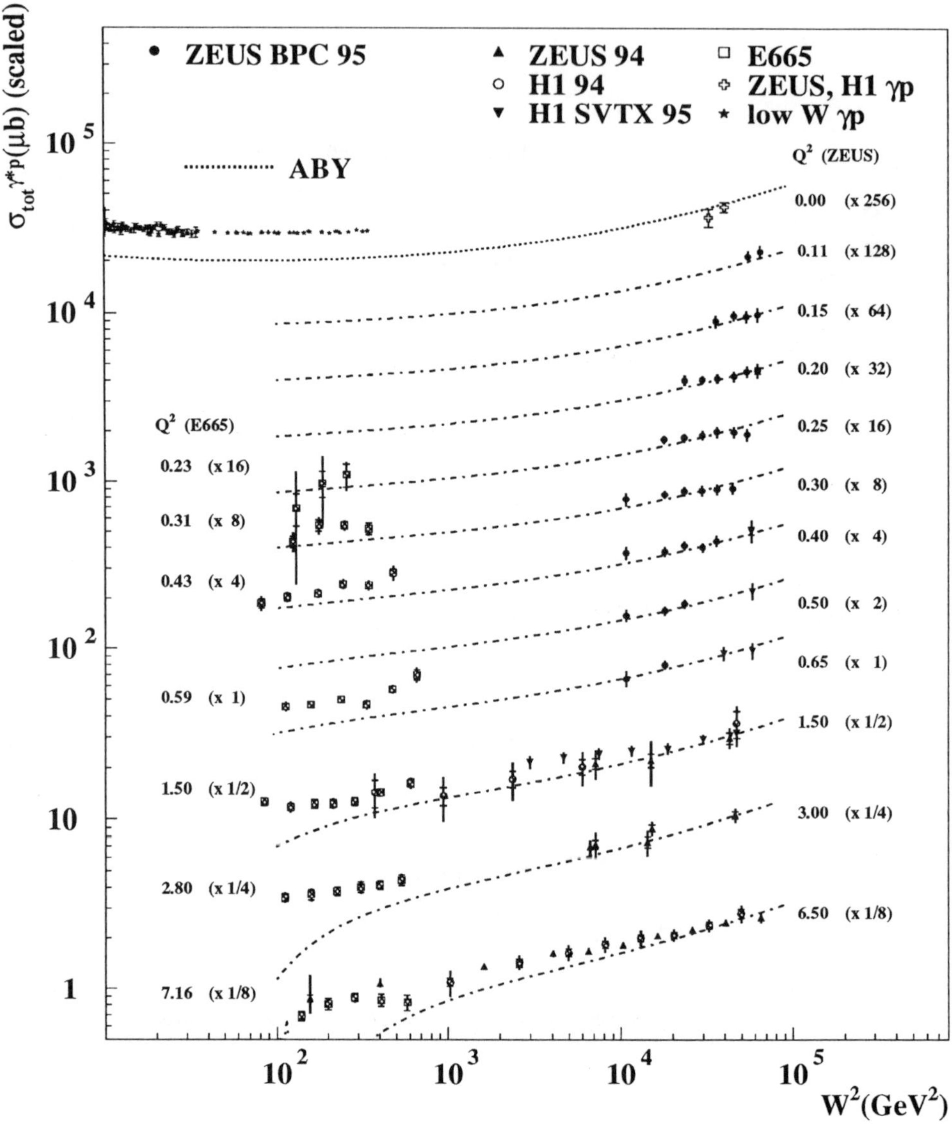

Figure 13: Fits to $\sigma_{tot}(\gamma^* - p)$ not including the P' trajectory

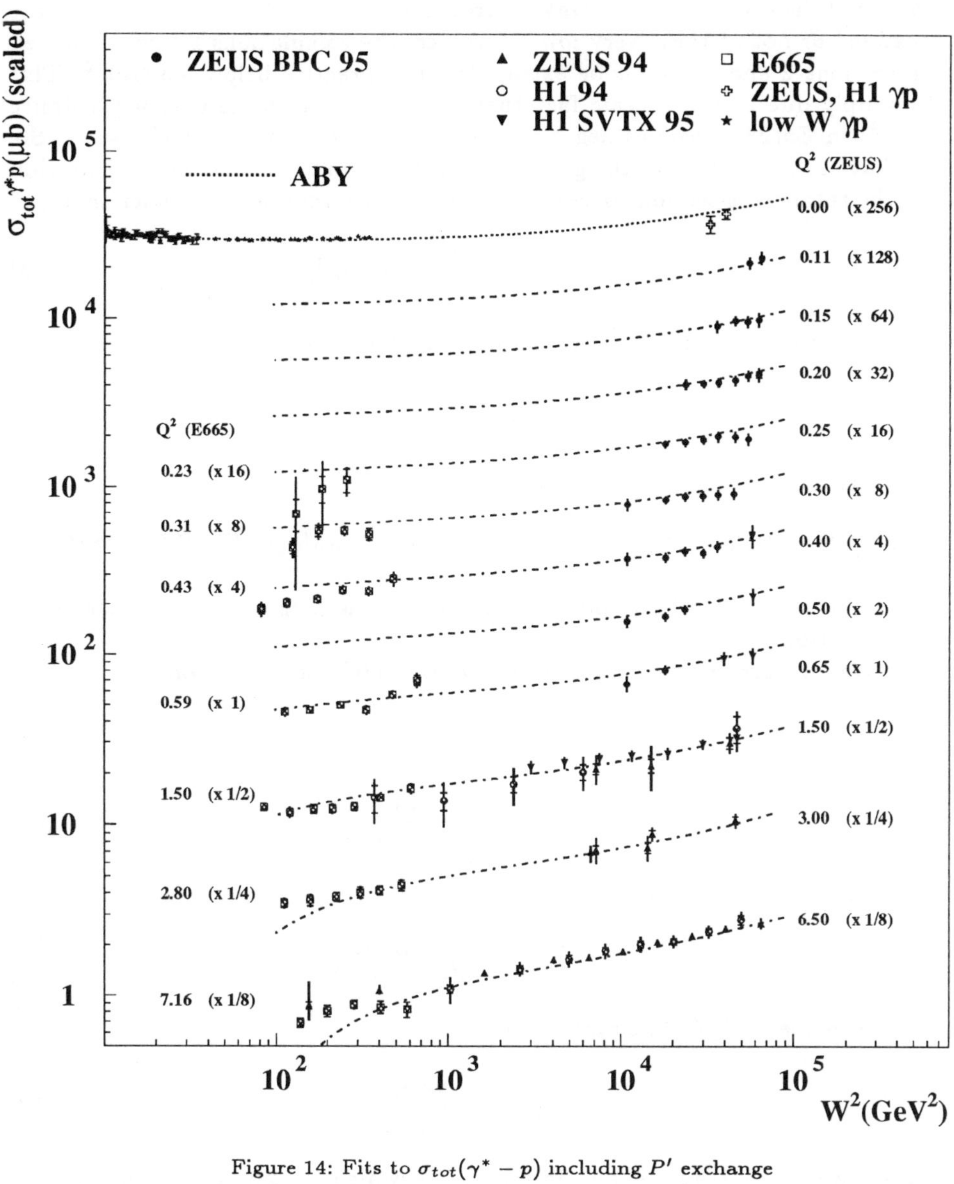

Figure 14: Fits to $\sigma_{tot}(\gamma^* - p)$ including P' exchange

From the discussion above, it seems that perturbative QCD works down to values of the order of $0.5-1\ GeV^2$. Irrespective of whether the predictions are justified or not at these very low Q^2 values, the calculations I have presented in sections 2 and 3 cannot be extrapolated in practice below $0.3\ GeV^2$. This is simply due to the prosaic fact that α_s becomes so large that for instance $\nu(Q^2)$ in Eq(28) becomes negative and enters into an unphysical region. So, we have to put a limit to the growth of α_s if we want to make some connection to the total cross-section for $\gamma - p$ scattering. We remind the reader that

$$\sigma_{\gamma^*-p}(s) = \frac{4\pi\alpha}{Q^2} \cdot F_2(x, Q^2) \tag{53}$$

and at very low x values we can approximately write

$$x = \frac{Q^2}{s} \tag{54}$$

The LO prediction for F_2 reads

$$F_2(x, Q^2) = B_S[\alpha_s(Q^2)]^{-d_+(1+\lambda_S)} \cdot x^{-\lambda_S} + C + B_{NS}[\alpha_s(Q^2)]^{-d_{NS}(1-\lambda_{NS})} x^{\lambda_{NS}} \tag{55}$$

with the first two (third) pieces representing the singlet (resp. non singlet) contribution.

We now assume that α_s saturates and modify its expression

$$\alpha_s(Q^2) = \frac{4\pi}{\beta_0 log \frac{Q^2}{\Lambda^2}} \tag{56}$$

as follows

$$\alpha_s(Q^2) = \frac{4\pi}{\beta_0 log \frac{Q^2+\Lambda^2_{eff}}{\Lambda^2_{eff}}} \tag{57}$$

which in the limit $Q\to 0$ simplifies to

$$\alpha_s(Q^2) = \frac{4\pi\Lambda^2_{eff}}{\beta_o \cdot Q^2} \tag{58}$$

where we have introduced an effective Λ_{QCD} which we expect to be of the order of the ρ mass. Accordingly the predictions for the singlet and non-singlet pieces of the proton structure function should be written as

$$F_{NS}(x,Q^2) = B_{NS}[\alpha_s(Q^2)]^{-d_{NS}(1-\lambda_{NS})} \cdot Q^{2\lambda\ NS} \cdot s^{-\lambda_{NS}} \tag{59}$$

$$F_S(x,Q^2) = B_S[\alpha_s(Q^2)]^{-d_+(1+\lambda_S)} \cdot Q^{-2\lambda_S} \cdot \quad s^{\lambda_S} + C\frac{Q^2}{Q^2+\Lambda_{eff}^2} \tag{60}$$

Since in order to obtain a finite total photon proton cross sections the powers of Q^2 in F_2 should match the $\frac{1}{Q^2}$ in Eq. (53) we obtain the following consistency relations

$$d_+(1+\lambda_S) = 1+\lambda_S \qquad \rightarrow \lambda_S = 0.47 \tag{61}$$

$$d_{NS}(1-\lambda_{NS}) = 1-\lambda_{NS} \qquad \rightarrow \lambda_{NS} = 0.52 \tag{62}$$

in very good agreement with the values obtained from LO fits in the perturbative regime and in agreement with known Regge phenomenology. Results of the fits to $\sigma_{tot}(\gamma^* - p)$ are shown in Fig. 12. Note that with this prescription we can predict the cross sections for virtual photon-proton scattering at low x as well as the HERA cross sections for real photon scattering on protons. However these fits fall too short of the photon-proton cross sections data at lower energies. They also fail to describe the tails of the virtual photon cross section data at low W^2 values. It is easy to see that the difference between our fits and the fixed target data falls as $\frac{1}{\sqrt{(s)}}$. We adscribe this effect to a term due to the P' trajectory which was not included in the singlet piece. We parametrise such a term as [41]

$$F_{S,P'} = B_{P'} \cdot x^{0.5}\sqrt{\frac{Q^2}{Q^2+\Lambda_{eff}^2}} \cdot (1 + \frac{A_{P'} \cdot Q^2}{Q^2+\Lambda_{eff}^2}) \tag{63}$$

with two extra parameters namely $B_{P'}$ and $A_{P'}$. This term is important at high x i.e. low W^2 values as exemplified in Fig. 13. The fixed target total photon proton cross section data is thus also properly described.

7 Conclusions

We have compared recent HERA data on structure functions at low x with QCD analytical calculations derived under standard DGLAP evolution equations. NLO predictions in QCD describe the rate of growth of the proton structure function F_2 in a wide Q^2 and x domain. The dominant term at high Q^2 and low x behaves like $x^{-\lambda_S}$ with $\lambda_S \sim 0.35 \pm 0.02$ independent of Q^2. This spread takes into account possible dependences on the number of excited quark flavours. If one wishes to extend the predictions to the complete kinematic range in the (x, Q^2) plane, one has to consider subleading terms. In the singlet piece, this amounts to including terms proportional to x^0 and $x^{0.5}$, associated to the classical soft pomeron and $P' - f_0$ trajectories. The non-singlet piece, is associated to the ρ trajectory.

Acknowledgements

Thanks are due to Prof. B. Adeva and his colleagues in Santiago for organizing an excellent meeting in La Toja.

References

1. H1 Coll., T. Ahmed *et al.*, *Nucl. Phys.* B **439**, 471 (a)nd I. Abt *et al.*, *Nucl. Phys.* B **407**, 515 (1993)
2. ZEUS Coll., M. Derrick *et al.*, DESY 95-193. and *Phys. Lett.* B **316**, 93 (412) and *Z. Phys.* C **65**, 379 (1995)
3. ZEUS Coll., M. Derrick *et al.*, *Phys. Lett.* B **345**, 576 (1995)
4. H1 Coll., S. Aid *et al.*, *Phys. Lett.* B **354**, 494 (1995)
5. A.D. Martin, W.J. Sterling and R.G. Roberts, *Phys. Lett.* B **354**, 1995 (155), and *Phys. Rev.* D **50**, 6734 (1994)
6. J. Botts et al., *Phys. Lett.* B **304**, 15 (1993)
7. M.Glueck, E. Reya and A.Vogt, *Phys. Lett.* B **306**, 391 (1993).
8. V.N. Gribov and L.N. Lipatov, *Sov. J. Nucl. Phys.* **15**, 438 (1972), G. Altarelli and G. Parisi, *Nucl. Phys.* B **126**, 298 (1977)
9. E.A. Kuraev *et al.*, JETP 45 (1977) 199 and Y.Y. Balitsky and L.N. Lipatov, *Sov. J. Nucl. Phys.* **28**, 822 (1978).
10. A. Mueller *Nucl. Phys.* B **18**, 125 (1992)
11. J. Bartels et al, *Z. Phys.* C **54**, 635 (1992) and K. Kwiecinski et al, *Phys. Rev.* D **46**, 921 (1992)
12. F.J. Ynduráin, FTUAM-96-12
13. A. de Rujula *et al.*, *Phys. Rev.* D **10**, 1649 (1974)
14. R.D Ball and S. Forte, *Phys. Lett.* B **335**, 77 (1994)
15. M. Glueck, E. Reya and M. Stratman *Nucl. Phys.* B **422**, 37 (1994)
16. S. Riemersma, J. Smith and W.L. van Neerven *Phys. Lett.* B **347**, 77 (1994)
17. E. Laenen et al., *Phys. Lett.* B **291**, 325 (1992)

TEV GAMMA-ASTRONOMY

V. FONSECA
Dept. Fisica Atomica, Fac. Ciencias Fisicas,
U. Complutense, 28040 Madrid, Spain
E-mail: fonseca@gae.ucm.es

After many years of unsuccessful efforts TeV Gamma ray (γ) astronomy has now revealed the first sources. Motivation for TeV γ astronomy, an overview of the observation techniques and detection problems and the first discoveries are presented. Also an outlook for the next years will be given.

1 Introduction

Information about our universe is retrieved through astronomical observation, mostly by detection of electromagnetic waves respectively electromagnetic particles at higher energy. Observation of the relativistic universe (often also called the nonthermal universe) is the field of γ astronomy. Charged cosmic rays (CRs) are many orders of magnitude more abundant than cosmic γs. The energy of charged CRs extends up to 10^{20} eV. Unfortunately charged CRs are useless to reveal the location of their origin. The weak galactic magnetic fields randomize completely the charged particles' direction such that they cannot be traced back to their origin. Only stable (long lived) neutral particles, such as gammas, neutrinos and ultra-high energies neutrons, can be traced back to their origin.

At energies higher than 10^{18} eV a neutron lives long enough to span just the distance from the center of our galaxy to the earth. For extragalactic observations only gammas and neutrinos are suitable. Neutrinos require very large detection volumes (order of km^3 of ice or water) and neutrino astronomy is in its infancy[a].

The current belief of the origin of cosmic rays is that they are somewhere accelerated in our universe[b]. The prevailing model of the origin of cosmic rays detected on earth is that they are predominantly generated in our galaxy up to about 10^{16} - 10^{17} eV. Up to that energy the weak galactic magnetic field confines them in our galaxy (protons up to about 10^{15} - 10^{16} eV, Fe up to a few times 10^{17} eV). At about 10^{18} eV cosmic particles generated in our galaxy escape and therefore one "sees" predominantly the extragalactic cosmic

[a](low energy) νs have been detected from only 2 objects, the sun and SN 1987A.
[b]It cannot be excluded that at least a fraction of cosmic rays originate from decays of ultra-heavy particles of the time of the big bang.

ray flux. The search for the location of the cosmic accelerators is the field of gamma astronomy.

In the keV - GeV energy regime γs are absorbed by the earth atmosphere. KeV - GeV γ astronomy has to be carried out by satellite or high flying balloon borne detectors. In the upper energy range, 100 MeV - 10 GeV, satellite borne detectors such as COSB and EGRET have detected in the last 2 decades more than 250 γ sources (end of 1997). Above 10 GeV the cosmic gamma flux becomes so low that gammas cannot be detected any more with detectors currently flown on satellites. Energetic gammas (and also cosmic rays) interact with the nuclei in the atmosphere and generate extensive air showers. For a few TeV showers only Cerenkov light can reach ground, while above 30-100 TeV the shower tail particles (mainly low energy e^+,e^- and γ) might reach ground and can be detected by appropriate detectors such as scintillation counters or tracking detectors.

Due to fluctuations and uncertainties in the shower development process in the atmosphere , most of the CR primary parameters such as the precise initial energy and primary particle nature are difficult to determine. Nevertheless Cherenkov light and secondary particles spread over a large area on ground such that a coarse sampled detector or even a single Cerenkov light detector has many orders of magnitude larger detection area than a satellite borne γ detector. As the γ flux is only a tiny fraction ($< 10^{-4}$) of the total CR flux one is confronted with the need of efficient gamma/hadron (γ/h) separation methods. While rather easy in satellite borne detectors, a high γ/h separation is very difficult to achieve with ground based detectors. This difficulty was the main reason why the nearly a century long search for the sources of the CRs remained unsuccessful until recently.

This paper has the following structure: In chapter 2 some consequences of the low and dominantly charged cosmic ray flux are very briefly discussed. In chapter 3 the production of γs and the transport losses in the universe will be mentioned while in chapter 4 an overview of the current observation techniques will be discussed. In chapter 5 a summary of the recent source detections will be given. An outlook and some future research projects are discussed in chapter 6.

2 The total cosmic ray spectrum

Cosmic rays were first detected in the year 1912 by Victor Hess who searched for the origin of background radioactivity in a series of balloon flights. In the following decades an intense search for the origin of cosmic rays has been carried out, albeit with negative results. Nowadays the situation is still unclear

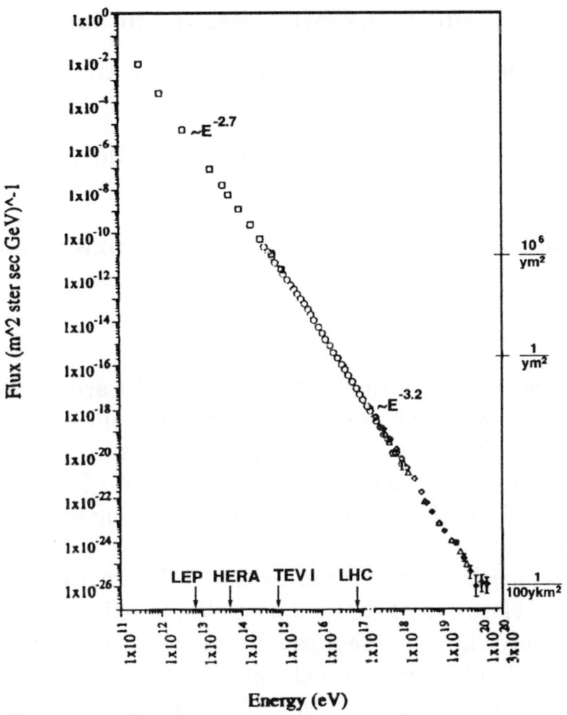

Figure 1: Differential cosmic ray flux highlighting the flux problem

but a relatively good knowledge of the total cosmic ray spectrum up to $3 \cdot 10^{20}$ eV has been collected. Fig. 1 shows the differential cosmic ray flux between 10^{11} to 10^{20} eV. The coarse charge composition is known up to about 10^{14} eV. The integrated flux of cosmic γs is still unknown, but it is estimated to be $< 10^{-4}$ times smaller than the total charged flux.

Also noted in Fig. 1 are some expected cosmic ray rates per year for a hypothetical detector area of 1 m^2 and 1 km^2, respectively. The steep flux drop highlights a general problem in cosmic ray experiments. Experiments can only be carried out over two or three decades in energy. The expected rates are many order of magnitude smaller than in typical accelerator experiments.

3 Generation of γs and transport losses in the universe

TeV γs need parent particles of higher energies (masses). Gamma generation processes are:

a) hadronic interaction resulting in $\geq$ TeV π^0 production and subsequent decay into $\gamma\gamma$,

b) Inverse Compton (IC) scattering of $\geq$ TeV electrons on low energy photons,

c) decays of supermassive particles.

Current models assume that either processes a) or b) are responsible for the generation of TeV γs while process c) is up to now speculative. Process a) would be a natural link to the observed charged CRs. Hadronic interactions will generate typically the same order of magnitude of νs as γs; therefore simultaneous observation of γs and νs would give strong evidence for "cosmic hadron accelerators".

The underlying mechanism to generate the TeV parent particles is assumed to be shockwave acceleration of charged particles, mostly protons or electrons. Shockwaves occur anywhere where a rapidly moving gas volume hits a slower (or stationary) gas cloud. Strong shocks are, for example, expected after supernovae (SN) explosions. The concept of shockwave acceleration has been originally developed by Fermi [1] (Fermi-acceleration of first and second order) and is constantly refined. Shockwave acceleration is quite efficient to generate energetic particles. It is predicted that in the follow-up of SN explosions, particles are accelerated up to at least 10^{15} eV. In case of hadrons these particles might interact with nuclei of the gas clouds (stellar wind) ejected before the supernova explosion or with interstellar gas and generate secondary particles, among them π^0s, which in turn decay into γs.

An important alternative of TeV γ generation occurs through electron acceleration. Electrons might equally be accelerated through shockwave acceleration or in time variable magnetic fields (betatron acceleration), e.g. in the strong, variable magnetic fields of pulsars. Part of the gained energy is lost by bremsstrahlung and synchrotron radiation, respectively. These low energy photons might be "upscattered" by the energetic electrons through inverse Compton scattering. The so-called SSC-model (Synchrotron Self Compton Scattering) from DeJager and Harding [2] is for the time being capable to explain all up to now observed TeV γ sources.

Our universe is not empty and γs might undergo interactions, mostly with low energy photons. There are about 10^9 times more photons in the universe

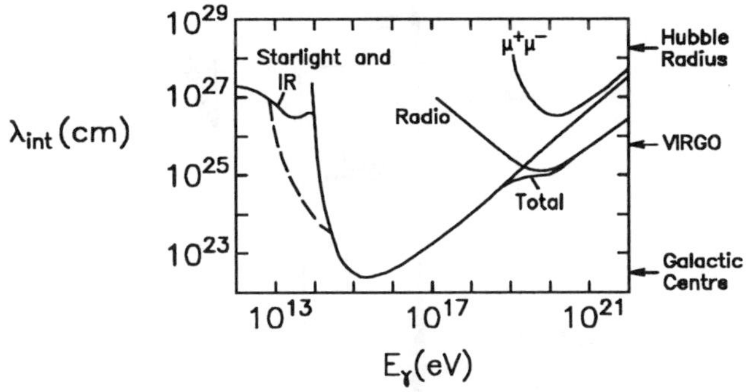

Figure 2: γ attenuation length as a function of energy. The dashed line indicates the possible effect from the IR background. The dominant effect is due to the 2.7° background. (Calculation from Wdowczyk and Wolfendale [3].)

than hadrons, therefore photon-photon interactions dominate although cross sections are small. The universe is filled with various low energy photon fields such as the 2.7° K microwave background, infrared radiation, starlight, radio waves, etc. These fields limit the visibility of high energy gammas. Fig. 2 shows a calculation of the γ attenuation as a function of energy. At about 10^{15} eV one can barely 'see" the center of our galaxy. The up to now unquantified IR background might, severely influence the prospects to search for distant AGNs in the 10^{13} - 10^{15} eV range. It also follows from Fig. 2, that the best window for γ astronomy of distant objects is below 10^{12} eV. It should be noted that one of the safest tests on the distance of the up to now unexplainable gamma ray bursts (GRB) could come from their spectral study in the 10^{12} - 10^{15} eV region.

4 Detection methods of TeV γs

In the MeV-GeV range γ astronomy can be carried out by detectors flown on balloons or satellites unaffected by atmospheric absorption. The many orders of magnitude larger hadronic background can be efficiently vetoed and the fluxes are generally high enough that detectors of less than 1 m^2 detection area can be used. Above a few GeV the γ flux from even the strongest sources becomes so low that observations would require detectors of many square meters area, but the costs for such large devices to be transported into orbit are prohibitively high.

The atmosphere is for electromagnetically and hadronically interacting particles an efficient absorber. The vertical air mass corresponds to 27 radi-

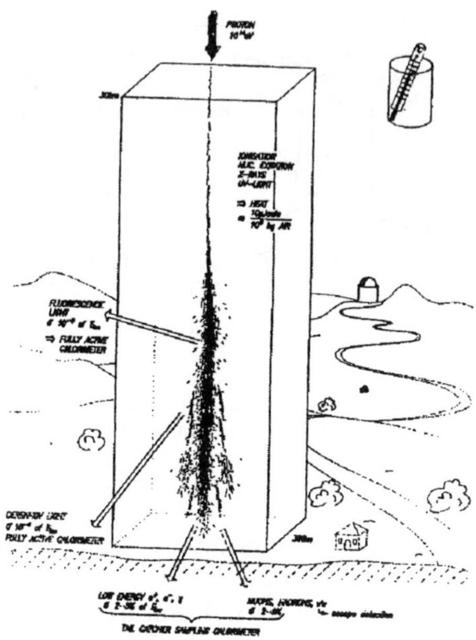

Figure 3: Example of a large air shower

ation lengths (rl) and ≈ 13 hadronic absorption lengths (X_0). Interaction of TeV CRs with the atmosphere leads to extensive air showers (EAS). Figure 3 shows an example of a large air shower (proton of 10^{14} eV) together with some observable quantities. Above a certain energy, the shower tail particles can reach the ground, where they can be detected by scintillation counters or similar detectors for charged particles. The threshold is typically between 30 to >100 TeV depending on the zenith angle and detector elevation. The advantage of the showering process is that a single particle can "spray" a large area on the ground with secondary particles, mostly low energy electrons and gammas and in the case of a primary hadron with a few muons and hadrons and also with Cerenkov light and photons from air fluorescence. Figure 4 shows a "flash photo" of the different shower components just before the shower hits the ground together with the principle of the main detection methods. In all cases only a sparse array of detectors is needed, i.e., it is possible to achieve a large sensitive area by only one (few) detector(s) but at the prize of reduced precision. The "workhorse" for many CR studies and γ astronomy is an array of scintillation counters that detects the charged particles in the shower tail.

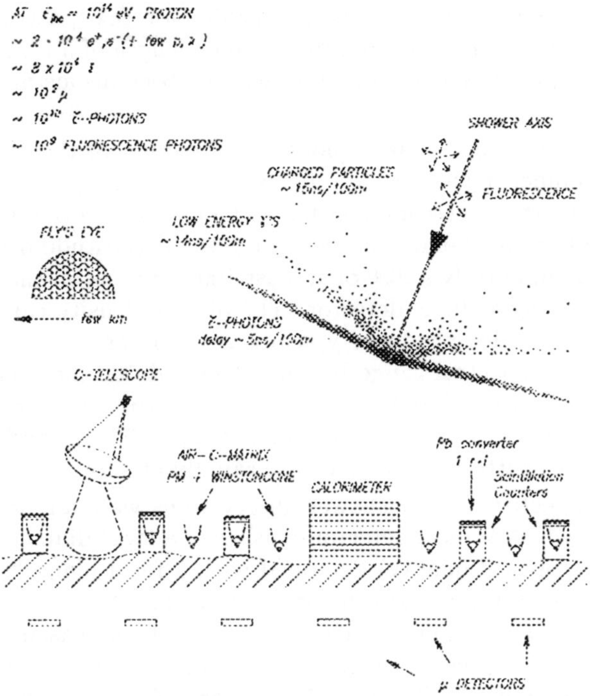

Figure 4: Flash photo of shower components + detection methods.

The primary's direction is determined by sampling the arrival time of the secondary particles while the energy is determined from a so-called NKG fit to the sampled particle density in the shower tail. The main advantages of scintillator arrays are their 24 hour up-time and their large angular acceptance (typically one sterad). Their main disadvantages are a high energy threshold and a poor γ/h separation power. The muon content in electromagnetic and hadronic showers differs by at least an order of magnitude and this difference could in principle be used for discrimination. The tiny fraction of muons combined with large fluctuations in hadronic showers make discriminations by muon counting very difficult. No sources have yet been discovered with scintillation arrays.

Cerenkov light and air fluorescence can only be detected during clear, dark nights, when the non-negligible night sky background light ($2 \cdot 10^{12}$ photons/m^2 sec sterad, at 300-500 nm wavelength) sets the lowest achievable detection threshold. Air fluorescence detectors are unsuited for TeV γ astronomy because the low light yield ($\approx$ 4 photons per meter track length of a minimum ionizing track) and isotropic emission. Air fluorescence observations are only feasible above 10^{17} eV.

Cerenkov photons are peaked along the direction of the shower development. The combined result of two effects: 1) the multiple scattering of the electrons (and the transverse momentum kick of secondaries in hadronic showers) and 2) the Cerenkov emission angle of 0.3-1.3° (altitude dependent) is to illuminate nearly uniformly a disc of at least 100 m radius on the ground where it can be detected by a single detector. The photon density can be quite high because $\beta = 1$ particles produce typically 24 photons/m track length close to the ground in a spectral range between 300 and 500 nm. The $1/\lambda^2$ emission spectrum of Cerenkov radiation favours detectors that are UV sensitive but there is atmospheric Ozone absorption below 290 nm and losses due to Rayleigh and Mie scattering that increase rapidly at shorter wavelengths.

Cerenkov detectors for air shower observations were already proposed in the 60's by Jelley [4]. A typical detector consists of a large diameter optical telescope(of modest optical quality) equipped with a focal plane "camera" made up of a matrix of fast photomultipliers, a so-called Air Cerenkov Telescope (ACT). Only recently a rapid development towards increased sensitivity has begun as ACTs are at present the instruments most likely to achieve a low energy threshold and a high γ/h separation power. ACTs with their limited field of view (a few degrees), can only be used for the study of preselected objects. The observation of Cerenkov light does not require that the charged shower tail reaches the ground, so low energy showers are detectable as long as their signal stands out against the night sky background light. As the ratio of signal to background increases only with the square root of the mirror area the tendency is to build large diameter dishes. The presently largest telescope is the Whipple one with a 10 m diameter mirror (Cawley [5]). The real breakthrough in the ACT approach came with the introduction of shower image analysis (Hillas [6]; Punch [7]). In an ACT of 3-4° camera diameter one might see a rather compressed (and sometimes a very complex) image of the entire shower. Fig. 5 shows simplified examples of a hadronic and an electromagnetic shower and their images in a telescope's focal plane. Discrimination of γs against hadrons is possible if the camera has sufficient image resolution. While γ and hadron showers are nearly indistinguishable at tail level the Cerenkov images of the entire showers are quite different in longitudinal and transversal structure.

When searching for γ showers from a point source, the telescope follows the source on the sky and the principal axis of the showers should point to the centre of the camera, while the directions of hadronic showers are randomly

Telescope	Whipple	CAT	HEGRA system	HEGRA CT1[a]	TA	CANGA.	Durham Mk6
Site	Mt.Hopkins	Themis	LaPalma	LaPalma	Dugway	Woomera	Narrabri
Longitude	$-110.5°$	$-2.0°$	$-17.8°$	$-17.8°$	$-130.0°$	$+136.8°$	$+149.8°$
Latitude	31.4° N	42.5° N	28.8° N	28.8° N	40.3° N	31.1° S	30.5° S
Elevation (m)	2300	1650	2200	2200	1600	160	200
# of telescopes	1	1	4	1	3	1	1 (3 dishes)
Mirror area (m^2)	74	17.5	4×8.4	5.0	3×6.0	11.3	3×42.0
# of pixels	151	548+52	4×271	127	3×256	256	109+19+19
Pixel ⊘ (°)	0.25	0.10	0.25	0.25	0.25	0.12	0.225
Threshold energy (GeV)	250	300	500	1500	600	1000	250
Sensitivity[b] ($cm^{-2}s^{-1}$) at the threshold	10^{-11}	10^{-11}	4×10^{-12}	3×10^{-12}	10^{-11}(?)	4×10^{-12}	4×10^{-11}

[a] For the old mirrors, i.e. up to Nov. 97.

[b] defined as the 5 σ limit for 50 h observation time.

Table 1: Main parameters of the leading ACTs.

oriented. Hadronic showers are in general slightly wider and their light is less concentrated as compared to γ showers. Experimenters have developed a series of parameters for image analysis such as the so-called Width, Length, Concentration, Miss, Alpha etc. γ/h discrimination power is intimately related to the telescope's optical quality and camera granularity. Table 1 summarizes the parameters of the leading air Cerenkov telescopes.

There exists another promising concept for air Cerenkov light detection. The Cerenkov light can be detected by a large matrix of open photomultipliers. The HEGRA AIROBICC detector is an example (Karle[8]). This detector has a large angular acceptance, excellent angular resolution and also γ/h separation capabilities when information about the shower tail are combined with observations from a scintillator array. The combination of the atmosphere as Cerenkov radiator and a ground based extended detector matrix basically forms a fully active calorimeter with good energy resolution. TeV γ showers produce about a factor of two more light than hadron showers of equivalent energy while the integral number of particles in the shower tail is the same within the measuring precision. Comparison of the total Cerenkov light yield and the number of particles in the shower tail therefore allow γ/h discrimination. Unfortunately, wide angle Cerenkov detectors have higher thresholds than ACTs because the former integrate the night sky light over large solid angles and are not yet suitable for γ astronomy below, say, 10 TeV.

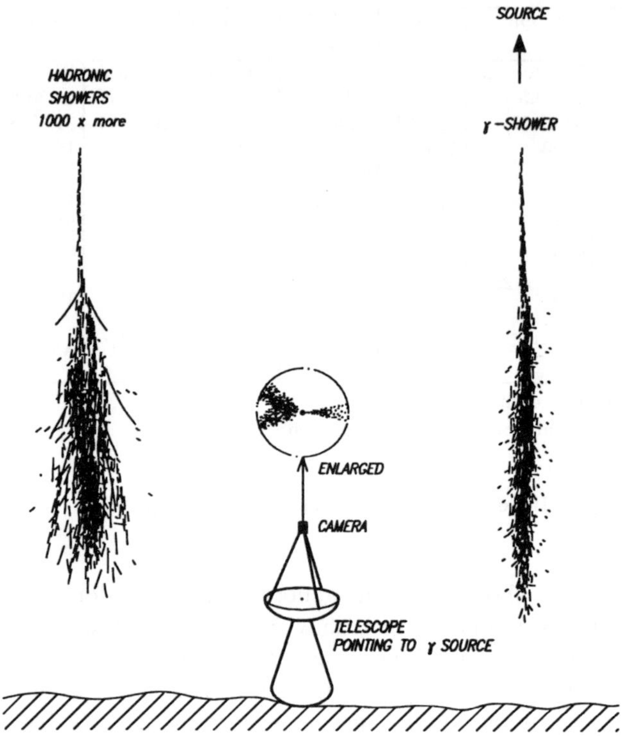

Figure 5: Example of an electromagnetic and a hadronic air shower seen by an ACT. Note that the images are right-left exchanged.

5 Overview of detected TeV γ sources

Potential TeV γ sources can be classified in

- galactic sources, such as SNRs, Plerions, X-Ray Binaries, cataclysmic variables , the center of our galaxy,
- extragalactic point sources, predominantly AGNs,
- exotic sources such as GRBs.

Other sources of γs might be diffuse such as γ emission from the galactic plane or large gas clouds (both illuminated by higher energy charged cosmic rays). Exotic sources might come from spontaneous breakdown of topological defects

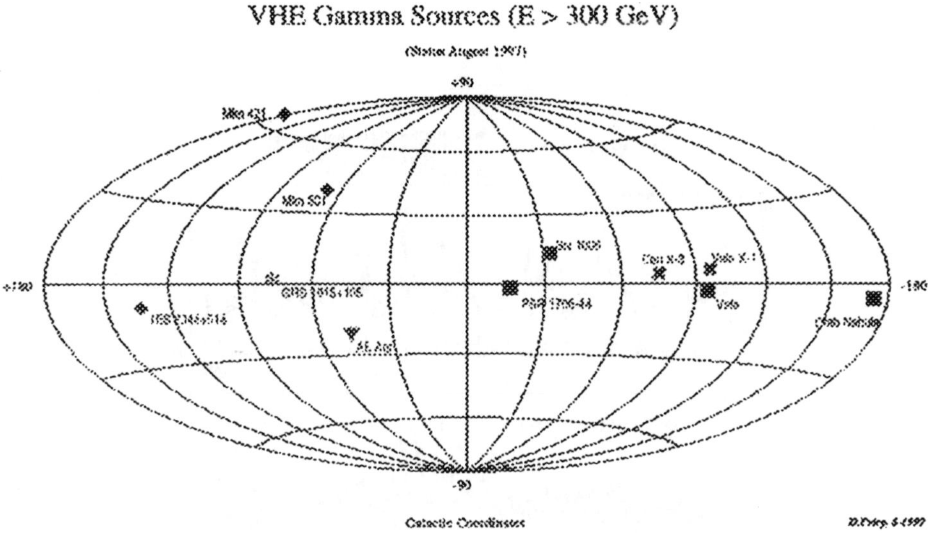

Figure 6: The galactic coordinates of point sources of Gamma-radiation of energy $E > 300$ GeV as observed by the various Cerenkov telescope observatories around the world.

or annihilation of the lightest supersymmetric particles with their conjugates, perhaps amassed around the center of our galaxy.

The first TeV γ source, the Crab nebula, was detected in 1989 by the Whipple group (Weekes [9]) with a significance of about 8 σ. Since then a total of 8 sources have been detected with high significance ($> 6\sigma$) and mostly confirmed by at least one other group. Fig. 6 shows the sky map with those sources together with some 5σ observations. Some sources, such as the Crab, Mkn 501 and Mkn 421 have significance levels exceeding more than 30 σ. All discoveries were made with Cerenkov telescopes but it should be mentioned that only a few % of the sky map has been scanned up to now. A guideline for possible objects can be found in sky map studies at lower energies, for example from EGRET that discovered over 250 sources in the 100 MeV - 10 GeV domain. One interesting aspect is that one of the TeV γ sources, Mkn 501, was not observed by EGRET. This is explainable as good Cerenkov telescopes have huge collection area in comparison to satellites, such that this large area ratio over-compensates for spectra following unbroken power law with a coefficient of < 2.

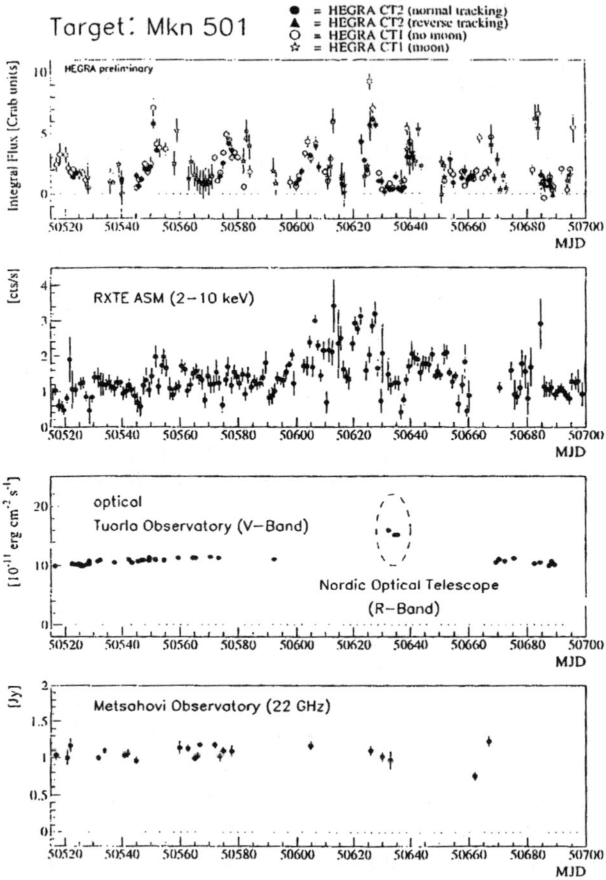

Figure 7: The light curves of the active galaxy Mkn 501 from March to September 1997 in the energy bands: > 1 TeV, 2 - 10 keV, optical and 22 GHz. Note that the encircled data points in the optical light curve were taken in a different filter band.

5.1 Galactic sources

TeV γs from the Crab nebula

As already mentioned, the detection of the first TeV γ source, the Crab nebula, by the Whipple group (Weekes 1989) opened the new window of TeV γ astronomy. The Crab nebula originates from a supernova explosion well documented by Chinese astronomers in 1054. In 1968 a pulsar was discovered in the central area of the nebula. Successively pulsed emission in the radio wave regime, optical domain, x-rays and MeV-GeV γ rays were discovered. It was found that the pulsar period was slowly increasing with a characteristic time of $\sim$ 1000 years. Because of the high keV-GeV luminosity the Crab was a prime candidate for TeV γ emission and many searches concentrated on this object. Eventually the first observations were made above 300 GeV with the 10 m Whipple ACT and showed an unpulsed γ emission. Due to our current understanding the TeV γ emission is not originating close to the pulsar but from the area of the shock wave acceleration in the nebula, where due to the slow acceleration mechanism (with respect to the pulsar period) and extended area the time structure is lost. It is believed that electrons are accelerated. These electrons produce high energy γs by inverse Compton scattering on low energy photons mainly generated by synchrotron radiation of the electrons during early acceleration in the strong and changing magnetic field around the pulsar. The SSC model [2] describes both the pulsed keV-GeV and unpulsed TeV γ emission up to at least 10 TeV with very little freedom. Between 500 GeV and 5 TeV the spectrum is well presented by a power law of

$$F = (3.2 \pm 0.7) \cdot 10^{-7} (E/\mathrm{TeV})^{-2.49 \pm 0.06} / \mathrm{m}^2 \ \mathrm{sec\ TeV}\,.$$

The Crab nebula is, for the time being, the strongest known galactic TeV γ source. Up to now no variability has been discovered. Due the these two facts the Crab nebula is also used as a calibration source for TeV γ astronomy.

Other galactic sources

Recently some more galactic TeV γ sources have been discovered. The sources PSR1706-44 (Kifune [10]) and Vela (Yoshikoshi [11]) have similar structure as the Crab nebula.

Supernova remnants (SNR) are considered to be the sources of the CRs up to about 10^{15} eV. Except for one case all searches in the TeV domain were negative. Only recently TeV γs have been observed from SN 1006 (Tanimori [12]). The observation of a strong non-thermal X-ray signal from the north-east rim of the supernova shell indicates at least dominance of electron acceleration. If the next generation of higher sensitivity instruments will not reveal TeV γ rays from SNRs then the model of SNRs being the source of the charged CRs has to be revised.

5.2 Extragalactic sources

Up to now observations of γ rays from three extragalactic sources have been reported, namely from Mkn 421, Mkn 501 and 1ES2344, the former two from multiple detection with very high significance. All three sources are low red shift ($z < 0.05$) AGNs of the BL-Lac type, i.e., large elliptical galaxies with a central black hole ($m \sim 10^6$ - 10^8 solar masses) and a relativistic jet pointing towards the earth. All three objects show high variability with short-term intense flaring of the γ emission. The most spectacular event occurred in 1997 where the source Mkn 501 could be observed nearly continuously for almost 8 months by many ACTs around the world. Fig. 7 shows a result from the HEGRA collaboration together with RXTE observation in the 2 - 10 keV domain and optical and radio wave observations. The most intense variability occurred in the TeV domain with flares exceeding the Crab flux by up to a factor 10. There is a significant correlation between the RXTE and TeV data giving strong evidence for electron acceleration. As predicted by the above mentioned SSC model the TeV intensity seems to increase/decrease in quadrature of the 2 - 10 keV intensity. The large abundance of nearly pure γ candidates will allow to determine with high precision the emission spectrum up to the multi-TeV domain and to carry out decisive tests on the limited transmission of our universe. Preliminary results indicate a rather low IR background, see section 3 and Fig. 2. High sensitivity studies of γ-emitting AGNs will be one of the most important areas of future γ astronomy.

6 Directions in the next years

Currently, the energy range between 10 GeV and 250 GeV is completely unexplored. The lower limit is given by the small detection area of satellite borne γ detectors while the present ACTs can observe only down to 300 GeV. This limit is basically a product of mirror area $\times$ quantum efficiency of the PMs (+ some smaller effects). Cerenkov light is produced in electromagnetically induced air showers down to below 1 GeV, albeit with a photon flux being reduced in first order proportional to the energy.

The exploitation of the above mentioned energy gap is of high importance because it is expected that various changes in the γ spectra should occur, such as cosmic accelerators reaching their endpoint in energy or important γ-absorption effects over cosmological distances. Also experiments in this gap

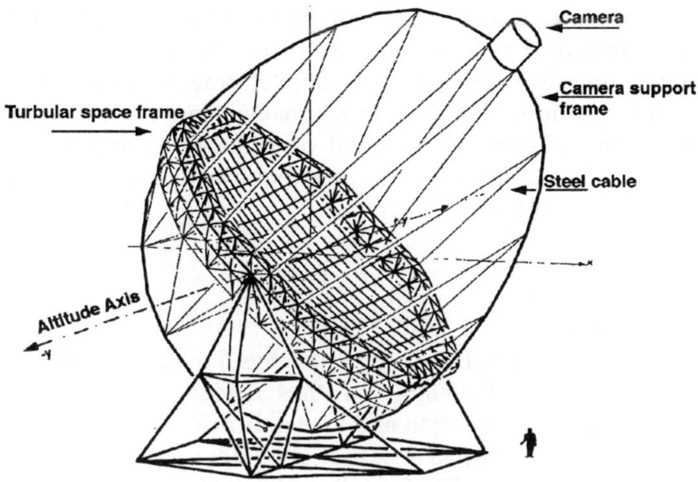

Figure 8: A model of the 17 m ⊘ MAGIC telescope.

will "bridge" low energy satellite experiments with ground based ultra high energy experiments.

In order to lower the threshold new ACTs are proposed following two approaches. The first one is to cluster improved telescopes of the "10 m ⊘" class. Observing showers from different direction improves the measuring precision and allows also to separate with high efficiency low energy shower images against the night sky light background. The HESS (High Energy Stereo System) collaboration [12] proposes to build up to 16 telescopes while the VERITAS collaboration [12] plans to build a system of 7 ACTs. Another approach, chosen by the MAGIC collaboration [12], is to build a 17 m ⊘ mirror and combine it with novel, high QE hybrid photomultipliers in order to reach a threshold of $\approx$ 15 GeV.Figure 8 shows the design of the 17 m ⊘ ACT. Other efforts,-such as the Celeste, Stacee, and Mini-GRAAL projects [12] -, concentrate on using existing solar power plants with very large mirror combined with photomulitipliers mounted on a tower.

It should be mentioned that a project of an advanced satellite borne γ detector, GLAST, is presently in its decision phase at NASA.

7 Conclusions

TeV γ astronomy is at present rapidly evolving. During the recent years the first sources of TeV γ emission have been found with high significance. These observations help us to understand high energy physics reactions in our universe. Due to technical progress and also the inflow of concepts developed for high energy physics experiments at accelerators we expect further rapid advancements in the next decades. In a broader view the entire field of high energy astroparticle physics research will very likely be intensified in the next years. A full understanding of our cosmos requires to explain the highest energy processes in our universe.

References

1. E.Fermi: Phys. Rev. 75, 1169 (1949).
2. O.C.DeJager and A.K.Harding: Astrophys.J.396, 161 (1992)
3. J.Wdowczyk and A.Wolfendale: J.Phys.5, 1419 (1972)
4. J.Jelley: Cerenkov Radiation and its Applications, Pergamon Press 1958.
5. M. Cawley et al.: 1990, Exp. Ast.1, 185 (1990).
6. A.Hillas: Proc. 19th ICRC La Jolla vol.3, 445 (1985).
7. M.Punch et al.: Proc. Towards a Major Atmospheric Cerenkov Detector II, Calgary, 163 (1993).
8. A.Karle et al.: Astroparticle Phys. 3, 321 (1995).
9. T.C.Weekes et al.: Astrophys. J.342, 379 (1989).
10. T.Kifune et al.: Astrophys. J. Lett. 91, 438 (1995).
11. T. Yoshikoshi et al.: Astrophys. J. Lett. 65, 487 (1997).
12. See Proc. Towards a Major Atmospheric Cerenkov Detector V, Kruger National Park (1997).

QCD AT LEP

KLAUS HAMACHER
DELPHI *Collaboration and Fachbereich Physik,*
Bergische Univ.-Gh. Wuppertal, Gaußstr. 20, 42097 Wuppertal, Germany
E-mail: Klaus.Hamacher@cern.ch

This lecture is a brief introductory to QCD at LEP. It is shown, how the basic properties of strongly interacting fields have been verified at LEP and how the strong coupling is measured from event shapes. Important properties of the hadronic final state, namely the energy evolution of fragmentation functions and of multiplicities as well as the importance of coherence effects are discussed. The connections between the different topics are shown and some recent results are included.

1 Introduction

Quarks, spin $\frac{1}{2}$ fermions, were introduced in the quark model of hadrons to explain the large number of observed strongly interacting particles. Quarks come in three different colours, they form colour triplets. The observed hadrons, however, are colourless conglomerates of quarks. Hadronic wave functions are symmetric under global SU(3) transformations in colour space.

The colour quantum number is interpreted as the "charge" of the strong interaction. The strong force is mediated by spin 1 "vector" gluons. In order to be able to couple quarks of different colour they need to carry colour themselves. In fact they carry colour and anti-colour. Therefore there are eight gluons [a].

A dynamical theory of the strong interaction, Quantum Chromodynamics[1] (QCD), has been constructed along the guide lines given by QED requiring invariance under local SU(3) colour transformation. This leads to a Lagrangian shown symbolically in EQ. 1:

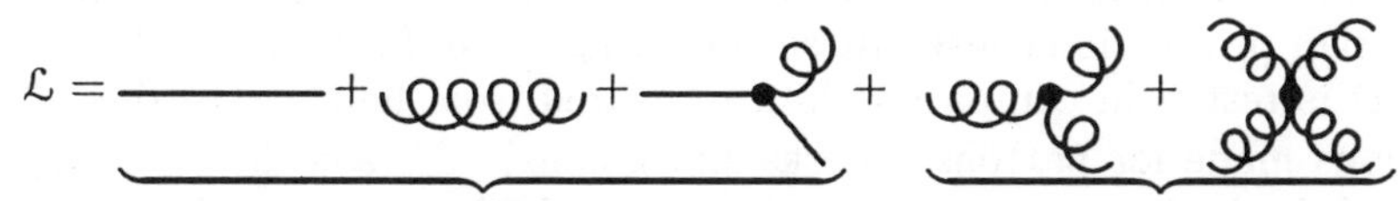

(1)

Besides the terms well known from QED now also the non-abelian gluon self-couplings, the so-called triple-gluon (TGV) and the four-gluon vertex exist. They are a direct consequence of gauge invariance under a non-abelian transformation group.

[a] One of the 3 · 3 colour combinations is a colour singlet which does not contribute.

The coupling for all vertices is universal, however, due to the different colour combinations possible, the individual vertices attain different appearant strength. The relative strength is determined by the so-called Casimir or Colour factors. They are $C_F = \frac{4}{3}$ for the quark-gluon vertex, $C_A = 3$ for the triple gluon vertex and $T_F = \frac{1}{2}$ for the production of a $q\bar{q}$ pair from a gluon.

The additional gluon self couplings lead to contributions to the loop diagrams describing the vacuum polarisation between colour charges. These terms act oppositely to the fermionic terms known from QED, they lead to an anti-screening of the colour charge, which in fact over-compensates the screening effect of the fermionic loops. So, contrary to QED, at small distances (corresponding to large momentum transfer, Q) the effective coupling, α_s , is small and at large distance (small Q) it is large. This is the reason why the strong interaction is "asymptotically free" and in the limit of high momentum transfer can be treated perturbatively using Feynman-graph techniques. At small momentum transfer typical for bound systems, however, these techniques fail leaving many strong interaction phenomena uncalculated. Presumably the increase of the effective strong coupling at small Q^2 is also connected to the phenomenon of confinement, which hides the colour charge in colour neutral hadrons.

In many respects e^+e^- annihilation is the optimal experiment for QCD studies. The simplest possible strongly interacting state, a $q\bar{q}$ pair is produced with a flavour composition known from electroweak theory. A special feature at LEP is the extremely high cross section at the Z pole. This results in a vast amount of events with negligible backgrounds. Also initial state photon radiation (ISR) at the Z pole is small. At higher energies the cross section falls off rapidly, by about three orders of magnitude. ISR in this case leads to a severe background. Often photons are radiated with energies such that the energy of the remaining system is close to the Z pole, where the cross section is large. These so-called "Z-return" events are strongly boosted opposite to the radiated photon(s). Also 2-γ events and at energies above 161 GeV especially W^+W^- pair production constitute severe backgrounds for many QCD studies.

QCD is tested in comparing data measured from hadrons to theoretical predictions made for partons (quarks and gluons). Always the troubling influence of the hadronisation, the so far uncalculable transition from partons to hadrons is present. As this influence for many observables vanishes as inverse powers of the centre-of-mass energy another advantage of LEP is the large energy compared to previous e^+e^- machines. The wide range of energies covered by LEP meanwhile even allows some experimental control of hadronisation.

The first part of this article, corresponding to the first lecture, verifies some of the basic ingredients of QCD using LEP results: Quarks have spin $\frac{1}{2}$ and Gluons have spin 1. The presence of the TGV follows from the angular dependence of the four-jet cross section. The determination of α_s from event-shapes and their hadronisation corrections are discussed. The running of α_s and of the b-quark mass is shown. The second part discusses some phenomena of the hadronic final state and their interconnection: scaling violations, the energy dependence of the multiplicity, coherence effects and, finally, a comparison of the properties of quark and gluon jets.

2 Basic Properties of Strongly Interacting Fields

2.1 Event Shapes and Jets

Jets, narrow bundles of hadrons are observed in the hadronic final state of e^+e^- annihilation, as well as in other high energy inelastic reactions. These jets are remnants of the initially produced quarks or gluons. The properties of these partons have to be derived from the observed hadrons. The simplest observable one may think of is the direction of the initial (back to back) $q\bar{q}$ pair. This is e.g. inferred by means of the so called thrust axis $\vec{n}_T$. This is the axis maximising the sum of the hadron momenta parallel to this axis:

$$T = \max_{\vec{n}_T} \frac{\sum |\vec{p}_i \cdot \vec{n}_T|}{\sum |\vec{p}_i|} \tag{2}$$

The so-called thrust, T, is the classical example of an observable characterising the event shape. Very two-jet like, "pencil" events will have large $T \sim 1$. Spherical events, however, will have small T, close to the kinematically possible lower limit $T = \frac{1}{2}$. Today most event shape variables (commonly denoted y) measure the amount of transverse momentum or mass in an event. It can be shown that $1 - T$ has a similar quality.

An important property of all shape observables used in QCD studies is that the particle momenta add linearly. The value of the event shape observable so is independent of the amount of hadrons seen. Only the distribution of energy or momentum in space matters. The observables become insensitive to details of hadron formation, decays of instable particles etc. Theoretical uncertainties entering due to the divergent cross sections for soft (infrared IR) or collinear gluon radiation also become unimportant. Therefore these observables are called IR and collinear safe.

More complicated topologies require the reconstruction of several partons. This is performed with so called jet-algorithms. The commonly used algorithms are of the so called JADE type[2]. They use a distance measure

$y_{ij} \sim (\text{inv.mass})^2/E^2_{CM}$ and a particle recombination scheme as basic ingredients. Firstly y_{ij} is calculated as a test variable for all hadron pairs (i,j). The pair with the smallest y_{ij} then is recombined, in the simplest case by adding the four-momenta of the hadrons and a pseudo-particle is formed which replaces the hadrons. So the particle multiplicity is reduced by one. This procedure then is iterated until the remaining y_{ij} are bigger than a predefined cut-off value y_{cut} or if a requested number of jets is reached.

The quality of the algorithm to reconstruct the underlying partonic state strongly depends on the choice made for y_{ij}. It turned out, that the so-called Durham or $k_\perp^2$ measure is a good choice:

$$y_{ij}^{Durham} = 2\left\{\frac{\min(E_i,E_j)}{E_{vis}}\right\}^2 \sin^2\frac{\theta_{ij}}{2} \;, \tag{3}$$

which is also well suited for theoretical calculations. Recently it has been suggested, to order the sequence of recombinations according to a separate ordering variable v_{ij} which is essentially the angle between (pseudo-) particles. Pairs are then recombined as predetermined by v_{ij} as long as $y_{ij} \leq y_{cut}$. It turns out that this strategy reduces the amount of "junk" jets which are formed accidentally. Furthermore soft particles at large angles may be left (freezed) out, as they may be emitted coherently from an ensemble of several partons (see SEC. 3.3). This algorithm is called Cambridge algorithm.

Alternate recombination schemes to the simple four-momentum sum have been invented (called E, E0, P, P0 etc.), mainly to leave the clustered jets massless as they are treated in theoretical calculations. As in general the directions of the jets are much better determined experimentally than their energies, it is often advantageous to rescale the jet-energies in order to fulfil energy conservation.

2.2 Angular Distributions, Spins and Colour Factors

By measuring the polar angle distribution of the thrust axis $\vec{n}_T$ it can be verified that indeed spin $\frac{1}{2}$ particles are produced in e^+e^- annihilation. It is known that the electroweak interaction conserves the fermion helicity. This causes the spin of the exchanged vector boson (γ or Z) to be aligned in the beam direction. Assuming spin $\frac{1}{2}$ particles one may add up all possible amplitudes of final states and finds that the distribution of the angle of the final state axis with respect to the beam, θ, is $\propto 1+\cos^2\theta$. In fact the "opposite" solution for spin 0 particles is much easier to explain: here the particles need to carry away the spin of the exchanged boson as angular momentum. So they need to be emitted primarily perpendicular to the beam. The angular

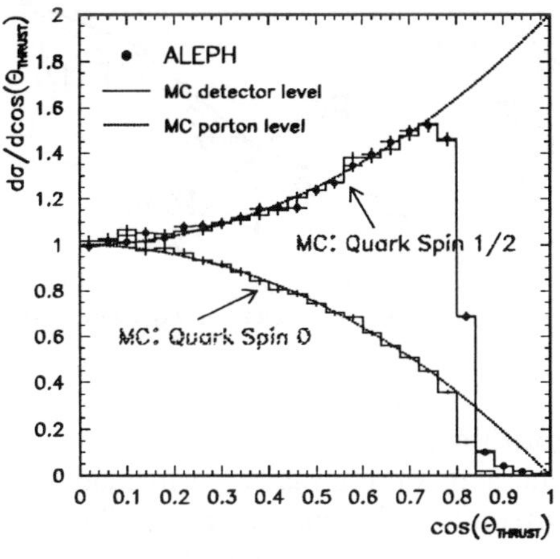

Figure 1. $\cos\theta$ distribution of the thrust-axis.

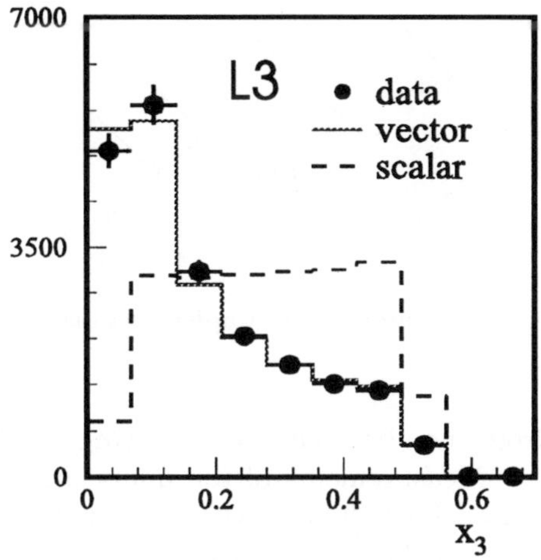

Figure 2. Scaled energy distribution of low energy (gluon) jets in three-jet events.

distribution in this case is $\propto \sin^2\theta$.

A result as given by ALEPH is shown in FIG. 1 [3]. The observed distribution follows the expectation for spin $\frac{1}{2}$ quarks. Deviations at large $\cos\theta$ are due to a limited geometrical acceptance of the apparatus.

The determination of the gluon spin is performed using three-jet events. The jets stem from the quark, the anti-quark and the gluon radiated at high transverse momentum. In terms of the scaled energy fraction of the quarks $x_{\mathrm{q}(\bar{\mathrm{q}})} = 2E_{\mathrm{q}(\bar{\mathrm{q}})}/E_{CM}$ the cross section can be written as:

$$\frac{d^2\sigma}{dx_{\mathrm{q}}\, dx_{\bar{\mathrm{q}}}} \propto \frac{x_{\mathrm{q}}^2 + x_{\bar{\mathrm{q}}}^2}{(1-x_{\mathrm{q}})(1-x_{\bar{\mathrm{q}}})} \tag{4}$$

Typical for the radiation of a spin 1 particle from the spin $\frac{1}{2}$ quarks are the so-called bremsstrahl poles $1/(1-x_{\mathrm{q}(\bar{\mathrm{q}})})$. They similarly appear in the QED case and cause the $1/E$ behaviour of bremsstrahlung. If gluons would be scalar particles these poles would not be present. FIG. 2 is part of a more general test of the three-jet matrix element [4] which similarly has also been performed by other experiments [5]. The scaled energy distribution of the least energetic jet is shown, which according to EQ. 4 most likely is the gluon jet. Mind, that from energy conservation $x_{\mathrm{g}} = 2 - x_{\mathrm{q}} - x_{\bar{\mathrm{q}}}$. The typical $1/E$ like behaviour is observed. Deviations at very small and large x_{g} are artifacts introduced due to the restrained possibilities to reconstruct parton properties

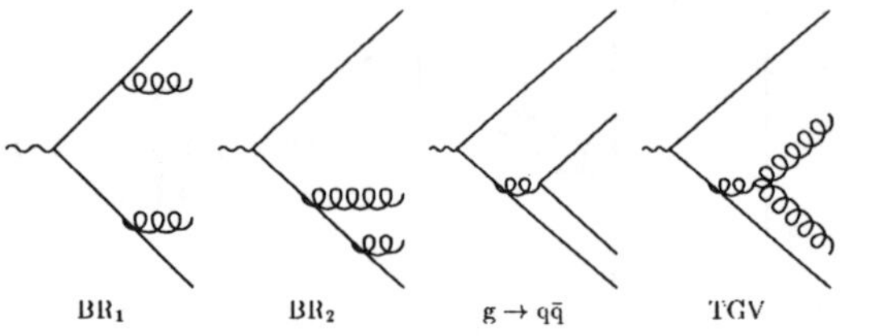

Figure 3. Lowest order 4-jet diagrams.

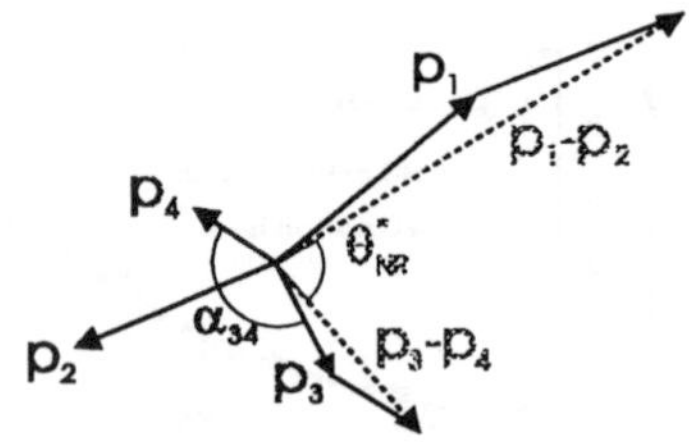

Figure 4. Def. of θ^*_{NR} and α_{34}

from the observed hadrons using jet algorithms.

Measurements of the quark and gluon spin were already performed long before LEP though with less evident outcome. The convincing verification of the TGV, however, which relies on a large statistic of four-jet events only became possible for the LEP experiments. Four jet production, to lowest order, proceeds via the graphs shown in FIG. 3. It turns out to be possible to split the cross section into three terms in a gauge invariant way:

$$\frac{d^2\sigma^{(4-jet)}}{d\theta^*_{NR}d\alpha_{34}} = \sigma_0\alpha_s C_F \cdot \{C_F R_{BR}(\theta_{NR}, \alpha_{34}) + C_A R_{TGV}(\theta_{NR}, \alpha_{34}) + n_f T_F R_{q\bar{q}}(\theta_{NR}, \alpha_{34})\} \quad (5)$$

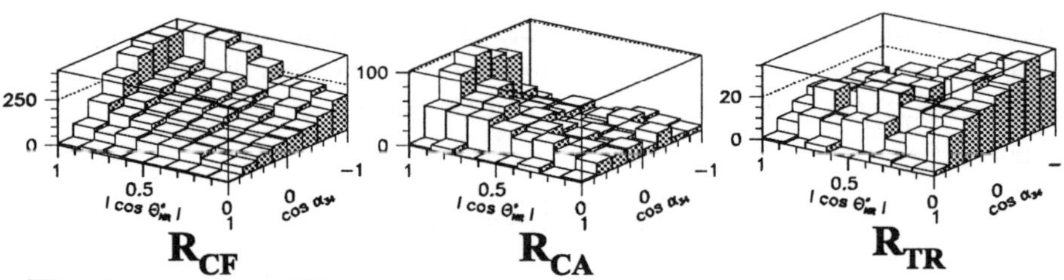

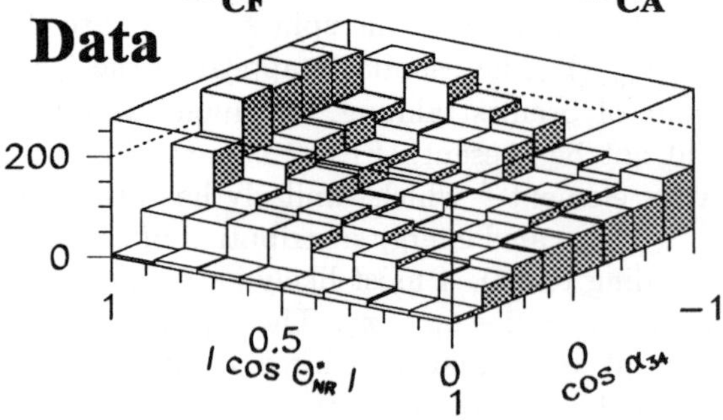

Figure 5. Upper plots: reference distributions corresponding to the different colour factor terms given in EQ. 5 as calculated with a second order ME model. Left plot: Four-jet angular distribution as measured by DELPHI.

This is a consequence of the helicity structure of the amplitudes and also holds for gauge theories in general. These three terms are directly proportional to the colour factors of gluon bremsstrahlung, C_F , the TGV, C_A , and q$\bar{\text{q}}$ pair production, T_F . T_F always appears multiplied with the number of flavours, n_f, which can be pair-produced at a given energy. C_F , the colour factor for gluon radiation from a quark occurs in all terms multiplied by α_s as gluons are not *a priori* present in e^+e^- annihilation.

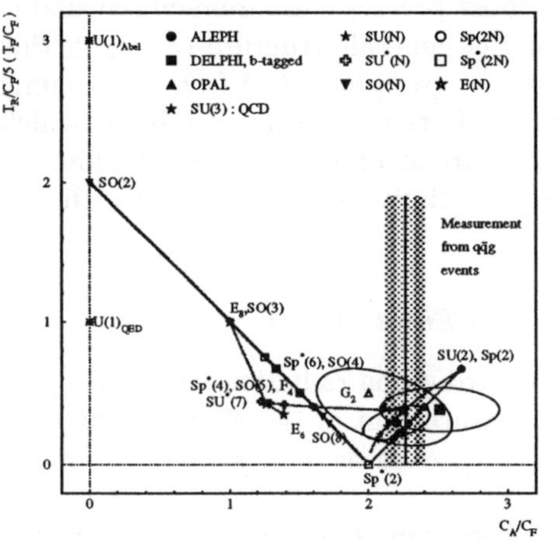

Figure 6. Measurements of C_A/C_F and T_R/C_F compared with expectations from numerous groups.

In principle five observables are needed to fully specify a four parton final state. It turns out, however, that again because of the helicity structure it is possible to perform the analysis as function of two observables only. Possible choices are the generalised Nachtmann-Reiter angle, θ^*_{NR}, and the angle α_{34} between the directions of the primary and secondary particles[6] (see FIG. 4). Most analyses use the jet energy to distinguish jets originating from the primary and secondary produced partons. Secondary partons are likely to be the least energetic ones.

Because of helicity conservation, in a rest system where the gluon is perpendicular to the radiating quark, the gluon will be longitudinally polarised. Angular momentum conservation then causes gluons originating from a splitting of this gluon (in a TGV) to be aligned predominantly along the initial quark direction. E.g. gluon jets from a TGV will predominantly have small θ^*_{NR}. In contrary a secondary q$\bar{\text{q}}$ pair, as both quark spins can be added up to give the gluon spin, will be emitted along the radiated gluon direction which is perpendicular to the initial quark. Thus q$\bar{\text{q}}$ pair production leads mainly to large θ^*_{NR}. Finally double gluon bremsstrahlung will happen dominantly along the direction of the quarks, leading mainly to large α_{34}. This behaviour can be read off from the reference distributions[7] shown in FIG. 5 predicted by a second order matrix element Monte Carlo model[8].

It should be noted that these reference cross sections strongly differ in size. R_{T_R} is about one order of magnitude smaller that R_{BR}. Especially for the determination of this term, which is sensitive to the production of so-called gluinos, the hypothetical super-partners of gluons in SUSY models, it is advantageous to try to really identify the origin of the final state jets. A possibility, which has been applied[7], is the identification of secondary vertices of heavy hadron decays in b quark jets. Because of their large mass b quarks are unlikely to be produced in a gluon splitting.

Only the colour factor ratios C_A/C_F and T_R/C_F can be determined in fits of the observed $(\theta_{NR}, \alpha_{34})$-distribution (like FIG. 5) as the total four-jet rate is $\propto \alpha_s C_F$. The results of the most precise measurements available [9,7,10] are shown in FIG. 6. Because of the general structure of EQ. 5 this measurement is able to pin down the gauge group of QCD from a dynamic measurement. It is observed, that possible alternative groups to SU(3) which also have three colours like $U(1)_3$ or SO(3) are incompatible with the four jet measurements. Some experiments [9] also give limits on light gluinos with this type of measurement.

2.3 Determining the Strong Coupling from Event Shapes

Intuitively, it is clear that the amount of hard gluon radiation will be proportional to the strong coupling. Thus a measurement of the share of three-jet events, $R_3(y_{cut})$, - in a well defined way - serves as a measurement of the strong coupling, α_s.

R_3, like any other (event shape) distribution, $R(y)$, may depend on the energy scale, Q, of the process. If the masses involved in the process are small with respect to the scale one may expect R to scale, i.e. to be constant. This is not so, however, in QCD as in other renormalisable field theories. Ultraviolet (UV) divergences appear in loop graphs. These need to be regularised by the introduction of some cut-off at the energy scale μ. The final result obtained from the theory should be μ-independent. In finite order perturbation theory, however, the result turns out to depend on the ratio Q^2/μ^2. Especially also the strong coupling will depend on the renormalisation procedure and evolves with the energy scale. α_s is said to run. It can be shown, that the scale dependence of any observable, R, can be expressed by the running of α_s which itself can be calculated from the so called renormalisation group equation (RGE). In second order this yields:

$$\alpha_s(Q^2) = \frac{1}{bL}\left\{1 - \frac{b'}{b}\frac{\ln L}{L}\right\}, \qquad L = \log\frac{Q^2}{\Lambda^2}, \tag{6}$$
$$b' = \frac{153 - 19n_f}{2\pi(33 - 2n_f)}, \qquad b = \frac{33 - 2n_f}{12\pi}.$$

Λ is a strong interaction scale parameter. Usually $\Lambda^{(5)}_{\overline{MS}}$ corresponding to the modified Minimal Subtraction renormalisation scheme and $n_f = 5$ is quoted. However, as this is a rather unphysical parameter, which takes scale values outside the validity of perturbation theory, since the start of LEP one chooses to quote the strong coupling at the Z mass scale, $\alpha_s(M_Z^2)$.

Complete matrix element predictions for event shape observables so far exist only in second order, $\mathcal{O}({\alpha_s}^2)$. These predictions (corresponding to a measured distribution $R(y)$), take the general form:

$$f(y) = \frac{1}{\sigma_0}\frac{d\sigma}{dy} = A(y) \cdot \bar{\alpha}(\mu^2) + \left[B(y) + A(y) 2\pi b \log\frac{Q^2}{\mu^2}\right] \bar{\alpha}^2(\mu^2) \quad (7)$$

With $\bar{\alpha} = \alpha_s/2\pi$. σ_0 is the total hadronic cross-section. $A(y)$ and $B(y)$ denote calculable coefficient functions. The physical scale Q only enters in the coefficient of the $\bar{\alpha}^2$ term. Therefore a measurement of α_s at a well defined scale becomes first possible in second order. The prediction for $f(y)$ depends on the unphysical cut-off scale μ via the renormalisation scale factor $x_\mu = Q^2/\mu^2$. This introduces the so called renormalisation scale uncertainty in the α_s measurement.

Although formally μ is a completely arbitrary parameter several so-called optimal scale choices have been proposed [11]. These were strongly debated and most experimentalists prefered to vary x_μ in a wide range. The central result mainly is given for $\mu^2 = Q^2$ ($x_\mu = 1$). This choice is motivated by the stability of the perturbative expansion. Growing powers of the $\log Q^2/\mu^2$ term in EQ. 7 are known to appear in higher orders. Thus for $x_\mu >> 1$ one may expect the perturbative expansion not to converge well.

An optimised value of μ has also been determined experimentally, by fitting α_s and μ simultaneously to the data. From an experimental point of view this definitely is an allowed (and successful) procedure. Theorists often feel uneasy about it, as μ is unphysical.

In $\mathcal{O}({\alpha_s}^2)$ calculations exist up to four final state particles (compare FIG. 3). Experimentally many final state hadrons are observed. The transition from partons to hadrons introduces further contributions to event shape distributions. They cannot be calculated perturbatively and are usually determined using so-called fragmentation models (see FIG. 7) which simulate the complete e^+e^- annihilation process into hadrons using Monte Carlo methods.

The initial $e^+e^- \rightarrow q\bar{q}$ (g) process is well known from electroweak theory, then in a perturbative QCD phase partons are generated according to a coherent parton shower process (see SEC. 3.2). Older models here alternatively employ a second order matrix element. At some cut-off scale $Q_0 \sim 1$GeV the parton shower terminates. Hadrons then are created using non-perturbative cluster[12] or string[13] models.

In the cluster model gluons from the perturbative phase first are split into $q\bar{q}$ pairs. Then colour neutral clusters are formed from nearby quarks and anti-quarks. This is easily possible as the parton splitting processes conserve colour (see for illustration FIG. 7). Colour and anti-colour therefore will be found in nearby regions of phase space. This feature is called preconfinement. The clusters then decay into (possibly instable) hadrons.

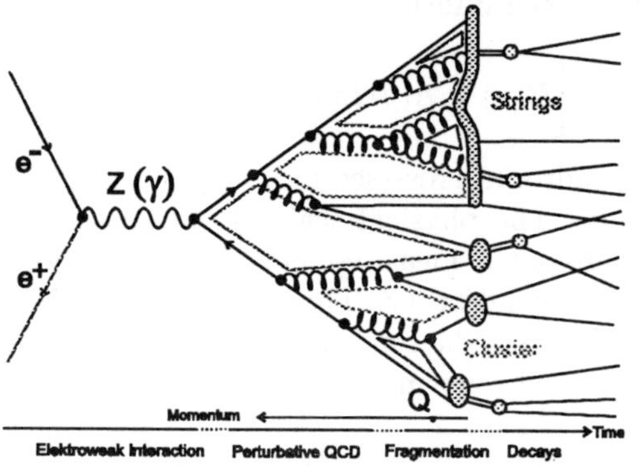

Figure 7. Cartoon of $e^+e^- \rightarrow$ hadrons.

The alternative string model stretches a colour string from the initial quark to the anti-quark via the (preconfined) chain of colour-anti-colour charges formed by the gluons emitted in the parton shower. The string evolves in space-time and breaks up into pieces. These pieces are then identified with hadrons. Finally after the creation of hadrons in both types of models instable hadrons decay and observable particles are formed.

In these models, event shape distributions at the partonic level and after stable hadron formation can be compared. The ratio of the partonic to hadronic distribution is taken as the effect of hadronisation:

$$C_{hadr} = \frac{f_{hadron}(y)}{f_{parton}(y)} \tag{8}$$

The product of this hadronisation correction and the perturbative prediction EQ. 7 then should directly represent the data distribution.

Clearly C_{hadr} depends to some extend on the model and the model parameters chosen. It turns out, however, that the discrepancies for the predicted C_{hadr} are small, provided the model represents the data distributions well. The most successful models describe the important distributions with uncertainties of a few percent only[14,15]. A remaining criticism still is that the amount of partons produced in the models (typically 6-10 at Z energies) is much higher than in the $\mathcal{O}(\alpha_s{}^2)$ calculation where it is at most four.

Experimentally also the influence of the limited acceptance and resolution of the detector needs to be corrected. Similar to the hadronisation correction this is done by comparing the distributions as calculated from hadrons generated by the fragmentation models to those found after tracking these hadrons through a detailed simulation of the detector. Uncertainties of this correction depend mainly on the quality of the detector simulation program and are in general small for the LEP experiments for most of the phase space.

As examples of event shape observables the thrust and a typical jet-algorithm have already been discussed. There are, however, many more event shape observables on the market. A study which is discussed below employs in total 18 observables (for detailed definitions see[16]). Instead of explaining these observables in detail, it is argued here that they can be ordered in four (or maybe three) classes of strongly correlated observables. The first class, (J), enfolds the jet-algorithms. Lets for simplicity assume a rather three-jet like event. Clearly here the jet resolution y at which the event "splits" from a two-jet into a three-jet event will be determined from the wide hemisphere of the event. So the three-jet rate, or equivalently the change of the two-jet rate, $D_2(y) = (R_2(y) - R_2(y - \Delta y))/\Delta y$, mainly depends on the wide side of the event. There is a second class, (W), of event shape observables which behave similar to the jet rates. The observable ρ_h is the invariant mass (squared) of the hadrons in the wide hemisphere of an event normalised to the centre-of-mass energy. Correspondingly the wide jet broadening B_{max} is the normalised sum of the transverse momenta with respect to the thrust axis. Also the jet cone energy fraction, $JCEF$, is calculated from the wide hemisphere.

A third class, (S), of event shape observables depends on particles of the whole event. The most prominent example here is the thrust, T. Other examples are the sum of the high and the low mass, $\rho_s = \rho_h + \rho_l$, the equivalent construction for the jet broadening, $B_{sum} = B_{max} + B_{min}$, the so-called C-parameter and the energy-energy-correlation, EEC. Some of these observables are highly correlated. Approximately the following expression holds: $\rho_s \approx (1 - T) \approx C/6$.

Finally there are observables (class D) which are calculated from properties of the wide event hemisphere minus the narrow hemisphere. These are the difference of the high and the low mass, $\rho_d = \rho_h - \rho_l$ and the asymmetry of the EEC, $AEEC$. Similarly this also applies to the oblateness, $O = T_{Major} - T_{Minor}$.

For a comparison of these different classes the emission of only one hard gluon is assumed (i.e. $\mathcal{O}(\alpha_s)$ is considered). Then all "perturbative" transverse momentum in the event is due to this gluon which is on the wide side of the event. It will contribute to all four classes. The non-perturbative contribution, however, will be different. To a rough approximation it is twice as important in observables summing both hemispheres compared to those measured from the wide hemisphere only. In fact this is not fully so as there are more particles on the wide side. Following this argument the hadronisation

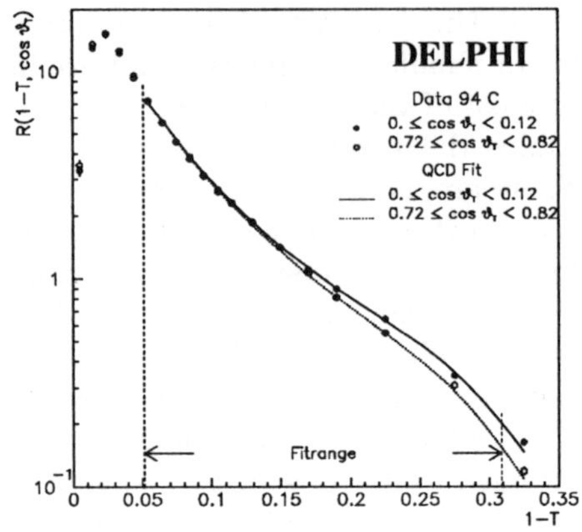

Figure 8. Thrust distribution for different $\cos\theta$. The lines represent a QCD fit.

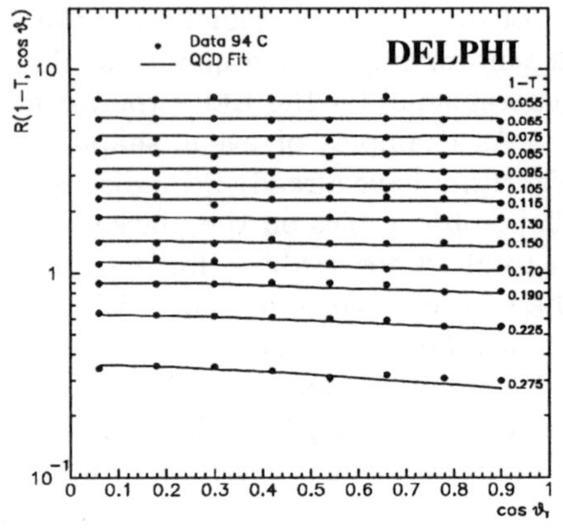

Figure 9. Thrust distribution as function of $\cos\theta$. The lines represent a QCD fit.

effect should almost vanish for the "difference" observables. In fact this is the reason why these observables have initially been introduced.

This classification is definitely oversimplified. Still it serves to order the multitude of observables and will allow to obtain some experimental control on the precision of the hadronisation correction using fragmentation models. It may be notable, that the above arguments for the influence of the hadronisation apply as well for the influence of higher order terms if interference effects are neglected.

DELPHI recently presented a detailed analysis of 18 event shape distributions[16]. The dependence of the distributions on the cosine of polar angle, $\cos\theta_T$, of the thrust axis of the events has been measured simultaneously. This measurement is explicitly sensitive to the varying probability of gluon emission on the polar angle and also allows improved experimental control of the data. As an example FIG. 8 shows the $1-T$ distribution for events almost perpendicular to (small $\cos\theta_T$) and close to the beam (large $\cos\theta_T$). In the first case the distribution fall off less steeply corresponding to more gluon emission. FIG. 9 directly shows the $\cos\theta_T$ dependence of the $1-T$ distribution.

A $\cos\theta$-dependent extension of EQ. 7 has been fitted to the data. Only a poor description of the data could be obtained for most of the observables if the traditional scale choice $\mu^2 = Q^2 = M_Z^2$ was made. For some observables the ratio of the $\mathcal{O}(\bar{\alpha}^2)$ to the $\mathcal{O}(\bar{\alpha})$ terms is of the order one, indicating a

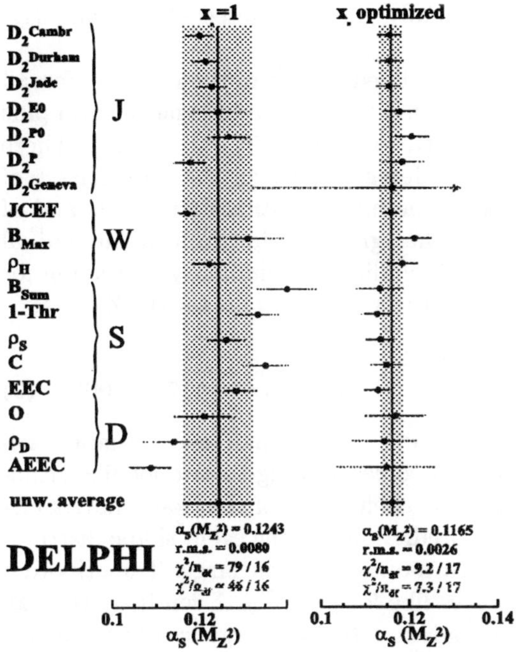

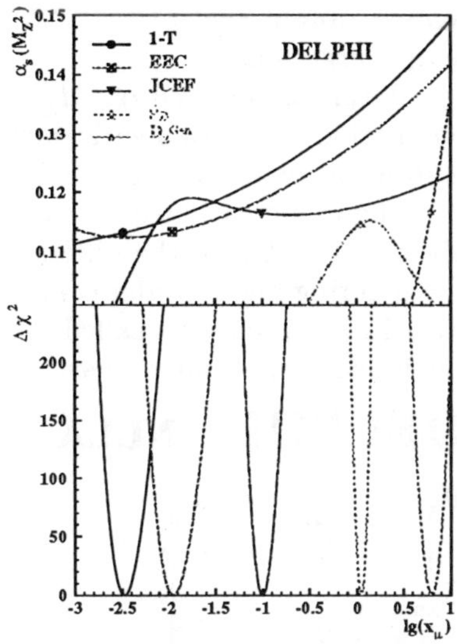

Figure 10. Upper plot: α_s dependence on x_μ for some observables. Lower plot: χ^2 dependence on x_μ.

Figure 11. Upper left: α_s values as measured from 18 event shape distributions. Left: $x_\mu = 1$, Right: x_μ optimised to the data. The full lines indicate the experimental plus hadronisation error. Dotted lines include an additional renormalisation scale error.

bad convergence of the perturbative series. This may naturally explain the large spread of α_s values obtained (left side of FIG. 11). The χ^2/N_{df} for the α_s values is 40/17 clearly demanding an additional (theoretical or scale) error to restore consistency among the individual results. This observation has similarly been made before in other α_s determinations from event shapes.

The situation drastically improves if also the renormalisation scale factor x_μ is adjusted to the data. Satisfactory fits ($\chi^2/N_{df} \sim 1$) are obtained for all observables in this case (see FIG. 8 and FIG. 9). Also the ratio of the $\mathcal{O}(\bar{\alpha}^2)$ to the $\mathcal{O}(\bar{\alpha})$ terms indicates a better convergence.

FIG. 10 shows the dependence of $\alpha_s(M_Z^2)$ and the change of χ^2/N_{df} as function of the scale factor x_μ. The fitted x_μ values for the different observables are well determined by the fit. They vary over a wide range of scales $-2.5 \lesssim \log_{10} x_\mu \lesssim 1$. It turns out that the summed observables (S) tend to prefer small x_μ, the wide side observables (W) and the prominent jet-rates mainly obtain central scale values. Finally the subtracted observables (D) adopt large scales.

Astonishingly the $\alpha_s(M_Z^2)$ values obtained at the optimum scale for all observables agree well with each other within the estimated experimental plus hadronisation errors (see right side of FIG. 11, here $\chi^2/N_{df} = 9.2/17$). There is a slight indication for a remaining difference between the four classes of observables. Given the different size of the hadronisation corrections and of the differing higher order contributions the agreement between the individual α_s values can only be called excellent. Including a preliminary correction for the influence of b-quark events yields for the strong coupling at the Z scale:

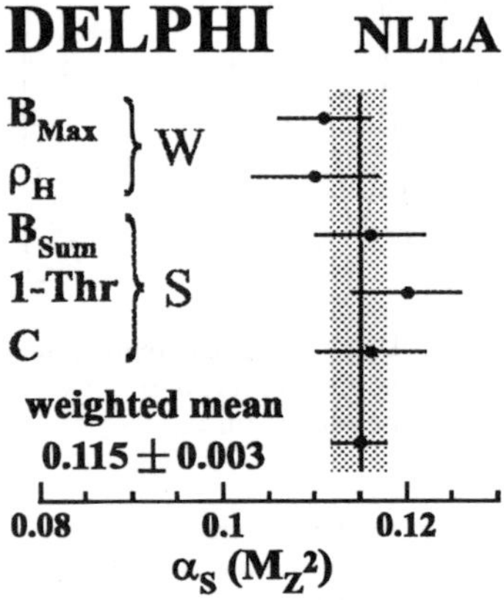

Figure 12. α_s results from an NLLA analysis. Experimental plus hadronisation errors are shown.

$$\mathcal{O}(\alpha_s^2): \qquad \alpha_s(M_Z^2) = 0.117 \pm 0.003. \tag{9}$$

In the limit of small event shape variables, y, corresponding to two-jet like situations, the coefficients of the exact matrix element predictions for the event shape distributions turn out to become large. The nth order behaves like $\log y^{2n-1}/y$. Even if α_s is small this leads to a bad convergence of the perturbative series. Therefore techniques have been developed to resum the leading and next to leading logarithmic terms to all orders. These NLLA methods provide reliable predictions in the small y limit. However, they are feasible only for some, so-called exponentiating observables.

The measurement of α_s from event shapes at small y is largely independent from the measurement described above as the data stem from a different range in y. The NLLA calculations match better to the calculations of the hadronisation corrections with parton shower models as multiple gluon emission is treated by the all orders resummation. The resulting α_s values from the NLLA analysis given in [16] are shown in FIG. 12. The y region of the fit has been chosen such that sub-leading terms are much smaller than "exponentiated" terms. The hadronisation uncertainty for these two-jet like situations is somewhat increased. Again, the individual results agree well with each other and, especially, the average strong coupling from this analysis:

$$NLLA: \qquad \alpha_s(M_Z^2) = 0.115 \pm 0.003, \tag{10}$$

is in good agreement with the $\mathcal{O}({\alpha_s}^2)$ result given in EQ. 9.

Conventionally the resummed and the matrix element calculation have been matched in order to obtain a prediction which is valid over a wide range of y. However, this introduces additional "matching" ambiguities[17]. Moreover the matched theory only gives a poor description of the data[16], as in this case no scale optimisation of the $\mathcal{O}({\alpha_s}^2)$ calculation is performed. This causes the α_s values determined with this method to be higher than the individual $\mathcal{O}({\alpha_s}^2)$ or NLLA results, which clearly should not be the case. Therefore α_s measurements obtained using matched predictions are not considered here.

2.4 Power Terms – Trying to Understand Hadronisation Corrections

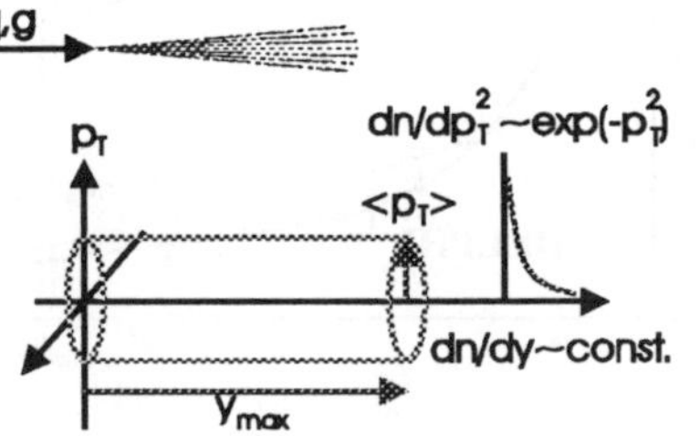

Figure 13. The tube model.

In the recent years it has been tried to understand the hadronisation effects better by studying so-called renormalon terms appearing in perturbation theory[18]. The discussion of renormalons is beyond the scope of this lecture. It turned out, however, that the renormalon predictions are similar to those of the longitudinal phase space model, which today is called tube model[19]. The tube model assumes that the transverse momentum of hadrons due to fragmentation is energy independent. The p_t^2 distribution may be Gaussian, the average p_t is constant. Then the maximum rapidity y_{max} of a particle grows logarithmically with the jet energy:

$$y = \frac{1}{2}\log\frac{E+p_{\|}}{E-p_{\|}} \quad \Rightarrow \quad y_{max} \simeq \log\frac{E}{\mu} \quad ,\text{with } \mu = \sqrt{p_t^2+m^2} \ . \tag{11}$$

Thus the phase space for particle production is a *tube* with radius $\langle p_t \rangle$ and length y_{max}(see FIG. 13). Then it is easy to derive predictions for the non-perturbative contributions to event shape means. For example the jet energy and the jet longitudinal momentum can be expressed as function of y_{max} by integrating over the tube. It follows that:

$$M_{jet}^2 = E^2 - p_{\|}^2 \simeq \lambda \cdot E_{jet} \Rightarrow \quad \langle \rho \rangle_{had} \simeq \frac{\lambda}{E} \quad \text{with } \lambda = \langle \mu \rangle \sim 500\text{MeV} \tag{12}$$

Thus the hadronisation correction vanishes like an inverse power of the centre-of-mass energy. Renormalon calculations for some observables predict higher powers of $1/E$. As EQ. 12 applies for each hemisphere independently it is

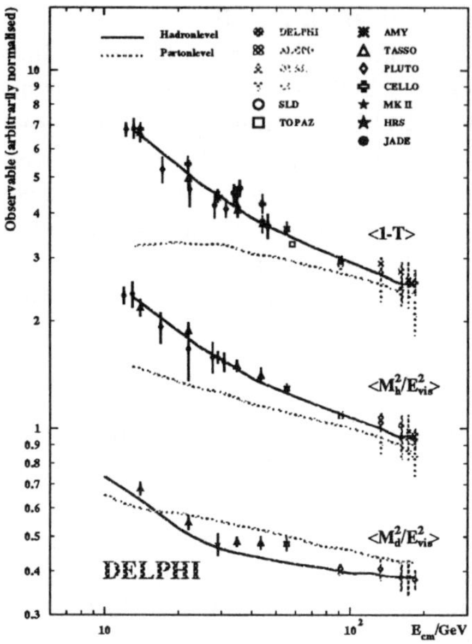

Figure 14. Energy dependence of $\langle 1-T \rangle$, ρ_h and ρ_d compared to JETSET.

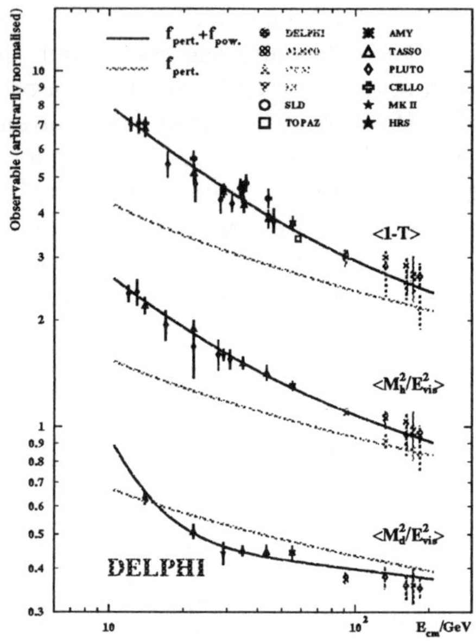

Figure 15. Energy dependence of $\langle 1-T \rangle$, ρ_h and ρ_d compared to fits of the ansatz EQ. 13.

clear that the hadronisation correction for ρ_s (or $T \simeq \rho_s = \rho_h + \rho_l$) is about twice as big as for ρ_h. Similar arguments apply for the jet broadenings.

FIG. 14 compares the energy dependence of $\langle 1 - T \rangle \sim \langle \rho_h \rangle + \langle \rho_l \rangle$, $\langle \rho_h \rangle$ and $\langle \rho_d \rangle = \langle \rho_h \rangle - \langle \rho_l \rangle$ with the expectation of a fragmentation model (right side) calculated on parton as well as on hadron level. FIG. 14 similarly compares this data to fits of the following ansatz adding perturbative and non-perturbative contributions incoherently:

$$\langle R \rangle = \langle f_{\mathcal{O}(\alpha_s{}^2)} \rangle + \frac{C_1}{E_{CM}} + \frac{C_2}{E_{CM}^2} \quad , \tag{13}$$

where $\langle f_{\mathcal{O}(\alpha_s{}^2)} \rangle$ denotes the $\mathcal{O}(\alpha_s{}^2)$ prediction for the event shape mean (compare EQ. 7). It is remarkable, that this simple ansatz reproduces the description of the mean event shapes given by the fragmentation model rather well. The power terms for $1 - T$ are observed to be roughly twice as big as for ρ_h (see TAB. 1)[20]. Also the leading coefficient for B_{sum} is bigger than for B_w. It is important to note that the subtraction made in ρ_d over-compensates the hadronisation correction. Here the perturbative expectation is above the full result (see also FIG. 14).

Obs.	C_1/GeV	C_2/GeV
$\langle 1-T\rangle$	1.09±0.03	-3.2±0.9
$\langle \rho_h\rangle$	0.67±0.12	-1.9±0.7
$\langle \rho_d\rangle$	-0.30±0.05	5.9±1.0
$\langle B_{max}\rangle$	1.27±0.09	-10.3±4.4
$\langle B_{sum}\rangle$	0.33±0.06	0.2±2.3

Table 1. Power terms for some observables.

Motivated by the study of power terms a model has been developed which describes the energy dependence of event shape means[21] as well as of full event shape distributions[22,25]. It is based on the notion of an effective universal strong coupling α_{eff} which differs from the perturbative coupling EQ. 6 at small scale. For event shape means this leads to the following prediction:

$$\langle R\rangle = \langle f_{\mathcal{O}(\alpha_s{}^2)}\rangle + \langle f_{pow}\rangle \tag{14}$$

$$\langle f_{pow}\rangle = c_R \frac{4C_F}{\pi^2}\mathcal{M}\frac{\mu_I}{E_{CM}}\cdot\left\{\alpha_0(\mu_I) - \alpha_s(\mu) - b\left(\log\frac{\mu^2}{\mu_I^2} + \frac{K}{2\pi b} + 1\right)\alpha_s^2(\mu)\right\}$$

c_R is a parameter depending on the observable R, $\mathcal{M}$ is a universal, so-called *Milan* factor and $\alpha_0(\mu_I)$ an effective non-perturbative coupling below a IR matching scale μ_I. This ansatz, as well as the corresponding calculation for full distributions has been applied to data[23]. Fits to the means of $1-T$ and ρ_h, for 30GeV $\leq E_{CM} \leq$ 184GeV yield a description of the data similar to the fits displayed in FIG. 15 and the following result for the strong coupling[24]:

$$\text{power law fit :} \qquad \alpha_s(M_Z^2) = 0.119 \pm 0.002 \pm 0.004. \tag{15}$$

The scale dependence of this result is reduced with respect to the $\mathcal{O}(\alpha_s{}^2)$ result as the divergence of the fixed order expansion is canceled due to the merging with the non-perturbative counter-term.

The successful application of the ansatz EQ. 14 implies that α_s evolves according to the QCD expectation EQ. 6. To quantify this running it is advantageous to determine $d\alpha_s^{-1}/d\log Q^2$ which is only slightly energy dependent. The DELPHI result from the measurements up to 183GeV is[24]:

$$\frac{d\alpha_s^{-1}}{d\log Q^2} = 0.70 \pm 0.17(stat.) \pm 0.08(syst.) \tag{16}$$

in good agreement with the QCD expectation $d\alpha_s^{-1}/d\log Q^2 \approx 0.635 \sim b$.

2.5 Status of α_s and its Running

In order to terminate the section on α_s determinations a brief overview of the important α_s measurements at different energies is given. Beside the

e^+e^- event shape measurements, which have mainly been performed in the energy range $22\text{GeV} \leq Q \leq 189\text{GeV}$, precise results come from the τ hadronic decay width, R_τ ($Q = 1.78\text{GeV}$), scaling violations of structure functions $\langle Q \rangle \sim 10\text{GeV}$, and from the Z hadronic width, R_Z ($Q = M_Z$). The theoretical calculations for R_Z and R_τ are performed in next to next to leading order.

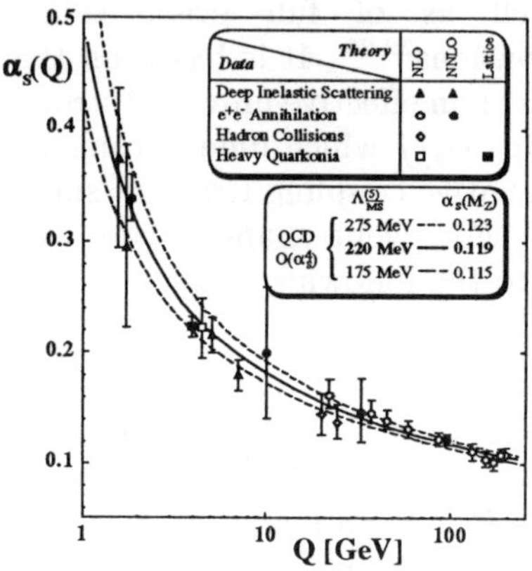

Figure 16. Running of α_s .

All measurements are consistent with the current (1998) world average[25,26]:

$$\alpha_s(M_Z^2) = 0.119 \pm 0.004.$$

This average is slightly higher than the measurements discussed mainly as the α_s results from matched $\mathcal{O}(\alpha_s^2)$ plus NLLA event shape measurements enter. Also the τ and Z results turn out to be slightly higher. The individual measurements are displayed in FIG. 16 as function of the energy scale. The energy evolution of the measurements is fully consistent with the QCD expectation and impressingly demonstrates asymptotic freedom.

2.6 The Running of the b-Quark Mass

Quark masses enter as free parameters in the QCD Lagrangian similarly to the coupling. In perturbation theory so-called mass singularities appear, which need to be regularised. As a consequence the renormalisation group equation needs to be extended to include also the quark masses. Similarly to the coupling this leads to a running (more specifically a decrease with increasing scale) of the quark masses. Using a running quark mass (again as in the case of a running coupling) to a large extend absorbs the influence of higher order effects compared to the case when the perturbative pole mass is used.

Recently NLO calculations of multi-jet production including mass effects[27] became available. These now allow to determine quark mass effects at a defined scale and so to assess the mass running. It turned out that the mass effect on the three jet rate is $\propto (m^2/Q^2)/y$, thus it is amplified by the factor $1/y$ with respect to the naive expectation $\propto m^2/Q^2$. This makes the running of the b quark mass a measurable effect at LEP ($(m_b^2/M_Z^2)/y \sim$ a few percent for small $y \sim 1/10$)[28].

The running of the b quark mass has first been verified by DELPHI[29] comparing the three-jet rate in b tagged events and light quark events. This comparison is advantageous as the influence of the strong coupling as well as of fragmentation effects cancel to a large extend. Residual fragmentation effects turned out to be small for the Durham algorithm. b events were identified by using large impact parameters of tracks due to the displaced decay vertices of long lived b hadrons.

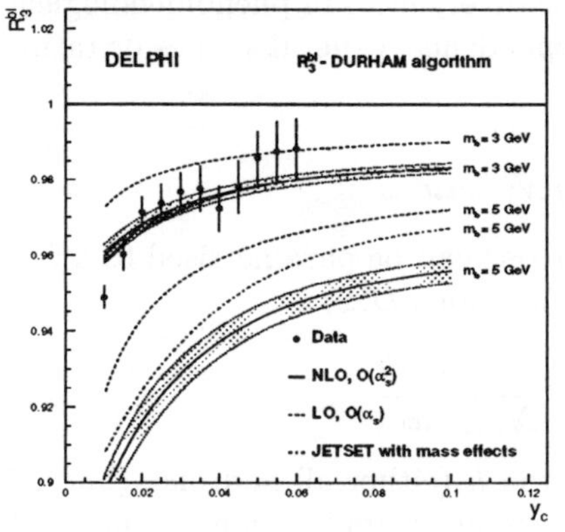

Figure 17. Ratio of the three-jet rate in b compared to light quark events.

The measured effect on the ratio of the three jet rates as function of the jet resolution y:

$$R_3^{bl}(y) = \frac{R_3^b(y)}{R_3^{light}(y)} \tag{17}$$

is compared to different LO and NLO calculations in FIG. 17. The expected increase at small y is observed. Note, however, that the individual data points are strongly correlated as the large y data are included in the low y points. From the measurement of $R_3^{bl}(y = 0.02)$ where the influence of higher jet rates is still negligible the b quark mass at the Z scale is measured to be:

$$m_b(M_Z) = 2.67 \pm 0.25(stat.) \pm 0.34(frag.) \pm 0.27(theo.)\text{GeV} \tag{18}$$

Besides the experimental errors and uncertainties due to the residual fragmentation correction theoretical errors are significant. These arise due to the renormalisation scale uncertainty and a so called mass-ambiguity[29]. When compared with the b mass determined from bound b states a mass running close to that expected from QCD is observed:

$$\Delta m_{\rm b} = m_{\rm b}(M_Z) - m_{\rm b}(M_\Upsilon/2) = -1.49 \pm 0.52\text{GeV} \tag{19}$$

This implies that at large energies, besides the inverse power suppression, b quark mass effects are further reduced due to the mass running. If the running of the b mass is taken for granted, the above measurement can also be interpreted as a test of the flavour independence of the strong coupling. From the measurement of $R_3^{bl}(y)$ it follows:

$$\alpha_s^b/\alpha_s^{light} = 1.007 \pm 0.005(exp.) \pm 0.008(theo.) \tag{20}$$

The equivalence of the strong coupling for other quark flavours has also been checked at LEP and SLC[30].

3 Energy Evolution of Soft Hadronic Quantities and Coherence

The second part of this talk covers measurements of quantities which are sensitive to the radiation of soft and collinear gluons. These quantities cannot be directly predicted by perturbative QCD as IR and collinear divergences occur. It is, however, often possible to absorb these divergences in phenomenological constants or functions. Only the energy dependence, general shapes or ratios are predicted by theory.

3.1 Fragmentation Functions and Scaling Violations

Hadron production from high energetic partons can be parametrised to a fair approximation by so-called fragmentation functions $D(x)$:

$$D_i^h(x) = \frac{1}{\sigma_{tot}} \frac{d\sigma^h}{dx} = \frac{1}{N_{Evt}} \frac{dN^h}{dx} \tag{21}$$

These are the probabilities to create a hadron h (either all hadrons, charged hadrons or of specified hadron type) with the momentum fraction $x = p_h/p_i$ of the parton momentum p_i. Theoretical papers prefer to use the energy fraction which can be calculated from the momentum provided the particle mass is known. As the hadron multiplicity in general is bigger than one $D(x)$ is not a properly normalised probability density but is normalised to the average number of hadrons of a given type:

$$\int_0^1 dx D_i^h(x) = \langle N_h \rangle \tag{22}$$

In the parton model [31] $D(x)$ is expected to scale, thus to depend on the momentum fraction x only. It will depend, however, on the type of parton "fragmenting" and also on the type of hadron. For light quarks and hadrons $D(x)$ falls about exponentially with x (see FIG. 18[3]). As heavier the hadron is as weaker will be the fall off. Heavy hadrons containing b or c quarks tend to pick up most of the momentum of the heavy quark. Still, as the heavy hadrons decay into many particles the observed stable hadron spectrum is softer (e.g. falls off more rapidly) than in light quark case. Gluon fragmentation, finally gives an even softer distribution (see FIG. 26). This is as gluons, contrary to quarks, are not valence particles of hadrons. These valence quarks first need to be produced in additional gluon splittings. The momentum of the gluon is shared by the quarks and hence $D_g^h(x)$ will be softer. Reversely the harder quark fragmentation function may be explained by the leading particle effect, namely the presence of the fragmenting quark in the produced hadron.

The overall charged hadron spectrum in e^+e^- annihilation is a superposition of the fragmentation functions of the individual quarks:

$$D^h(x|\sqrt{s}) = \sum_i \epsilon_i(\sqrt{s}) D_i^h(x) \tag{23}$$

ϵ_i is the probability to create a quark of type i in e^+e^- annihilation. Due to the varying importance of γ and Z exchange and their interference the ϵ's are strongly depending on the centre-of-mass energy $\sqrt{s}$.

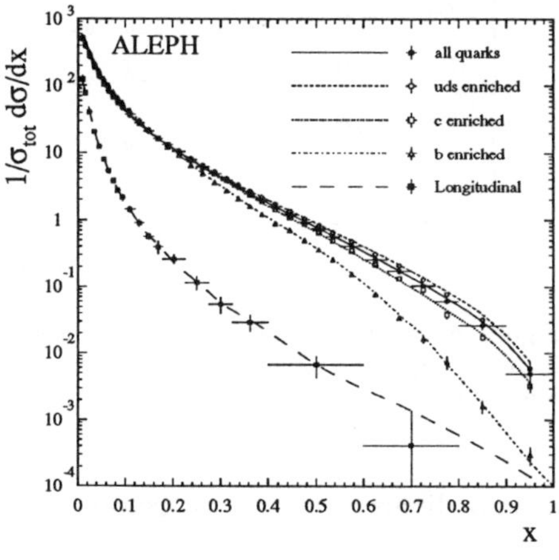

Figure 18. Several fragmentation functions.

The initial partons as they carry colour charge will emit soft gluon bremsstrahlung. The amount of radiation is proportional to the coupling and thus the "colour charge" of the parton. Furthermore this radiation will grow strongly with energy due to the increase of phase space. The emitted gluons on the one hand reduce the energy of the remaining parton and, on the other hand themselves give rise to the production of soft, low energetic (low x) hadrons. The fragmentation function thus with increasing energy falls off at large x and rises at small x. Bremsstrahlung leads to a twist of the fragmentation function about a "central" value of x (see FIG. 19 and FIG. 20). This process is called scaling violation and similarly occurs for structure functions in deep-inelastic scattering. Scaling violations are described by the so-called DGLAP integro-differential equations[32]:

$$\frac{d}{d\log s} D^h(x,s) = \sum_j \int_x^1 \frac{dy}{y}\frac{\alpha_s}{2\pi} P_{ji}(y) D(\frac{x}{y}, s) \tag{24}$$

Here the so called splitting functions $P_{ji}(y)$ describe the splitting processes of the partons:

$$P_{gg}(y) = 2C_A \frac{(1-y(1-y))^2}{y(1-y)} \qquad P_{gq}(y) = C_F \frac{1+y^2}{1-y}$$
$$P_{q\bar{q}}(y) = n_f T_F (y^2 + (1-y)^2) \tag{25}$$

The two bremsstrahlung kernels P_{gq} and P_{gg} show the typical infra-red ($\propto 1/y$) divergence. It appears twice for the gg case as the gluons are in-

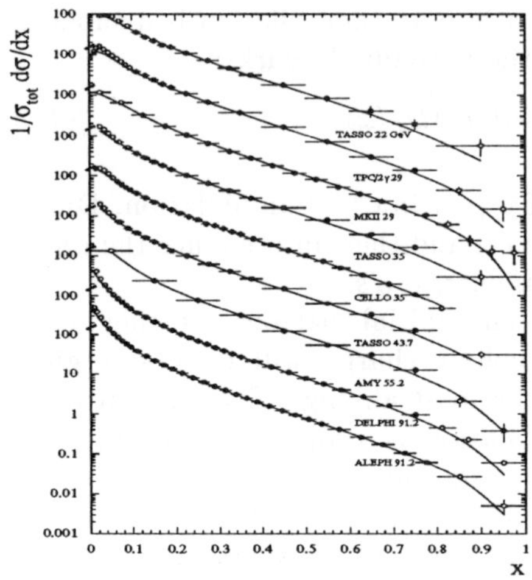

Figure 19. Energy dependence of the charged hadron fragmentation function in e^+e^- annihilation.

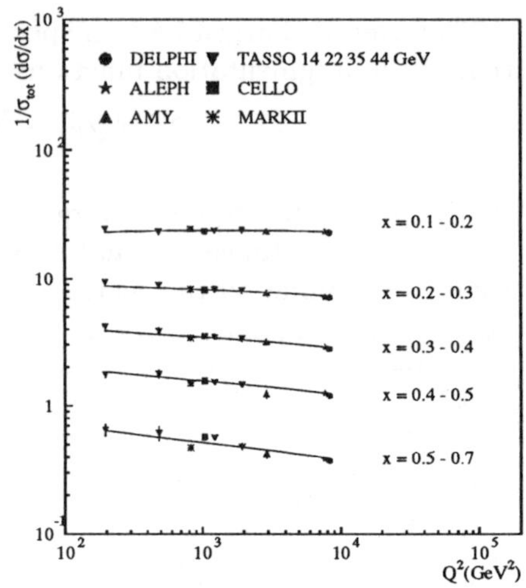

Figure 20. Scaling violations of the charged hadron fragmentation function in e^+e^- annihilation.

distinguishable. The $g \to q\bar{q}$ kernel shows no divergence and therefore is of reduced importance.

As is evident from EQ. 24 the logarithmic slope of $D(x)$ directly measures α_s . Experimentally, this requires, however, to precisely control the varying contributions of the different flavours j with energy $\sqrt{s}$. The fragmentation functions for the different quarks have been measured (see FIG. 18). Theoretically, beside the usual renormalisation scale uncertainty a further so-called factorisation scale uncertainty enters. It accounts for the barely understood interplay of the perturbative and non-perturbative domains. Measurements have been presented by DELPHI[33] and ALEPH[34] and yield the following result for the strong coupling:

$$\alpha_s(M_Z) = 0.124 \pm 0.005(exp.) \pm 0.010(theo.) \qquad (26)$$

Fragmentation functions can also be decomposed according to the contributions originating from exchange bosons (γ or Z) polarised transversely and longitudinally with respect to the direction of the produced hadron. Transverse polarisation leads to the well known $(1+\cos^2\theta)$ polar angle distribution, longitudinal polarisation to a $\sin^2\theta$ dependence:

$$\frac{1}{\sigma}\frac{d^2\sigma}{dx d\cos\theta} = \frac{3}{8}(1+\cos^2\theta)F_T(x) + \frac{3}{4}\sin^2\theta F_L(x) \qquad (27)$$

The fragmentation functions F_T and F_L can be written as product of coefficient functions times parton fragmentation functions D_i. The leading coefficient for F_L is $C_F\alpha_s/2\pi$. It turns out to mainly measures the gluon fragmentation function at small x:

$$F_L = C_F\frac{\alpha_s}{2\pi}\int_x^1 \frac{dz}{z}\left\{F_T(z) + 4(\frac{z}{x} - 1)D_g(z)\right\}. \tag{28}$$

Thus a measurement of the polar angle distribution in e^+e^- annihilation gives access to the gluon fragmentation function without identification of a gluon. It is, however, to be noted that F_L represents an integral over a gluon spectrum. A longitudinal fragmentation function is included in FIG. 18[3].

Integrating over x and $cos\theta$ shows that the total longitudinal cross section $\sigma_L \propto \alpha_s$. Thus this cross section as well as the total cross section $\sigma_0 = \sigma_T + \sigma_L$ can be used to measure α_s . In fact, as special techniques can be applied, the total cross section is calculated to $\mathcal{O}({\alpha_s}^3)$ and leads to the theoretically best defined α_s determination. This important topic has been, however, left aside in this lecture.

3.2 Angular Ordering and Multiplicities

QCD is a quantum theory, thus gluons will be emitted coherently. In order to depict an important consequence of this coherent emission, which is common to all gauge theories, the so-called Chudakov effect is considered. It concerns the radiation of a soft photon from an e^+e^- pair originating in a pair creation (see FIG. 21). Let $\Theta_{e^+e^-}$ be the opening angle of the pair and θ the angle of the emitted γ. E_i and $\vec{p}_i$ or $\vec{k}$ are the energies and 3-momenta of the electrons and of the photon, respectively. All angles are assumed to be small and all energies large compared to the electron mass, m. The ability of the emitted photon to resolve the e^+e^- pair is proportional to its transverse wavelength:

$$\lambda_\perp = \frac{\lambda}{\theta} = \frac{1}{k\theta} = \frac{1}{k_\perp} \tag{29}$$

The energy imbalance ΔE before the emission of the photon can be used to determine the approximate time before the photon is emitted:

$$\Delta E = \sqrt{{\vec{p}_1}^2 + m^2} + |k| - \sqrt{(\vec{p}_1 + \vec{k})^2 + m^2} \approx k\theta^2 = k_\perp\theta = \frac{\theta}{\lambda_\perp}. \tag{30}$$

The approximate time, Δt, until the photon is emitted and the separation of the e^+e^- pair during this time, Δd, is:

$$\Delta t = \frac{1}{\Delta E} = \frac{\lambda_\perp}{\theta} \qquad\qquad \Delta d = \Theta_{e^+e^-}\Delta t = \lambda_\perp\frac{\Theta_{e^+e^-}}{\theta} \tag{31}$$

Now it is clear that a photon can only be emitted if the photon can resolve the electron from the positron, thus is $\lambda_\perp < \Delta d$. Otherwise it would be as if a neutral object (the e^+e^- pair) would emit a γ, which clearly is impossible. The above condition can only be fulfilled if $\theta < \Theta_{e^+e^-}$, thus no radiation is possible outside the cone defined by the e^+e^- pair.

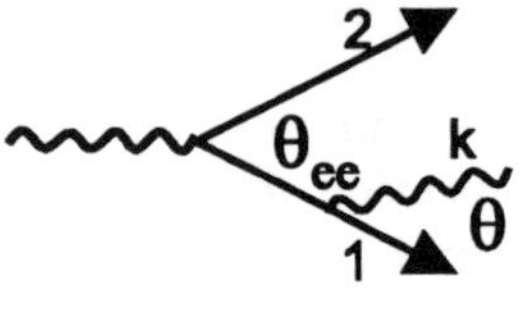

Figure 21. γ radiation off an e^+e^- pair.

The same result also holds in QCD. Here it is complicated by different colour factors of the radiating partons ($q\bar{q}$, qg or gg). In QCD many successive radiations occur which all respect the coherence condition. This coherent "shower" of emitted partons can be efficiently described by successive emission which occur at a reduced angle in each step. This is called "angular ordering".

One important consequence of angular ordering concerns the phase space available for soft gluon radiation. It is proportional to the available energy times the maximum allowed opening angle. Thus the relevant scale in soft radiation processes is not (only) the energy but the product $E\theta$. For larger opening angles it turns out that $\kappa = 2E\sin\theta/2$ is a better choice. κ^2 is similar to y_{cut} and one of the reasons why these measures are used in the jet algorithms.

If the effect of coherence is taken into account by introducing the appropriate scale κ^2, the DGLAP equation (EQ. 24) can be used to also predict the energy evolution of the multiplicity, which is the integral of the fragmentation function (EQ. 22). A mathematically simple solution is given in [35]:

$$\langle N_{charged}(s)\rangle = C\exp\left\{\frac{1}{b}\sqrt{\frac{6}{\pi\alpha_s(s)}} + (\frac{1}{4} + \frac{5n_f}{54b\pi})\log\alpha_s(s)\right\} \tag{32}$$

An equivalent result using special functions is given in[36]. They express the energy dependence of the multiplicity by an α_s dependence and a non-perturbative scale factor C to absorb the effect of hadronisation. Both predictions describe the experimental data similarly well (see FIG. 22). This is an important observation as it shows that coherence effects, which necessarily occur in gauge theories, survive the hadronisation and can indeed be observed in the hadronic final state.

The value of $\alpha_s(M_Z) \sim 0.12$ obtained when fitting EQ. 32 to the data is quite reasonable. This could, however, be accidental as the underlying calculation does not fix the renormalisation scale. Also α_s is changed if non-perturbative contributions to the multiplicity are considered. However, these non-perturbative contributions are expected due to the leading particle effect.

The logarithmic slope of the fragmentation function and as well the increase of the multiplicity with energy is proportional to the product of the coupling times the colour factor (see Eq. 24). Therefore it is expected that the multiplicity in gluon jets is increased compared to quark jets by the colour factor ratio C_A/C_F . Although this basic prediction dates back more than 20 years [37] it has only been verified recently[38].

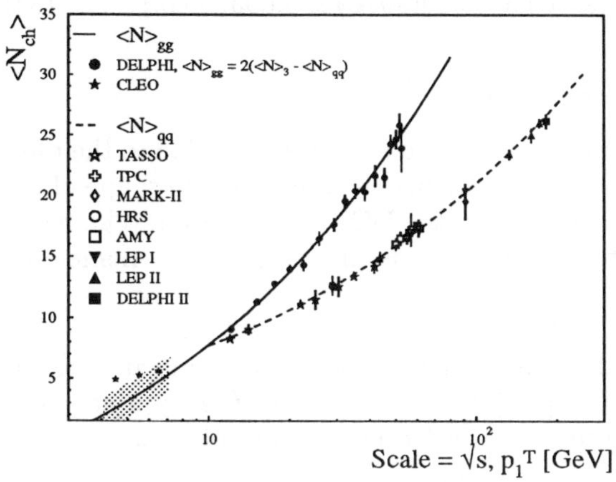

Figure 22. Multiplicity in $e^+e^- \to$ hadron events and in gg events as function of $\sqrt{s}$

The result is based on the measurement of the change of the gluon and quark multiplicities with scale, not on the measurement of the multiplicities itself. This is equivalent to the original prediction but avoids a direct dependence on non-perturbative effects influencing the observed multiplicity. Explicitly the multiplicity for symmetric three-jet ($q\bar{q}g$) events is measured as function of the opening angle. It turns out that the centre-of-mass energy of the $q\bar{q}$ pair, $2E^*$, the relevant scale for the $q\bar{q}$ system, is almost independent of the opening angle. The scale of the gluon jet $p_1^\perp$, however, which is proportional to the gluon transverse momentum increases proportional to the opening angle. Thus the observed multiplicity change is almost only due to the gluon. The original prediction for the three-jet multiplicity [39]:

$$N_{q\bar{q}g} = N_{e^+e^-}(2E^*) + r(p_1^\perp)\left\{\frac{1}{2}N_{e^+e^-}(p_1^\perp) - N_0\right\} \quad , \tag{33}$$

needs to be modified by an additional non-perturbative term, N_0. This term absorbs non-perturbative (leading particle) differences in the fragmentation of the quarks and the gluon. $r(E)$ is the ratio of the multiplicity in gluon to quark jets as predicted in[40].

The extracted gluon multiplicity (identified with the second term in Eq. 33) is compared to the $q\bar{q}$ multiplicity (taken from all hadronic e^+e^- events) in Fig. 22. It is clearly seen that the increase with scale is about twice as big for gluons compared to quarks. It should however be noted, that the absolute multiplicity of the gluon jet is not well determined, as the non-perturbative term, N_0, in EQ. 33 is a property of the full three-jet event and not of the gluon. The comparison of the three-jet multiplicity to the hadronic multiplicity in e^+e^- leads to the most precise measurement of the colour factor ratio C_A/C_F obtained so far[38] (see also FIG. 5):

$$\frac{C_A}{C_F} = 2.246 \pm 0.062(stat.) \pm 0.080(syst.) \pm 0.095(theo.) \tag{34}$$

3.3 Fragmentation at Small Momentum

As hadron production at small momentum is to a large extend due to the radiation of soft gluons, it is possible to also predict the approximate shape of the fragmentation function in this regime from the DGLAP evolution. The result usually is expressed in a variable which expands the range of low momenta:

$$\xi = \log 1/x = -\log x \tag{35}$$

Using again the scale κ defined above, which takes into account angular ordering and solving the DGLAP equation for small momenta by a so-called inverse Mellin-transform yields to lowest order a Gaussian solution:

$$x \cdot D(x) \propto \exp\left\{\frac{\xi - \xi^*}{2\sigma^2}\right\} \tag{36}$$

Here ξ^* is the peak position and σ the width of the Gaussian which are predicted to be:

$$\xi^* = \frac{1}{2}\log\frac{E_{CM}}{\Lambda_{eff}} \qquad\qquad \sigma \propto \log^{\frac{3}{2}}\frac{E_{CM}}{\Lambda_{eff}} \tag{37}$$

Λ_{eff} is an effective scale parameter. Also higher order corrections in terms of the skewness and the kurtosis of the Gaussian have been calculated[41]. An alternative calculation based on the so-called modified leading log approximation (MLLA) leads to numerically similar results [42].

Clearly the QCD calculations are made for partons but measurements are done with stable hadrons, thus a connecting ingredient is needed. This is the so-called Local Parton Hadron Duality (LPHD). It assumes that the spectrum predicted for partons directly applies to the observed hadrons, except for an overall normalisation factor. For a single number like the average multiplicity this is likely to be correct – for momentum dependent spectra, however, more caution is appropriate. Experimentally it is known, that more than half of the initially produced hadrons are instable and decay[14]. The decay products share the momentum and the energy of the primary particle. Decays thus shift

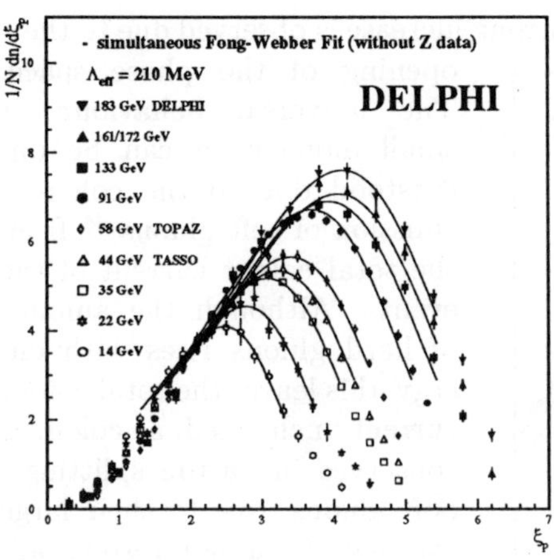

Figure 23. ξ dependence of the fragmentation functions for different $\sqrt{s}$.

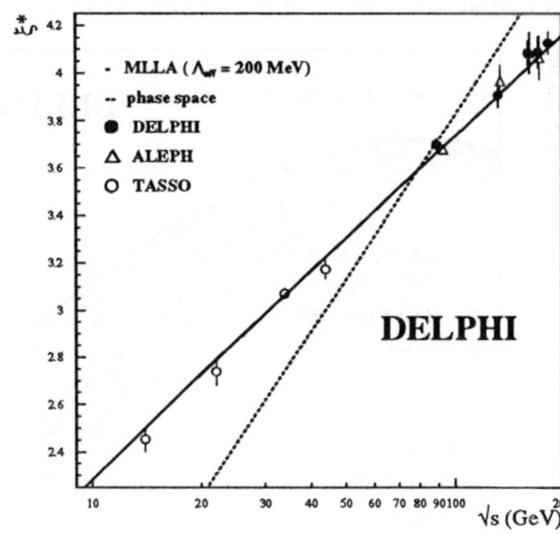

Figure 24. $\sqrt{s}$ dependence of ξ^*

particles to smaller momenta. So it would be false to take LPHD literally. However, the QCD calculations include a cut-off parameter which determines at which energy scale the showering process stops. An appropriate (small) choice of this parameter will mimic the effect of the decays. Still, it should be kept in mind that the interpretation of this scale in the context of QCD is questionable.

FIG. 23 shows the charged hadron fragmentation function as function of ξ for a wide range of centre-of-mass energies. It is seen, that a common fit of the skewed Gaussian prediction describes the data well. The MLLA prediction works even better in the range of small ξ (large x). The peak position ξ^* has been measured separately for each centre-of-mass energy (see FIG. 24[43]). Note that the factor $\frac{1}{2}$ in the expression for ξ^* in EQ. 37 is a direct consequence of the angular ordering (i.e. an effect of coherence). If hadron production would take place incoherently the peak is expected to move like $\log E/m$. m here is a typical hadron mass. Thus the slope in case of incoherent production would be about twice as big (dashed line in FIG. 24) as for the coherent case (full line), which describes the data well. This observation is among the best evidence for the coherence of gluon emission we have.

Still, there is a more intuitive look to this coherence phenomenon. Looking directly to the momentum spectrum observed in e^+e^- annihilation, FIG. 25[43], it is observed that hadron production seems to be almost universal (within ~20%) at small momentum $p < 1$GeV. In contrary at large momentum, which is the scaling region if expressed in x, a strong increase is observed due to the opening of the phase space. The universal behaviour at small momentum can be understood due to the coherent emission of soft gluons [44] from the total colour current of the event. Although the amount of hard gluons rises with energy, this leaves the total colour current unchanged, as colour is conserved in parton splittings. Soft gluons, due to their large wavelength cannot resolve additional hard gluons. In consequence soft gluon radiation and the amount of soft hadrons produced is approximately constant. The calculation given by [44] describes the data in a wide energy range $14\text{GeV} < \sqrt{s} < 183\text{GeV}$ well (see FIG. 25).

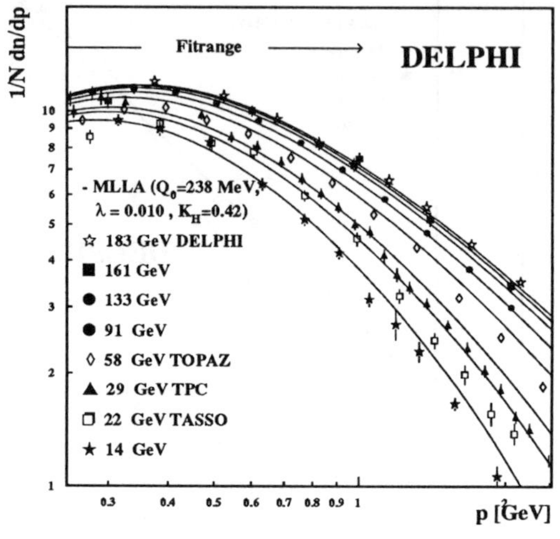

Figure 25. E_{CM} dependence of the charged hadron fragmentation function at small momentum.

3.4 Comparing Gluon and Quark Jets

High energy gluons as well as quarks give rise to jets. The characteristic differences between both types of jets have been studied intensively at LEP. In e^+e^- annihilation gluon jets appear first in three-jet ($q\bar{q}g$-) events (see SEC. 2.2). Originally the two low energy jets in mirror-symmetric three-jet events were compared[45,46]. The high energy jet most likely is a quark jet and has been left out. Among the two low energy jets the quark jet then has been tagged as a heavy quark jet, the remaining jet then must be a gluon jet. This tagging mainly has been performed using large impact parameters of tracks originating in a decay vertex of a long lived heavy hadron inside a jet. The corresponding information on a light quark jet then has been obtained from a mixed (1:1) gluon and light quark sample. This is again obtained from mirror-symmetric three-jet events, however, depleted from heavy quark events. Recently this technique has also been extended to non-symmetric event topologies[47]. In this case one relies on the q/g composition as predicted by the three-jet matrix element. However, this strongly improves the statistics and also extends the range of energy scales available. Again the scale κ

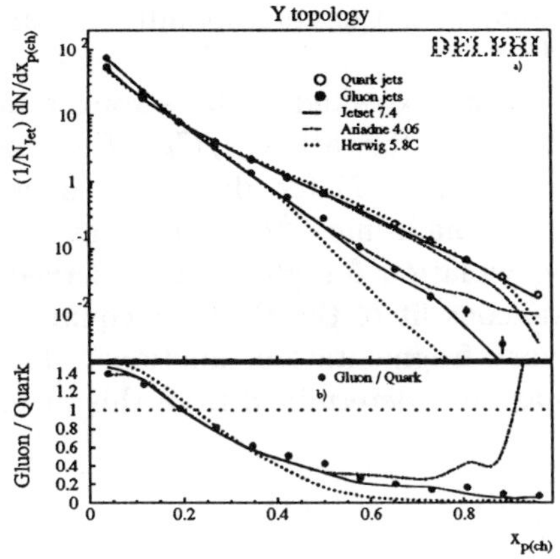

Figure 26. Comparison of the quark and gluon fragmentation function.

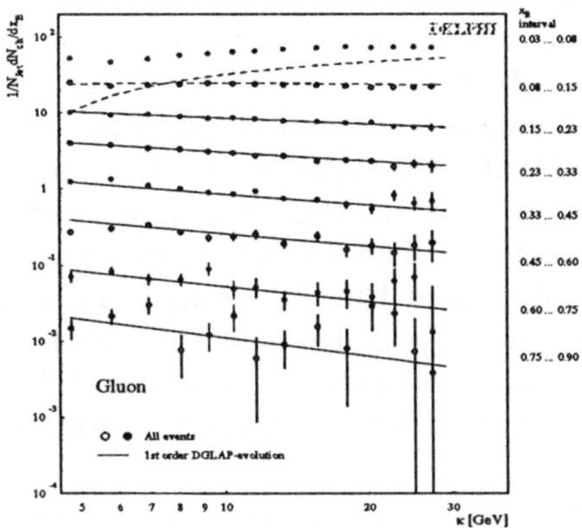

Figure 27. Scale dependence of the gluon fragmentation function.

motivated in SEC. 3.2 turns out to be a reasonable scale for the dynamical evolution of the jets. The results obtained even allow experimental tests of the correctness of this scale.

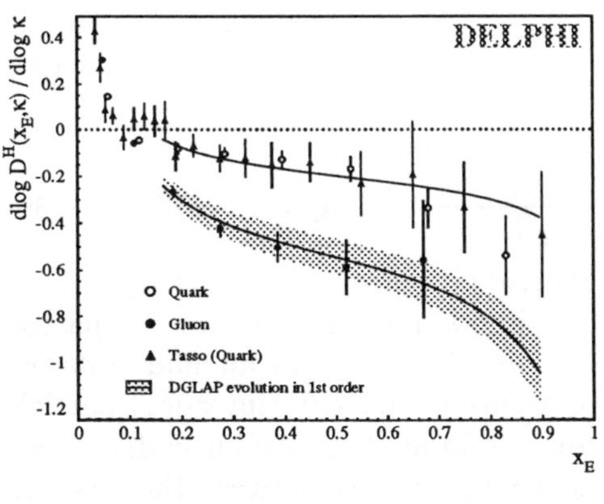

Figure 28. Comparison of the scaling violation in quark and gluon jets.

FIG. 26 compares the gluon and quark fragmentation function to charged hadrons obtained in symmetric three-jet (Y) events of 60° opening angle. Similarly to the longitudinal fragmentation function (see FIG. 18) the fragmentation function of gluons falls off more rapidly than the one of quarks. At small x the gluon fragmentation function is higher corresponding to a higher multiplicity in gluon jets (see SEC. 3.2). The scale (κ) dependence of the gluon fragmentation function is shown in FIG. 27. The typical pattern of scaling violation, an increase at small x and a depletion at large x, is observed. A corresponding result has also been observed for light quark jets in three-jet events. Here the measured fragmentation function and the scaling violation (see FIG. 28) agree well with the corresponding result obtained in low energy e^+e^- annihilation[48].

According to EQ. 24 scaling violations are expected to be stronger in gluon jets compared to quark jets by the colour factor ratio C_A/C_F . FIG. 28 compares the logarithmic slopes obtained from FIG. 27 for gluon jets to the corresponding result for light quarks and all hadronic e^+e^- events at low energy[48]. The expected, stronger scaling violation for gluons is observed. The lines in FIG. 28 result from a simultaneous fit of the DGLAP equation to the κ dependence of the gluon and quark fragmentation functions. The colour factor ratio $C_A/C_F = 2.44 \pm 0.21$ has been determined from this fit in good agreement with the expectation.

3.5 *More on Coherence, the String Effect*

The coherence of gluon radiation, beside the effect on the multiplicity and the momentum spectra at low energies also strongly influences the shape of the particle flow in space. This clearly shows that multi-parton events should not be represented as an incoherent sum of the contributions of the individual partons. The flow of soft gluon radiation originating from a $q\bar{q}g$ ensemble has been calculated[42] to be:

$$\frac{8\pi dN_{q\bar{q}g}}{d\Omega_{\vec{n}}} = \frac{2}{a_+}N_q'(Y_{q^+}, Y_q) + \frac{2}{a_-}N_q'(Y_{\bar{q}^-}, Y_{\bar{q}}) + \frac{2}{a_1}N_g'(Y_{g^1}, Y_g)$$
$$+ 2\left[I_{1+} + I_{1-} - (1 - \frac{2C_F}{N_c})I_{+-}\right] N_g'(Y), \quad (38)$$

where

$$Y_i = f(E_i, \lambda, a_i) \qquad I_{+-} = \frac{a_{+-} - a_+ - a_-}{a_+ a_-} \qquad a_i = 1 - \vec{n}\vec{n_i}. \quad (39)$$

The Y_i denote energy scales for the individual jets, $(+, -, 1) \rightleftarrows (q, \bar{q}, g)$, $\vec{n}$ is the direction in which the gluon flow is observed and $\vec{n_i}$ are the parton directions. The expression EQ. 38 includes the effects of angular ordering inside each jet as well as the collective nature of the soft "inter-jet" particle flow. The first three (N) terms describe the individual (collinear divergent) jets. The last non-divergent term gives the interference between the jets. An important observation here is that the gluon leads to a destructive interference in the angular range between the quark and the anti-quark jet (e.g. opposite to the gluon). The radiation here is in fact smaller than in the case where a photon is emitted in the gluon direction.

This interference was first observed qualitatively by the JADE experiment at PETRA and is called string effect[42]. The first quantitative tests here comes from the DELPHI experiment[46]. The particle flow in mirror symmetric $q\bar{q}g$ events and in $q\bar{q}\gamma$ events (Y events, opening angle $\sim 60°$) in the direction opposite to the radiated bosons was compared (see FIG. 29[46]). The particle flow was projected into the event plane. The quark jets were identified as heavy quark jets using impact parameter information. The polar angle in the event plane, Ψ, is always counted from the highest energy quark jet and oriented in the direction of the second quark jet. Integrating the particle flow in the range $35° \leq \Psi \leq 115°$ (opposite to the bosons) yields for the ratio of particle flows in $q\bar{q}g$ to $q\bar{q}\gamma$ events:

$$R_{g\gamma} = \frac{N_{q\bar{q}g}}{N_{q\bar{q}\gamma}} = 0.56 \pm 0.06 \tag{40}$$

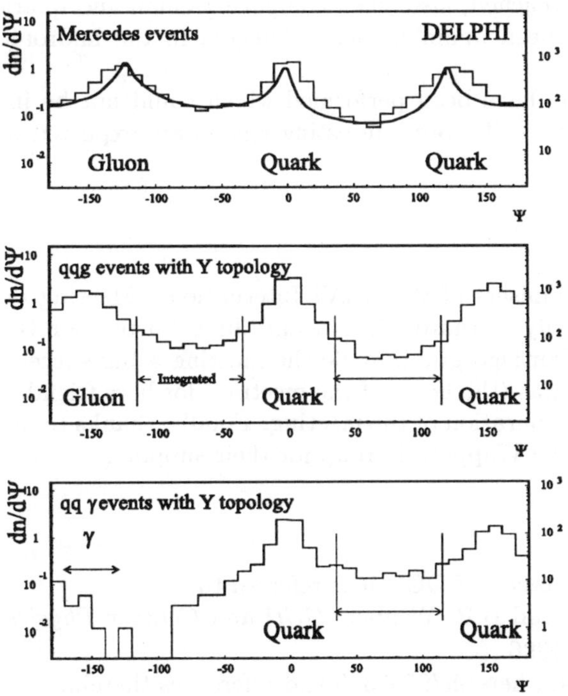

Figure 29. Angular particle flow for $q\bar{q}g$ Mercedes events, Y events and Y events with a hard γ.

Thus the destructive interference is clearly observed and in good agreement with the expected value $R^{th}_{g\gamma} \sim 0.6$.

It is instructive to also measure the particle flow in fully symmetric "Mercedes" events. The advantage of this event topology is that these events are unbiased by kinematical boosts [b]. The prediction EQ. 38[50] is normalised to the data in FIG. 29 (upper plot)[46] and describes the data well in the regions of the valleys[50]. Discrepancies in the region of the quark jets are due heavy particle decays. A comparison of the particle rates in the intervals $50° \leq \pm\Psi \leq 70°$ yields:

$$R_{g-q\bar{q}} = 2.23 \pm 0.37 \ , \tag{41}$$

[b]Note that originally the string effect has been explained by a boost of the colour string in the Lund string model[49].

which is again in good agreement with the expectation $R^{th}_{g-q\bar{q}} \approx 2.4$. This measurement clearly shows that the string effect is not a kinematical artifact but a genuine coherence effect.

4 Summary

Beside the success of LEP for studies of the electroweak sector of the Standard Model it has been proven to be an excellent laboratory for QCD studies. Precise measurements of the quantum numbers and couplings of the strongly interacting elementary fields have been obtained, including the most precise results available of the strong coupling, α_s . A better understanding of the hadronisation and of the energy evolution of fragmentation functions and the hadron multiplicity has been reached, also when comparing gluon and quark jets. Coherence effects have been unambiguously detected in the hadronic final state.

Many more measurements have been performed which could not be included in this brief lecture and still more interesting results are expected in the coming years.

Acknowledgements

I would like to thank the organisers of the XXVI International Meeting on Fundamental Physics, especially Bernardo Adeva and Juan Fuster, for the invitation to La Toxa. The Santiago group made the meeting a big success. Special thanks to Mercedes Fatas (the heart of the meeting) for here friendly, uncomplicated and efficient preparation of the meeting. Finally, thanks to my colleagues of DELPHI and of the Wuppertal group for their support.

References

1. For a comprehensive discussion of QCD it is referred to: R.K. Ellis, W.J. Stirling and B.R. Webber, *QCD and Collider Physics*, Cambridge Univ. Press, 1996.
2. see Yu.L. Dokshitzer et al., hep-ph/9707323 and references therein.
3. ALEPH Coll., *Phys. Rep.* **294**, 1 (1998).
4. L3 Coll. *Phys. Lett.* B **263**, 551 (1991).
5. ALEPH Coll., see [3]; OPAL Coll., *Z. Phys.* C **52**, 543 (1991); DELPHI Coll. *Phys. Lett.* B **274**, 498 (1992).
6. O. Nachtmann and A. Reiter, *Z. Phys.* C **16**, 45 (1982), G. Rudolph in *Physics at* LEP, CERN 86-2, Vol II, p.150.

7. DELPHI Coll., *Phys. Lett.* B **414**, 401 (1997),*Z. Phys.* C **59**, 368 (1993), *Phys. Lett.* B **255**, 466 (1991).
8. T. Sjöstrand, Comp. Phys. Comm. **82**, 74 (1994).
9. ALEPH Coll., *Z. Phys.* C **76**, 201 (1997),*Phys. Lett.* B **284**, 151 (1992).
10. OPAL Coll., *Z. Phys.* C **65**, 367 (1995).
11. P.M. Stevenson, *Phys. Lett.* B **100**, 81 (1981); S.J. Brodsky et al., *Phys. Rev.* D **28**, 2916 (1983).
12. G. Marchesini and B.R. Webber, *Nucl. Phys.* B **283**, 1 (1984); B.R. Webber, *Nucl. Phys.* B **283**, 492 (1984).
13. X. Artru, G. Mennessier, *Nucl. Phys.* B **70**, 93 (1974); B. Anderson et al., *Phys. Rep.* **97**, 33 (1983)
14. DELPHI Coll., *Z. Phys.* C **73**, 11 (1996).
15. I. Knowles et al., *QCD Event generators*, hep-ph/9601212.
16. DELPHI Coll., contr. paper **142** to the XXIX ICHEP, Vancouver (1998).
17. S. Catani et al., *Nucl. Phys.* B **407**, 3 (1993)
18. for an introduction see: G. Altarelli, CERN-TH/95-309
19. B.R. Webber, plenary talk at the ICHEP, Glasgow (1994), hep-ph/9410268.
20. D. Wicke, DELPHI Coll., priv. communication.
21. Yu.L. Dokshitzer and B.R. Webber,*Phys. Lett.* B **352**, 451 (1995).
22. Yu.L. Dokshitzer and B.R. Webber,*Phys. Lett.* B **404**, 321 (1997).
23. DELPHI Coll.,*Z. Phys.* C **73**, 229 (1997); K. Hamacher, *Nucl. Phys.* B (Proc. Supp) **54A**, 34 (1997); O. Biebel, *Nucl. Phys.* B (Proc. Supp) **64**, 22 (1998); D. Wicke, *ibid.* p.27.
24. DELPHI Coll., α_s *from* DELPHI *Measurements at* LEP *2*, to be published.
25. Yu.L.Dokshitzer, plenary talk at the ICHEP 1998, Vancouver, hep-ph/9812252.
26. S. Bethke, talk at the IV$^{\rm th}$ Int. Symp. Rad. Corr., Barcelona (1998).
27. G. Rodrigo, *Nucl. Phys.* B (Proc. Supp) **54**, 60 (1997); W. Bernreuther et al., *Phys. Rev. Lett.* **79**, 189 (1997); G. Rodrigo et al., *Phys. Rev. Lett.* **79**, 193 (1997); P. Nason, C. Oleari , *Phys. Lett.* B **407**, 57 (1997).
28. M. Bilenky et al.,*Nucl. Phys.* B **439**, 505 (1995).
29. DELPHI Coll., *Phys. Lett.* B **418**, 430 (1998).
30. ALEPH Coll., *Phys. Lett.* B **355**, 381 (1995); OPAL Coll., *Z. Phys.* C **65**, 31 (1995); SLD Coll., hep-ex/9808017.
31. R.P. Feynman, *Photon Hadron Interactions*, W.A. Benjamin (1972).
32. V.N. Gribov, L.N. Lipatov, Sov. J. Nucl. Phys., 15:438 and 675, 1972; G. Altarelli, G. Parisi, Nucl. Phys. **B126** (1977) 298; Yu.L. Dokshitzer, Sov. Phys. JETP **46** (1977) 641.

33. DELPHI Coll., *Phys. Lett.* B **311**, 408 (1993), *Phys. Lett.* B **398**, 194 (1997).
34. ALEPH Coll.,*Phys. Lett.* B **357**, 487 (1995), Errat.*Phys. Lett.* B **364**, 247 (1995).
35. B.R. Webber, *Phys. Lett.* B **143**, 501 (1984).
36. Yu. L. Dokshitzer et al.,*Int. J. Mod. Phys.* A **7**, 1875 (1992).
37. S.J. Brodsky and J.F. Gunion, *Phys. Rev. Lett.* **37**, 402 (1976).
38. DELPHI Coll., *The Scale Dependence of the Multiplicity in Quark and Gluon Jets and a Precise Determination of* C_A/C_F , to be published.
39. Yu.L. Dokshitzer et al., Sov. J. Nucl. Phys. 47 (1988) 881.
40. J. B. Gaffney and A.H. Müller, *Nucl. Phys.* B **250**, 501 (1985).
41. C.P. Fong and B.R. Webber,*Phys. Lett.* B **229**, 289 (1989).
42. Yu.L. Dokshitzer et al., *Basics of Perturbative QCD*, Ed. Frontières (1991).
43. DELPHI Coll., contr. paper 138 to the XXIX ICHEP Vancouver, (1998).
44. V. Khoze at al.,*Phys. Lett.* B **394**, 179 (1997),*Phys. Lett.* B **386**, 451 (1996).
45. OPAL Coll., *Phys. Lett.* B **265**, 462 (1991), *Phys. Lett.* B **302**, 523 (1993), *Phys. Lett.* B **388**, 659 (1996), *Z. Phys.* C **69**, 543 (1996).
46. DELPHI Coll., *Z. Phys.* C **70**, 179 (1996).
47. M. Siebel, Diplomarbeit, Bergische Univ. - GH Wuppertal, WUD 97-43.
48. DELPHI Coll., contr. paper 147 to the XXIX ICHEP, Vancouver, 1998.
49. B. Andersson et al., *Phys. Lett.* B **94**, 211 (1980).
50. V. A. Khoze and W. Ochs, *Int. J. Mod. Phys.* A **12**, 2949 (1997).

SUPERSYMMETRY SEARCH AT LEP

S. NAVAS
IFIC, Valencia-CSIC, and D.F.A.M.N., U.de Valencia, Av. Dr.Moliner 50
E-46100 Burjassot (Valencia) SPAIN
E-mail: navas@evalo1.ific.uv.es

A review of the main results obtained by the four LEP experiments on the search for new particles predicted by supersymmetric theories is reported, based mainly on the recent data collected during the beginning of the '98 high energy run of LEP. The analyses presented in this note are carried out in the framework of the *Minimal Supersymmetric Standard Model (MSSM)*, where the LSP is assumed to be the lightest neutralino, and of the *Gauge Mediated Supersymmetry Breaking Model (GMSB)*, where the LSP is the gravitino. *R-parity* conservation is assumed in both cases. No evidence of such signal is found, although candidate events consistent with the expectations from Standard Model processes are observed. Limits at 95 % C.L. on the production cross–sections and on the particle masses obtained by the four LEP experiments, and bounds on the parameters of the Minimal Supersymmetric Standard Model are reported.

1 Introduction

This note presents an overview of the main analyses performed at LEP [1] on the search for new particles predicted by supersymmetric theories. The results obtained by the four LEP experiments [2] are also presented. They are based mainly on the analysis of the 1997 data and on the ~40 pb^{-1} of data taken during the beggining of the last high energy run of LEP (early 1998), but some results combining all LEP energies are also reported. A copy of the transparencies presented in the conference can be found in Ref. [3].

Supersymmetry [4] extends the Standard Model with a new type of symmetry between bosons and fermions in such a way that both types of particles play a similar role in the theoretical description of the known interactions. This is achieved by increasing the ordinary field multiplets with additional fields differing by half a unit of spin. The resulting particle spectrum contains several new states: *gauginos*, associated with the ordinary gauge bosons; *higgsinos* with the Higgs bosons; *sleptons*, *sneutrinos* and *squarks* with the ordinary matter fermions. In this way, supersymmetric theories predict for each elementary particle a *superpartner* with the same quantum numbers but with a spin that differs by $\frac{1}{2}$ from that of the ordinary particle. In addition to its evident theoretical appeal the idea of supersymmetry is supported by a series of phenomenological clues that makes of great interest its experimental study, in particular the search for the particles that it predicts and which have not been found yet.

Ordinary particles and s-particles are distinguished by their *R-parity* [a], a multiplicative quantum number, which is assumed to be conserved to ensure leptonic and barionic number conservation. This has three extremely important phenomenological consequences: the lightest supersymmetric particle (LSP) must be absolutely stable and escapes detection; each s-particle must eventually decay into a state which contains an odd number of LSPs; and, in collider experiments, s-particles can only be produced in even numbers (usually two at a time).

The searches presented in this note have been performed assuming the universality of the parameters at the high mass scale, typical of Grand Unified Theories (GUT). Under this assumption, the mass eigenstates and couplings are dominantly governed by four parameters: the masses M_1 and M_2 of the gaugino sector [b] (which are assumed to satisfy the GUT relation $M_1 = \frac{5}{3}\tan^2\theta_W \cdot M_2 \approx 0.5 \cdot M_2$ at the electroweak scale); the universal mass m_0 of the scalar lepton sector particles (masses at the electroweak scale are derived using the renormalization group equations [5]), which enters mainly via the sneutrino mass; the Higgs mass parameter μ; and the ratio $\tan\beta$ of vacuum expectation values of the two Higgs doublets.

As discussed later, in the MSSM model the LSP is usually the lightest neutralino $\tilde{\chi}^0_1$. With unification of gaugino mass terms, in the region where $M_2 >> |\mu|$ the lightest chargino and neutralino have large higgsino components; this is referred to as the *higgsino region*. Here, both particles are generally close in mass (*degenerate*). In the opposite case, the region where $M_2 << |\mu|$ is referred to as the *gaugino region*; here $\tilde{\chi}^0_1$ has usually a mass roughly 0.5 times the mass of the $\tilde{\chi}^0_2$ and $\tilde{\chi}^\pm_1$, provided that the GUT relation between M_1 and M_2 holds. The intermediate region is called *mixed region.*

Since no supersymmetric partners have been found up to now, *supersymmetry must be a broken symmetry.* There are two main competing proposals for what the mediating SUSY breaking interactions might be.

• The first is that they are gravitational. In this *gravity-mediated supersymmetry breaking scenario* [4], the lightest supersymmetric particle (LSP) is assumed to be either the lightest neutralino ($\tilde{\chi}^0_1$) or the sneutrino ($\tilde{\nu}$). In this scenario, the gravitino mass ($m_{\tilde{G}}$) is usually comparable to the masses of the MSSM particles (at least 100 GeV/c^2 or so). Its interactions will be of gravitational strength, so the gravitino will not play any role in LEP physics.

---

[a] The *R-parity* is defined as $R = (-1)^{2J+3B+L}$ where J, B and L are the spin, the barionic and the leptonic numbers of the particle, respectively.

[b] M_1, M_2 and M_3 are the gaugino masses associated to the U(1), SU(2) and SU(3) gauge groups, respectively.

• The second main possibility is that the flavour-blind mediating interactions for supersymmetry breaking are the ordinary electroweak and QCD gauge interactions. In this *gauge-mediated supersymmetry breaking models (GMSB)*[6], the gravitino is almost certainly the LSP, and all of the MSSM s-particles will eventually decay into final states that include it.

A very light gravitino may be relevant for LEP phenomenology, because it contains as its longitudinal component the goldstino, which has a non-gravitational coupling to all s-particle–particle pairs ($\tilde{X}$, X). The decay rate of any s-particle into its Standard Model partner $\tilde{X} \to X\tilde{G}$ is usually not fast enough to compete with the other decays of s-particles, unless $\tilde{X}$ is the next-to-lightest supersymmetric particle (NLSP). Since the NLSP has no competing decays, it should always decay into its superpartner and the LSP gravitino.

The most obvious guess is that *the lightest neutralino ($\tilde{\chi}_1^0$) is the NLSP* in light gravitino models. Since $\tilde{\chi}_1^0$ contains a mixture of the photino (the linear combination of bino and neutral wino whose superpartner is the photon), it should then decay into [c] photon+gravitino with a width [7]:

$$\Gamma(\tilde{\chi}_1^0 \to \gamma\, \tilde{G}) = 2 \times 10^{-3} \cdot \kappa_{1\gamma} \cdot (\frac{m_{\tilde{\chi}_1^0}}{100\ GeV})^5 \cdot (\frac{\sqrt{|F|}}{100\ TeV})^{-4}\ eV, \qquad (1)$$

where $\kappa_{1\gamma}$ is the photino content of the $\tilde{\chi}_1^0$ in terms of the neutralino mixing matrix and $|F|$, defining the supersymmetry-breaking scale $\Lambda_S = |F|^{1/2}$, is related to the gravitino mass by: $|F| = \sqrt{3} \cdot M_p \cdot m_{\tilde{G}}$, $M_p \simeq 2.4 \times 10^{18}\ GeV/c^2$.

Note that the decay width is larger for smaller $|F|$, or equivalently for smaller $m_{\tilde{G}}$, if the other masses are fixed.

If $m_{\tilde{\chi}_1^0}$ is of the order of 100 GeV/c^2 and $m_{\tilde{G}}$ less than roughly 1 keV/c^2, then the decay $\tilde{\chi}_1^0 \to \gamma\, \tilde{G}$ can occur quickly enough to be observed in signatures which we will discuss further. In fact, from equation (1), the mean decay length for a $m_{\tilde{\chi}_1^0}$ with energy E in the lab frame is:

$$c\tau = 9.9 \times 10^{-3} \cdot \frac{1}{\kappa_{1\gamma}} \cdot (E^2/m^2_{\tilde{\chi}_1^0} - 1)^{\frac{1}{2}} \cdot (\frac{m_{\tilde{\chi}_1^0}}{100\ GeV})^{-5} \cdot (\frac{\sqrt{|F|}}{100\ TeV})^4\ cm\ , \quad (2)$$

which could be anywhere from microns to kilometres depending on the scale of supersymmetry breaking $|F|$.

Another likely candidate for *the NLSP*, in light gravitino models, *is the lightest stau* $\tilde{\tau}_1$. A slepton NLSP can decay as $\tilde{l} \to l\, \tilde{G}$ with a width and decay length just given by equations 1 and 2, with the replacements $\kappa_{1\gamma} \to 1$ and

---

[c] If $m_{\tilde{\chi}_1^0} > m_Z$, the decay $\tilde{\chi}_1^0 \to Z^0\, \tilde{G}$ may be non-negligible.

$m_{\tilde{\chi}_1^0} \to m_{\tilde{l}}$. So, just as for the neutralino NLSP case, the decay $\tilde{l} \to l\,\tilde{G}$ can be either fast or very slow, depending on the form of supersymmetry breaking.

Importance of ΔM in s-particle search

The parameter ΔM defined as the mass difference between the s-particle under study and the LSP ($\equiv m_{s-part} - m_{LSP}$) plays an essential role from an experimentally point of view since the typical missing transversal momentum ($\not{P}_t$) and visible mass grow with ΔM. Figure 1 (left) shows the contour levels of the total detection efficiency (in percentage) in the $(m_{\tilde{\chi}_1^\pm}, m_{\tilde{\chi}_1^0})$ plane assuming a stable $\tilde{\chi}_1^0$ (MSSM scenario). The highest efficiency is obtained when $M_{\tilde{\chi}^\pm} \sim 2M_{\tilde{\chi}^0}$, so for moderate values of ΔM (~ 40 GeV/c^2); due to the selection criteria against background from two-photon interactions (WW events) the efficiency decreases for low (high) values of ΔM. Figure 1 (right) shows the similar result assuming the neutralino decay $\tilde{\chi}_1^0 \to \gamma\,\tilde{G}$ (GMSB scenario). In this case, the dependence of the efficiency on $m_{\tilde{\chi}_1^\pm}$ and $m_{\tilde{\chi}_1^0}$ is weaker than in the previous case, thanks to the presence of isolated photons. Higher efficiencies are obtained for heavy neutralinos, which ensure the presence of highly energetic photons, easy to detect.

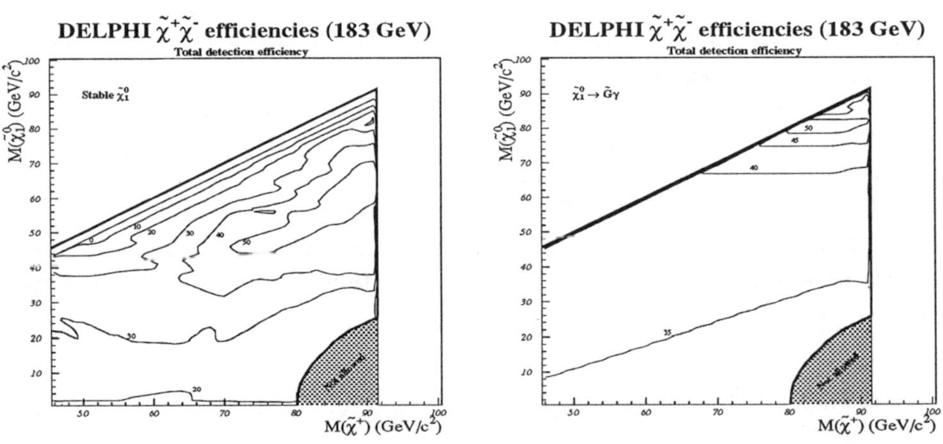

Figure 1: Contour levels of the parametrised total efficiency (in percentage) computed by DELPHI in the chargino search analysis, as a function of $m_{\tilde{\chi}_1^\pm}$ and $m_{\tilde{\chi}_1^0}$ in the neutralino LSP scenario (left plot) and in the gravitino LSP scenario (right plot).

Another phenomenological variable which depends on ΔM is the average length of flight of the searched s-particle. Different searches are optimised for different ranges of ΔM: if 3 GeV/$c^2 < \Delta M$ the s-particle decays promptly and the standard search is applied. The main signature is the missing energy and different selection criteria are optimised for different ΔM regimes; if 500 MeV/c^2<ΔM<3 GeV/c^2 the secondary vertex of the s-particle decay is not yet detectable. Only events with ISR photons are selected in order to reduce the high cross-section of two-photon interactions background; if 150 MeV/c^2<ΔM<500 MeV/c^2 the decay length is short and the s-particle still decays inside the detectors. Main signatures are the high impact parameter and the presence of kinks; if ΔM<150 MeV/c^2 the s-particle decays outside the detector and stable heavy charged particles are searched.

To optimize the signal detection efficiency, a large part of the SUSY analyses apply different selection criteria to different ΔM regions. Efficiencies are estimated from the Monte Carlo simulation of a huge range of s-particle masses and ΔM values and then parametrised and extended to all MSSM space.

2 Backgrounds at *LEP-II*

The characteristic feature of any SUSY production is the *large missing energy*, because of the undetected LSP. This is why, to be sensitive to such processes the detectors must be *as hermetic as possible*. Therefore, any Standard Model process which manifests missing energy can be a source of background for SUSY searches.

Two-photon events ($e^+e^- \rightarrow \gamma\gamma \rightarrow e^+e^-X$) are characterized by large missing energy and low visible mass. However, $\not{P}_t$ is small and the jet or lepton system tends to be coplanar with the beam direction, and there are often jets in the forward-backward region of the detector.

In $\mathrm{f\bar{f}}$ events with initial state radiation there is a high energy photon which is usually lost in the beam pipe. Such events are characterized by large visible mass ($\sim m_\mathrm{Z}$) and large missing momentum aligned with the beam direction. The background from $\mathrm{f\bar{f}}$ events with initial state radiation emitted at larger polar angles can be reduced by vetoing events with energetic isolated photons in the detector.

$\mathrm{W^+W^-}$ events are one of the most dangerous sources of background at centre–of–mass energies above 161 GeV. Leptonic topologies are usually characterized by the presence of high energetic isolated electrons or muons. Events can be irreducible when both W bosons decay into taus. Purely hadronic $\mathrm{W^+W^-}$ events can also constitute an irreducible background for some SUSY channels, like $\tilde{\chi}^+\tilde{\chi}^-$ topologies, when $m_{\tilde{\chi}_1^\pm} \sim m_W$ and $m_{\tilde{\chi}_1^0} \sim 0$. For large $\Delta M \sim m_\mathrm{W}$, the need to reject $\mathrm{W^+W^-}$ background translates in a loss of efficiency for these topologies.

$\mathrm{W\,e\,\nu_e}$ events are characterized by high values of the visible mass, $\not{P}_t$ and missing mass, since there is an energetic neutrino at high angles and an energetic electron ($E \sim E_{beam}$) at low angles which escape the detection.

3 Experimental results on the MSSM scenario

After the analysis of all the data collected at LEP, several candidates for charginos, neutralinos, sleptons and squarks have been observed in data by the four LEP experiments. Unfortunately, all these events are in agreement with the Standard Model expectations at the different energies. Therefore, no evidence of a signal has been found. Nevertheless, previous limits on s-particle masses have been significantly extended with the new statistics collected at E_{cm} ~189 GeV.

3.1 Neutralinos and charginos

Direct production of $\tilde{\chi}_2^0\tilde{\chi}_1^0$ and of $\tilde{\chi}^+\tilde{\chi}^-$ has been searched at LEP. Since the characteristics of neutralino and chargino events mainly depend on ΔM, different selection criteria have been applied. Model–independent upper limits are obtained at 95% C.L. for the production cross–section. Exclusion limits are determined from the observed number of events, the signal detection efficiencies and their uncertainties, and the number of background events expected and their uncertainties. To obtain a limit on a given ($M_{\tilde{\chi}_2^0}$,$M_{\tilde{\chi}_1^0}$) or ($M_{\tilde{\chi}_1^\pm}$,$M_{\tilde{\chi}_1^0}$) point, the independent analyses performed for different ΔM regimes are combined.

Figure 2 (left) shows the OPAL limit for $\tilde{\chi}_2^0\tilde{\chi}_1^0$ production assuming that $\tilde{\chi}_2^0$ decays dominantly to $\tilde{\chi}_1^0$ $Z^{0(*)}$. The best cross–section limit is ~0.14 pb. The worst case happens for low values of ΔM (<5 GeV/c^2) where the best limit is of the order of 1 pb.

Similar model–independent upper limits are obtained on $\tilde{\chi}_1^+\tilde{\chi}_1^-$ production cross–section. The best limit is ~0.15 pb, obtained for $M_{\tilde{\chi}_1^\pm} \sim 2M_{\tilde{\chi}_1^0}$ and $M_{\tilde{\chi}_1^\pm} >$ 70 GeV/c^2, assuming that Br($\tilde{\chi}_1^\pm \to \tilde{\chi}_1^0$W)=100%. The worst case occurs for low values of ΔM, where a limit of ~3 pb is reached for ΔM~3 GeV/c^2.

The model–independent limits can be translated into mass limits following the MSSM assumptions introduced in Section 1. These limits depend tightly on ΔM and on $M_{\tilde{\nu}}$ (so on m_0). For $m_0 > 300$ GeV/c^2 and ΔM>10 GeV/c^2 the minimal non excluded chargino mass ranges from ~92 GeV/c^2 for a chargino dominantly gaugino to the kinematic limit for a chargino dominantly higgsino. For an arbitrary value of m_0 the lower mass limit decreases to 57.2 GeV/c^2, due to a lower cross–section. For lower values of ΔM (<10 GeV/c^2) the chargino is purely higgsino and the mass limit is independent of m_0, but in this case the efficiency is lower due to the selection criteria used to reduce background from two–photon interactions. The lower mass limit obtained by DELPHI for ΔM=5 GeV/c^2 is 89.9 GeV/c^2. More sophisticated analyses are needed for lower values of ΔM as explained in Section 1. Figure 2 (right) shows the chargino mass limit from DELPHI as a function of ΔM obtained by the different analyses, in different regions of the (μ,M_2) plane, for different relations between M_1 and M_2, and for different values of $M_{\tilde{\nu}}$. In conclusion the lower chargino mass limit, valid when the chargino is dominantly gaugino, the $\tilde{\nu}$ is heavy and ΔM>0.5 GeV/c^2 is 62.2 GeV/c^2.

The results from the four LEP experiments at E_{cm}=183 GeV were combined by the LEP SUSY working group [8]. Mass limits are extended by 2–6 GeV/c^2 for 3 GeV/c^2 $\leq \Delta M \leq$ 5 GeV/c^2.

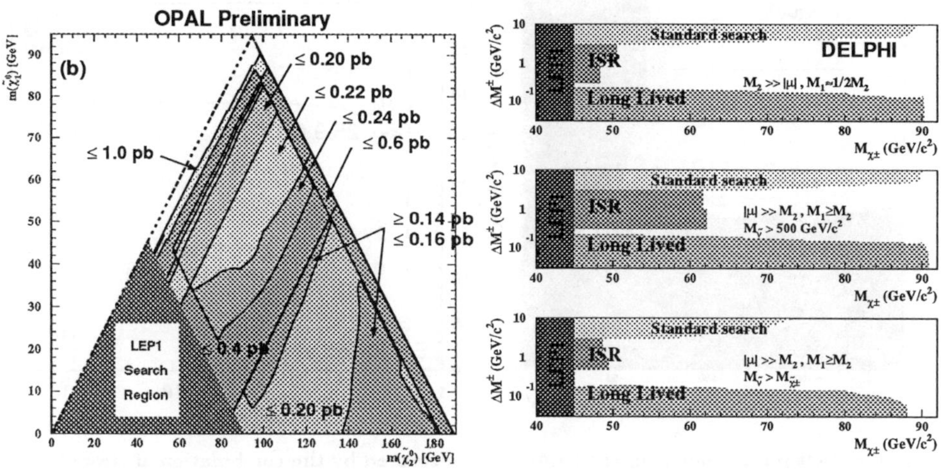

Figure 2: Left plot: The contours of the 95% C.L. upper limit for $e^+e^- \rightarrow \tilde{\chi}_2^0\tilde{\chi}_1^0$ production cross-sections at E_{cm} = 188.7 GeV are shown assuming Br($\tilde{\chi}_2^0 \rightarrow \tilde{\chi}_1^0 Z^{0(*)}$)=100%. The region for which $M_{\tilde{\chi}_2^0}+M_{\tilde{\chi}_2^0}<M_Z$ is not considered in this analysis. Right plot: Regions in the plane ($M_{\tilde{\chi}_1^\pm}$,ΔM) excluded at 95% C.L., for E_{cm} =189 GeV, using: the standard search for high ΔM charginos; the search for soft particles accompanied by ISR; the search for long lived charginos. The result is shown in three SUSY scenarios.

3.2 Lower mass limit on the LSP

In order to obtain an absolute lower mass limit on the LSP ($\tilde{\chi}_1^0$) it is necessary to combine chargino, neutralino and slepton searches. The commonly used method consists in converting the negative results from searches for charginos and neutralinos into exclusion regions in (μ,M_2) plane for different $\tan\beta$ values and determine the minimal allowed LSP mass as a function of $\tan\beta$. The exclusion regions obtained from chargino and neutralino searches depend however on the sneutrino and selectron mass values, respectively. The searches for sleptons are therefore used to set a minimal allowed sneutrino mass value. The scan over the sneutrino mass is performed starting from the value allowed by the slepton searches. The GUT relation with a common m_0 is used to relate selectron and sneutrino masses and to obtain a minimal allowed sneutrino mass for given $\tan\beta$ and M_2 values. The LEP1 search for $\tilde{\chi}_2^0\tilde{\chi}_1^0$ production with $\tilde{\chi}_2^0 \rightarrow \tilde{\chi}_1^0\gamma$ was also included in order to cover regions of the (μ,M_2) plane not excluded by the LEP2 gaugino searches (for $M_2 \sim |\mu|$, $\mu < 0$ and $\tan\beta<1.5$).

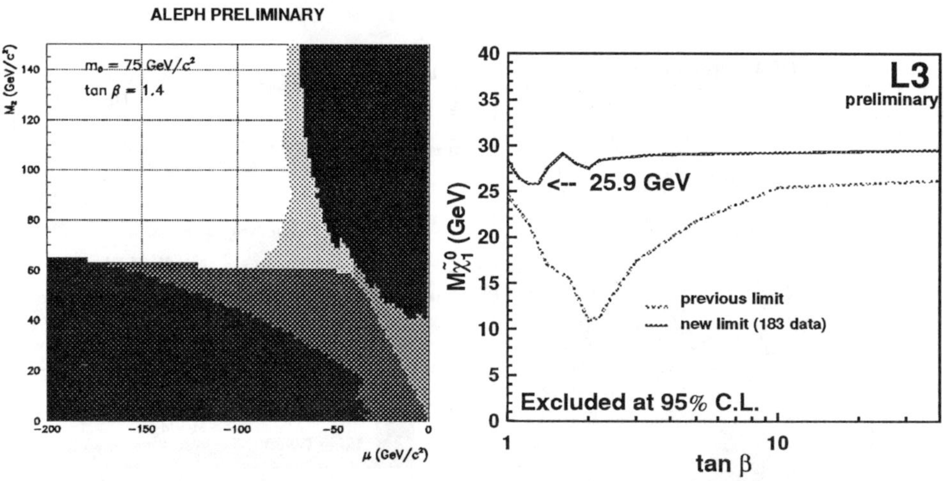

Figure 3: Left plot: Regions in the (μ,M_2) plane covered by the combination of (from light to dark gray) neutralino search at LEP2, neutralino search at LEP1, selectron search and chargino search at LEP2, for $\tan\beta$=1.4, m_0=75 GeV/c^2 and E_{cm}=189 GeV. Right plot: Lower limit on $M_{\tilde{\chi}_1^0}$ as a function of $\tan\beta$ and for any value of m_0, at E_{cm}=183 GeV.

An example of the exclusion regions resulting from searches by ALEPH for charginos, neutralinos and sleptons are shown in Figure 3 (left). The lower limit on the LSP mass obtained by L3 is shown in Figure 3 (right), as a function of $\tan\beta$, for any value of m_0, and at E_{cm}=183 GeV. The absolute minimum is 25.9 GeV/c^2. Similar values on the lower mass limit on the LSP are obtained by the other LEP experiments and they are summarised in Table 1.

3.3 Sleptons

Cross section limits and exclusion limits at 95% C.L. are determined in the same way as for neutralinos and charginos, but in the $(M_{\tilde{l}},M_{\tilde{\chi}_1^0})$ plane. Figure 4 (left) shows the right–handed selectron exclusion mass limit obtained by ALEPH at E_{cm}=189 GeV/c^2. The excluded limit is the best for $\tilde{\chi}_1^0$

masses smaller than 40 GeV/c^2 and becomes worse when the selectron is mass-degenerate with the LSP.

The numerical mass bounds on the slepton masses set by the four LEP experiments are listed in Table 2, for different values of ΔM and $\tan\beta$.

Table 1: Lower mass limits on the LSP ($\tilde{\chi}_1^0$).

	E_{cm}	Any m_0	m_0 >200 GeV/c^2
ALEPH	189 GeV	26.0 GeV/c^2	32.0 GeV/c^2
DELPHI	189 GeV	23.4 GeV/c^2	30.6 GeV/c^2
L3	183 GeV	25.9 GeV/c^2	30.1 GeV/c^2
OPAL	189 GeV	25.4 GeV/c^2	32.8 GeV/c^2

Table 2: Lower mass limits on sleptons masses at E_{cm}=183 GeV. The results obtained at E_{cm}=189 GeV are shown in italic.

	$M_{\tilde{e}_R}$	$M_{\tilde{\mu}_R}$	$M_{\tilde{\tau}_R}$
ALEPH	*84* GeV/c^2 for ΔM>15 GeV/c^2, $\tan\beta$=2	*78* GeV/c^2 for ΔM>15 GeV/c^2	65 GeV/c^2 for ΔM>15 GeV/c^2
DELPHI	81.4 GeV/c^2 for $M_{\tilde{\chi}_1^0}$<35 GeV/c^2, $\tan\beta$=1.5	70.3 GeV/c^2 for $M_{\tilde{\chi}_1^0}$<35 GeV/c^2	*70* GeV/c^2 for ΔM>15 GeV/c^2
L3	80 GeV/c^2 for ΔM>15 GeV/c^2, $\tan\beta$=1.41	69 GeV/c^2 for ΔM>15 GeV/c^2	
OPAL	78 GeV/c^2 for ΔM>2 GeV/c^2, $\tan\beta$=1.5	*73* GeV/c^2 for ΔM>3 GeV/c^2	*69.4* GeV/c^2 for ΔM>10 GeV/c^2

The results from the four LEP experiments at E_{cm}=183 GeV were combined by the LEP SUSY working group [8]. For a fixed centre–of–mass energy, mass limits are extended by 5–6 GeV/c^2 for $\tilde{e}_R$ and $\tilde{\tau}_R$ and by 2–3 GeV/c^2 for $\tilde{\mu}_R$. These results are superseded by the mass limits obtained by a single LEP experiment at E_{cm}=189 GeV.

3.4 Squarks

No evidence for $\tilde{t}_1\bar{\tilde{t}}_1$ and $\tilde{b}_1\bar{\tilde{b}}_1$ production has been observed in data. In order to calculate mass limits, the number of signal events passing the event selections is determined as a function of $M_{\tilde{q}_1}$, $M_{\tilde{\chi}_1^0}$ and the mixing angle $\theta_{\tilde{q}}$. The results obtained with all centre–of–mass energies data (E_{cm}=130–189 GeV) were used to calculate these limits.

Figure 4 (right) shows the OPAL 95% C.L. excluded regions in the ($M_{\tilde{b}_1}$,$M_{\tilde{\chi}_1^0}$) plane, for different values of the mixing angle between the left and right components. There is a good agreement between the expected and the obtained exclusion limit. The numerical $\tilde{t}_1$ and $\tilde{b}_1$ mass bounds for the four LEP experiments are listed in Table 3. The results from the four LEP experiments at E_{cm}=183 GeV were combined by the LEP SUSY working group [8]. For a fixed E_{cm}, mass limits are extended by 5–6 GeV/c^2.

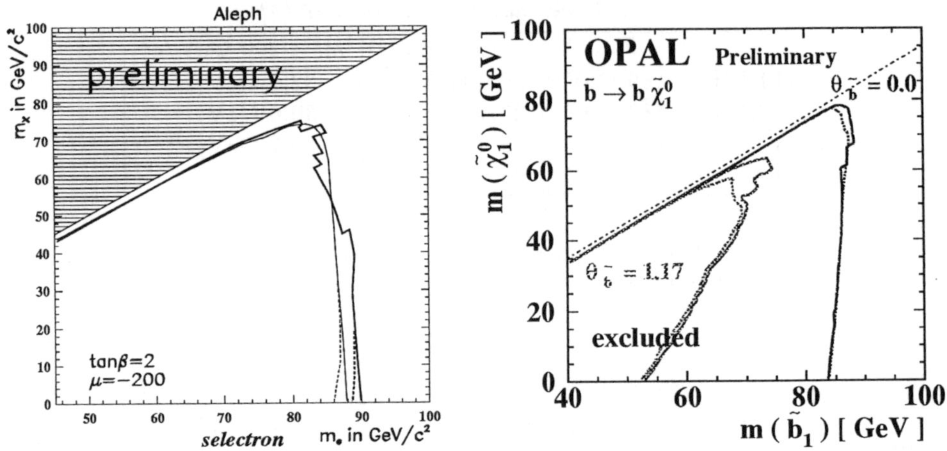

Figure 4: Left plot: The $\tilde{e}_R$ mass limit at E_{cm}=189 GeV. The actual (thick line) and expected (thin line) mass limits are shown for slectrons assuming Br($\tilde{e}_R \rightarrow e\tilde{\chi}_1^0$)=100% (full lines). The dashed curves show the effect of cascade decays assuming no efficiency for these decays. Right plot: The 95% C.L. excluded regions in the $(M_{\tilde{b}_1}, M_{\tilde{\chi}_1^0})$ plane, at E_{cm}=189 GeV, for a mixing angle of the $\tilde{b}_1$ assumed to be 0.0 and 1.17 rad, assuming that the $\tilde{b}_1$ decays into $b\tilde{\chi}_1^0$. The solid lines show the actual limits and the dotted lines show the limits calculated only with the expected number of background events.

4 Experimental results on the GMSB scenario

4.1 Chargino and neutralino search results

As already discussed, in GMSB with a $\tilde{\chi}_1^0$ NLSP, SUSY particles will decay to $\tilde{\chi}_1^0$, which in turn decays via $\tilde{\chi}_1^0 \rightarrow \gamma\tilde{G}$ with nearly a 100 % branching fraction. The final chargino and neutralino states would then include jets plus isolated photons, jets with an isolated lepton plus isolated photons, or two isolated leptons plus isolated photons, all with significant missing transverse momentum from the undetected gravitinos. Therefore, searches for such events are made in a way similar to the searches in the $\tilde{\chi}_1^0$ LSP scenario, but requiring the presence of isolated energetic photons.

As mentioned, the $\tilde{\chi}_1^0$ decay can occur near the interaction point, at an observably macroscopic distance, or outside the detector (see equation 2). This

Table 3: Lower mass limits on stop and sbottom masses at E_{cm}=183 GeV, for $\theta_{\tilde{q}}=0$. The results obtained at E_{cm}=189 GeV are shown in italic.

	$M_{\tilde{t}}$ $\tilde{t} \to c\tilde{\chi}$	$M_{\tilde{t}}$ $\tilde{t} \to bl\tilde{\nu}$	$M_{\tilde{b}}$ $\tilde{b} \to b\tilde{\chi}$
ALEPH	*84* GeV/c^2 for ΔM>10 GeV/c^2	*88* GeV/c^2 for ΔM>10 GeV/c^2	*84* GeV/c^2 for ΔM>10 GeV/c^2
DELPHI	78.4 GeV/c^2 for ΔM>10 GeV/c^2	82.0 GeV/c^2 for ΔM>15 GeV/c^2	78.6 GeV/c^2 for ΔM>10 GeV/c^2
L3	81 GeV/c^2 for ΔM>10 GeV/c^2		80 GeV/c^2 for ΔM>20 GeV/c^2
OPAL	*88.3* GeV/c^2 for ΔM>10 GeV/c^2	*87.3* GeV/c^2 for ΔM>10 GeV/c^2	*85.5* GeV/c^2 for ΔM>10 GeV/c^2

is why it is particularly desirable not to require that both photons be observed in the analysis, therefore extending the detection efficiency to larger decay lengths. For very long lifetimes, neither photon will be observed, and the final state will be identical to that from the gravity–mediated scenario.

As an example, let us consider the selection made by the OPAL experiment at E_{cm} =184 GeV. In the case of $\tilde{\chi}_1^+ \tilde{\chi}_1^-$ pair production with a 100 % branching fraction for $\tilde{\chi}_1^\pm \to \tilde{\chi}_1^0 W^\pm$, production cross–section and limit contours are shown in Figure 5 (left). The case of arbitrary $\tilde{\chi}_1^0$ lifetimes has been also considered. This is of particular interest since the lifetime is, for all practical purposes, arbitrary, and a long-lived $\tilde{\chi}_1^0$ may have escaped previous detection. If one of the $\tilde{\chi}_1^0 \to \gamma \tilde{G}$ decays occurs outside the detector, its missing energy will affect all event quantities (visible energy, transverse momentum, etc), and we must rely on the ability to find the single remaining photon. If both $\tilde{\chi}_1^0 \to \gamma \tilde{G}$ decays occur outside the detector volume, we rely on the gravity–mediated chargino–neutralino search. In Figure 5 (right), limits on the chargino pair production cross–section for three examples of ΔM values with $m_{\tilde{\chi}_1^\pm} = 90$ GeV/c^2 are shown. Included in that figure at large $c\tau$ are the limits for the gravity–mediated chargino search, and thus include all $0 < c\tau < \infty$.

4.2 Indirect search: Anomalous production of photon pairs

Some indirect searches for supersymmetry can also be performed at LEP. Let us consider, as an example, the search for anomalous production of photon pairs with missing energy made by the DELPHI experiment.

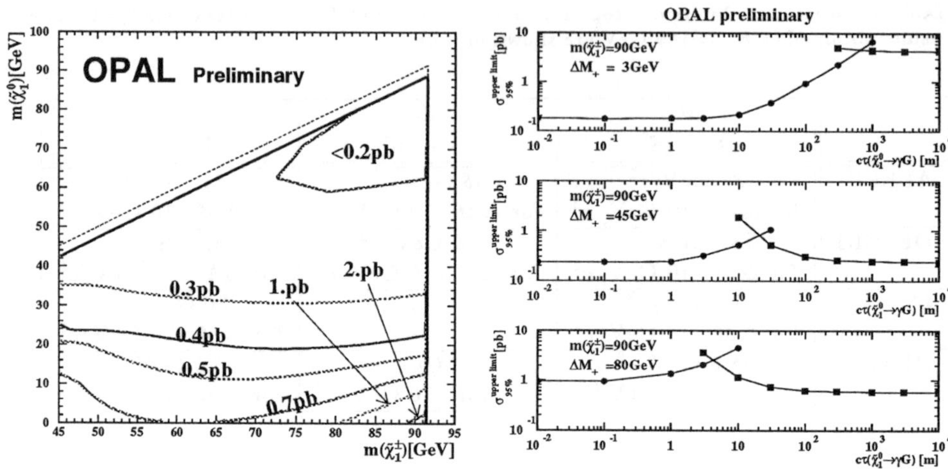

Figure 5: 95 % C.L. upper limits set by OPAL on the $\tilde{\chi}_1^+ \tilde{\chi}_1^-$ production cross–section, assuming both decay via $\tilde{\chi}_1^\pm \to \tilde{\chi}_1^0 W^\pm$. Left plot: Prompt $\tilde{\chi}_1^0 \to \gamma \tilde{G}$ decays considered. Right plot: $\tilde{\chi}_1^0 \to \gamma \tilde{G}$ decays, as a function of the $\tilde{\chi}_1^0$ lifetime considered. The large lifetime limits (connected squares) come from the gravity–mediated SUSY breaking chargino search.

Pairs of photons with missing energy are expected from reactions like $e^+e^- \to Y\,Y$ with the subsequent decay $Y \to X\,\gamma$, where X is a neutral, weakly interacting, stable particle. Such processes are expected in supersymmetry models, but also in the Standard Model:

– In supersymmetric models with gauge mediated supersymmetry breaking (*GMSB*) the Y particle can be the lightest neutralino ($\tilde{\chi}_1^0$), decaying to a photon and a gravitino (with masses of order $\sim$1 eV/c^2 or lower). The reaction looked for is thus $e^+e^- \to \tilde{\chi}_1^0 \tilde{\chi}_1^0 \to \tilde{G}\gamma\,\tilde{G}\gamma$ which produces a pair of photons accompanied by two invisible particles. The decay $\tilde{\chi}_1^0 \to \tilde{G}\gamma$ is assumed to take place in the proximity of the beam interaction point.

– Alternatively, in more conventional *MSSM models* with $\tilde{\chi}_1^0$ the LSP, the process $e^+e^- \to \tilde{\chi}_2^0 \tilde{\chi}_2^0 \to \tilde{\chi}_1^0\gamma\,\tilde{\chi}_1^0\gamma$ can give rise to similar final states, where the particle X, here identified as $\tilde{\chi}_1^0$, can be relatively heavy. The decay $\tilde{\chi}_2^0 \to \tilde{\chi}_1^0\gamma$ is assumed to occur in the proximity of the beam interaction point.

– Within the *Standard Model*, two–photon final state are expected mainly from the reaction $e^+e^- \to \nu\bar{\nu}\gamma\gamma$, where the photons are emitted from the initial state electrons and the neutrinos are produced via Z-exchange in the *s-channel.* This is the most relevant background in the search for anomalous production of events with photons and missing energy, although other potential sources of background can come from the QED process $e^+e^- \to \gamma\gamma(\gamma)$, and from $e^+e^- \to e^+e^-\gamma(\gamma)$.

The results obtained by DELPHI in the search for the process $e^+e^- \to \tilde{\chi}_2^0\tilde{\chi}_2^0 \to \tilde{\chi}_1^0\gamma\tilde{\chi}_1^0\gamma$ are shown in Figure 6 (left) in terms of 95 % C.L. upper limits on the signal cross–section. In this case the limits are shown as a function of the mass difference between $\tilde{\chi}_2^0$ and $\tilde{\chi}_1^0$. The signal–cross–section is below ~0.2 pb, provided that $m_{\tilde{\chi}_2^0} - m_{\tilde{\chi}_1^0}$ is larger than 10 GeV/c^2.

Similarly, for the case of a massless X-particle or, equivalently, the reaction $e^+e^- \to \tilde{\chi}_1^0\tilde{\chi}_1^0 \to \tilde{G}\gamma\tilde{G}\gamma$, the limits obtained are shown in the right plot of Figure 6. These are expressed in terms of 95 % C.L. upper limits on the signal cross–section $\sigma(e^+e^- \to \tilde{\chi}_1^0\tilde{\chi}_1^0)$ as a function of the $\tilde{\chi}_1^0 \equiv Y$ mass. The limit ranges from 0.16 pb and 0.105 pb, depending on the $\tilde{\chi}_1^0$ mass.

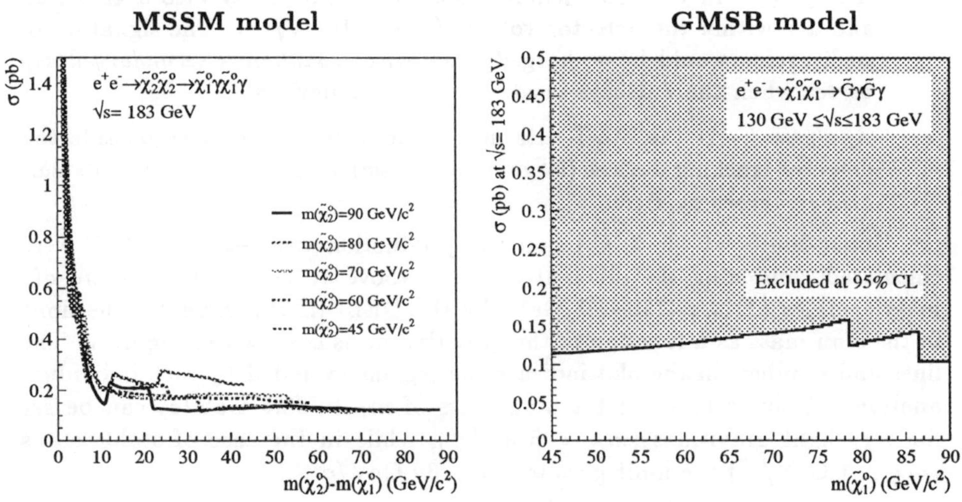

Figure 6: Upper limits at 95 % C.L. from DELPHI on the cross–section for: (left plot) process $e^+e^- \to \tilde{\chi}_2^0\tilde{\chi}_2^0 \to \tilde{\chi}_1^0\gamma\tilde{\chi}_1^0\gamma$ as a function of the mass difference between $\tilde{\chi}_2^0$ and $\tilde{\chi}_1^0$ for some given values of $m_{\tilde{\chi}_2^0}$; (right plot) process $e^+e^- \to \tilde{\chi}_1^0\tilde{\chi}_1^0 \to \tilde{G}\gamma\tilde{G}\gamma$ as a function of the $\tilde{\chi}_1^0$ mass. The limits are obtained by combining all the data taken at centre–of–mass energies from 130 GeV to 183 GeV. The limits are equally valid for a generic process of the type $e^+e^- \to YY \to X\gamma X\gamma$ where X is an undetected particle.

4.3 Stau search results

As already mentioned, although most of the attention has been focused in the case in which the neutralino is the NLSP in *GMSB* models, it is equally possible that the NLSP is a charged slepton and, in particular, a stau. In this case, its decay width $\Gamma(\tilde{\tau} \to \tau\tilde{G})$ and decay length are just given by equations (1) and (2). Therefore, different analyses are applied for the very distinctive signatures observed depending on $m_{\tilde{\tau}}$ and $m_{\tilde{G}}$:

- *Analysis 1*: At one extreme, if the decay length is too short to be detected ($m_{\tilde{G}} \sim 0$), the final state will be identical to that of the MSSM $\tilde{\tau}$ (two acoplanar taus) in the limit of vanishing neutralino mass ($m_{\tilde{\chi}_1^0} \to 0$).

- *Analysis 2*: If the $\tilde{\tau}$ just fly a few centimeters (small $m_{\tilde{G}}$) and the decay is too short to allow for the reconstruction of its track, only the decay products of the τ will be seen in the detector, but the tracks will be slightly displaced from the main interaction vertex and the search has to be based on the *measurement of track impact parameters*.

- *Analysis 3*: In the intermediate case, the $\tilde{\tau}$ can decay into a gravitino and a τ within the detector volume ($m_{\tilde{G}} \sim 10$ eV/c^2). The signature of such events would be a charged track with a *kink* or a *secondary decay vertex* when the $\tilde{\tau}$ decays inside the tracking devices.

- *Analysis 4*: At the other extreme, if the decay takes place outside the detector tracking devices (large $m_{\tilde{G}}$), the signature will be that of a *stable heavy charged particle*.

Upper limits at 95 % C.L. on the production cross–section of $\tilde{\tau}$ were calculated combining the different analyses above at the different centre–of–mass energies (E_{cm} = 130–183 GeV) by the DELPHI experiment. The limit in the stau mass as a function of the gravitino mass is shown in Figure 7. The lines and numbers in the plot indicate the regions excluded by each individual analysis. A lower limit on the stau mass of $m_{\tilde{\tau}} > 68.5$ GeV/c^2 can be set for very light gravitinos ($m_{\tilde{G}} < 5$ eV/c^2), while in the case of stable staus ($m_{\tilde{G}} > 1$ GeV/c^2) the limit goes to $m_{\tilde{\tau}} > 80$ GeV/c^2.

Acknowledgements

I would like to thank the conference organization committee their invitation to the conference and the four LEP experiments ALEPH, DELPHI, L3 and OPAL for the results and plots they provided to me to be shown in the conference.

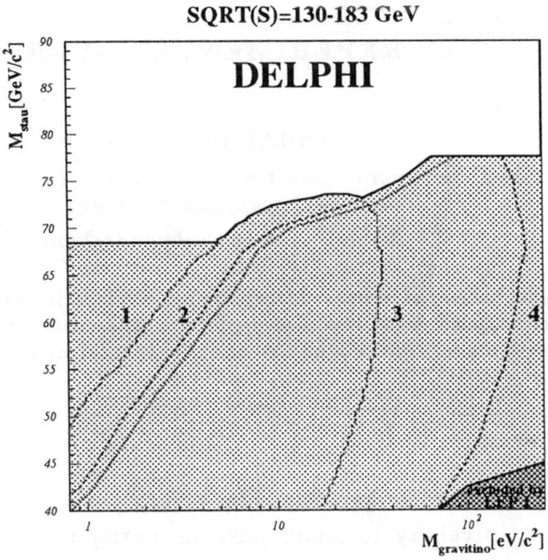

Figure 7: Region in the $(m_{\tilde{\tau}},\, m_{\tilde{G}})$ plane excluded by DELPHI at 95 % C.L. by combining Analyses 1, 2, 3 and 4 described in the text in the energy range E_{cm} =130–183 GeV.

References

1. "LEP Storage Ring", *Technical Notebook* **CERN** (1989)
2. ALEPH Coll., *Nucl. Inst. and Meth.* **A360** (1995) 481;
   DELPHI Coll., *Nucl. Inst. and Meth.* **A378** (1996) 57;
   L3 Coll., *Nucl. Inst. and Meth.* **A289** (1990) 35;
   OPAL Coll., *Nucl. Inst. and Meth.* **A305** (1991) 275
3. Talk transparencies at *http://ific.uv.es/~navas*
4. P.Fayet and S.Ferrara, *Phys. Rep.* **32** (1977) 249;
   H.E.Haber and G.L.Kane, *Phys. Rep.* **117** (1985) 75;
   M.Chen, C.Dionisi, M.Martinez and X.Tata, *Phys. Rep.* **159** (1988) 201
5. L.E.Ibañez and C.Lopez, *ibid.* B**233** (1984) 511;
   J.Ellis and F.Zwirner, *Nucl. Phys.* B**338** (1990) 317
6. P.Fayet, *Phys. Lett.* B**86** (1979) 272;
   J.Ellis, K.Enqvist and D.Nanopoulos, *Phys. Lett.* B**151** (1985) 357;
   S.Ambrosanio et al., *Phys. Rev.* D**56** (1997) 1761;
   S.Dimopoulos et al., *Phys. Rev.* D**54** (1997) 3283;
   J.L.Lopez and D.Nanopoulos, *Mod. Phys. Rev.* **D55** (1997) 4450
7. A.Brignole, F.Feruglio and F.Zwirner preprint *hep-ph/9711516*
8. LEP SUSY working group notes *LEPSUSYWG/98-02.1*, *LEPSUSYWG/98-03.1* and *LEPSUSYWG/98-04.1*.

POLARIZED DIS: PAST AND FUTURE
AN EXPERIMENTAL OVERVIEW

ABHAY DESHPANDE

Physics Department, Yale University, P.O.Box 208121, New Haven CT 06520-8121, USA

E-mail: Abhay.Deshpande@yale.edu

After a brief history of Polarized Deep Inelastic Scattering experiments I will review the present experimental status of g_1 data emphasizing the open questions. I will then present a short outlook on future experimental efforts.

1 History

Interest in physics with spin variables began with Madame Wu's experiments at Columbia University to study possible parity non-conservation in β-decay. The discovery[1] attracted a lot of attention to helicity effects in particle physics. This was around 1957/58 when internal structure of the proton was neither probed nor known. Efforts to produce a polarized electron beam, which could be used as polarized lepton probes, began soon after. Two methods tried were: a) Mott scattering of unpolarized electrons off high Z targets, and b) photoionization of polarized alkali atomic beams. The first method failed to produce significant polarizations in electron beams, while the second was successful[2], and produced average polarizations of 58%. Interestingly, the motivation for producing an electron beam with high polarization was not well focussed. It was believed that once a polarized electron beam was produced, it was sure to have many applications.

In 1968 at the Rochester conference in Vienna, P. Panofsky reported on the Deep Inelastic Scattering (DIS) of electrons from protons and the first experimental evidence for the internal structure of protons. Immediately after that J. Kuti and V. W. Hughes seriously considered an experiment which would scatter polarized electron beam off a polarized stationary proton target[2]. Experimental deep inelastic *polarized* scattering was thus born.

In 1966 J.D.Björken wrote down a relation[3] between the first moments of the proton and neutron spin structure function and the weak interaction constant g_A/g_V: $\Gamma_1^p - \Gamma_1^n = \frac{1}{6}\frac{g_A}{g_V}[\Delta u - \Delta d]$. It was derived using current algebra. The only assumption made in the derivation was that the weak currents are the *same* for quarks and lepton sectors. This relation became, and has remained, one of the most important relation in particle physics to be tested experimentally. It motivated the experimental effort then, as it has until today. The first polarized scattering experiment (E80) was approved[2] at SLAC in October 1971 and the Yale-SLAC experiment E130 followed.

2 Introduction

The Yale-SLAC experiments took data with polarized electron beams scattering off stationary proton targets. This was followed by the European Muon Collaboration's (EMC) experiment at CERN which probed deeper into the proton. In 1989 EMC published its results on spin structure functions which created tremendous excitement in the physics community. What were those results? What was their significance?

Quantum Chromodynamics (QCD), a field theory that describes interaction between quarks and gluons was in its infancy then and was largely untested. Now we know that the spin of the proton should have contributions from the valence quarks ($\Delta\Sigma$), the gluons (Δg), and the orbital angular momenta associated with the quarks (L_q) and gluons (L_g):

$$\frac{1}{2} = \frac{1}{2}\Delta\Sigma + \Delta g + L_q + L_g \tag{1}$$

The accepted theory for the spin of the proton before EMC was based on the Quark Parton Model (QPM). In QPM there are neither gluons nor any angular momenta associated with the internal components of the proton. The proton spin is carried by the valence quarks i.e. $\Delta\Sigma = 1$. With some relativistic corrections and assumption of no sea quark contribution ($\Delta s = 0$), the value of $\Delta\Sigma$ was expected to be about 0.6[4]. The EMC measurements[5] of $\Delta\Sigma = 0.12 \pm 0.17$ and $\Delta s = -0.19 \pm 0.06$ violently contradicted the predictions of the Quark Parton Model. Thus indicating (although with large uncertainty) that the valence quarks hardly contribute to the proton spin, while the sea quarks are polarized in a direction *opposite* to the proton's spin! A need for more precise experiment was felt. This resulted in a series of experiments E142 to E155 at SLAC, and the 5 year running of the Spin Muon Collaboration's experiment NA47 at CERN starting in 1992. The HERMES experiment at DESY was approved a little later and has started taking data since 1995.

In this paper, I will give an overview of the published measurements on g_1 structure function from the CERN, SLAC and DESY experiments. With the availability of more accurate data sophisticated theoretical techniques to analyze them have been developed in the last three years. Results from such analyses have led to new questions which can only be answered by future experiments in this field. I will discuss them very briefly towards the end.

3 What motivated SMC, SLAC and HERMES Experiments?

The first moment of the spin structure function for the proton (neutron) is given in terms of perturbative QCD coefficient functions C_S and C_{NS} for the singlet and nonsinglet sector by

$$\Gamma_1^{p(n)} = \left[\pm\frac{1}{12}a_3 + \frac{1}{36}a_8\right] C_{NS} + \frac{1}{9}a_0 C_S \qquad \text{where} \tag{2}$$

$$a_3 = \Delta u - \Delta d; \quad a_8 = \Delta u + \Delta d - 2\Delta s; \quad a_0 = \Delta u + \Delta d + \Delta s \tag{3}$$

where $\Delta q = q^{\uparrow} - q^{\downarrow}$ with the understanding that for any quark type q, the superscript $\uparrow$ ($\downarrow$) indicates that the quark is aligned (anti-aligned) w.r.t. the nucleon spin. In QPM the C_S and C_{NS} coefficient functions reduce to 1 without any dependence on the value of $\alpha_s(Q^2)$, the strong coupling constant.

Björken Sum Rule

In the framework of pQCD the Björken sum rule can be written as $\Gamma_1^p - \Gamma_1^n = \frac{1}{6}a_3 C_{NS}$ where $a_3 = |g_A/g_V^{n\to p}| = 1.2601 \pm 0.0025$[6] determined from neutron β decay assuming SU(2) symmetry. It is a fundamental result of QCD and the coefficient function C_{NS} is known to $\mathcal{O}(\alpha_s^4)$. Because of the fundamental nature of this sum rule (discussed in section 1), experimental test was of paramount importance in particle physics. Test of this sum rule needs measurements with proton as well as neutron targets.

Ellis-Jaffe Sum Rules

Assuming $\Delta s = 0$ in the equations for the first moments of the proton and neutron (above) and flavor SU(3), Ellis and Jaffe calculated[4] the values of $\Gamma_1^{p,n}$. This sum rule is less fundamental compared to the Björrken sum rule, but historically, suggested a simple and clean test of the quark parton model as we knew it at the time. Once the first moment for the proton or neutron spin structure function is experimentally determined, its straight forward to evaluate the value of $a_0 = \Delta\Sigma$ using the Eq.2,3 in the QPM. EMC's results indicating violation of the Ellis-Jaffe sum rule attracted so much attention that it was called the "spin crisis" in the theoretical world. The value of $a_8 = 3F - D = 0.579 \pm 0.025$[7] has been determined from the hyperon weak decay in the baryon octet. Perturbative QCD corrections to this rule have been estimated to $\mathcal{O}(\alpha_s^3)$.

4 Experimental Measurement of Spin Structure Function

In this section the general method of measurement of spin structure functions is described. This is followed by some specifics about the experiments by the SMC, SLAC, and HERMES collaborations.

Spin dependent structure function measurement needs polarized beams and polarized targets. The experiments measure the cross section asymmetries $A_{\parallel} = \frac{d\sigma^{\uparrow\downarrow} - d\sigma^{\uparrow\uparrow}}{d\sigma^{\uparrow\downarrow} + d\sigma^{\uparrow\uparrow}}$ and $A_{\perp} = \frac{d\sigma^{\uparrow\rightarrow} - d\sigma^{\downarrow\rightarrow}}{d\sigma^{\uparrow\rightarrow} + d\sigma^{\downarrow\rightarrow}}$. The arrow-doublets in the equation indicate the relative orientation of the the beam and target's spin orientation. Since the virtual asymmetries (A_1 and A_2) are related to the interaction of photon and the nucleon (and not between the beam and target nucleon), one has to take into account the depolarization factors involved. The relations between the $A_{\parallel}, A_1$, and g_1 (the spin structure function) are given in the set of equations: $A_1 = \frac{\sigma_{1/2} - \sigma_{3/2}}{\sigma_{1/2} + \sigma_{3/2}}$; $A_{\parallel} = D(A_1 + \eta A_2)$; and $A_1 = \frac{g_1 - \gamma^2 g_2}{F_1}$ where D, d are the depolarization factors mentioned before, and η, η' are other factors which also depend on the kinematics alone. The value of $\gamma^2 = 2Mx/Q^2$ where M stands for the target nucleon mass, is typically very small for SMC and moderately small for the other experiments. Normally in the measurement of the longitudinal structure functions g_1 the corrections due to the A_2 (which has been measured and found to be very small) are ignored, and an extra uncertainty is included in the total systematic uncertainty. The bottom line for the g_1 measurement is:

$$g_1 \approx A_1 F_1 = \frac{1}{D} A_{\parallel} \frac{F_2}{2x(1+R)} \tag{4}$$

where F_2 is the unpolarized structure function measured in unpolarized DIS experiments, and $R = \sigma_L/\sigma_T$ ratio which is typically very small. The x, Q^2 dependence of all the structure functions mentioned in this section is assumed but not explicitly stated. For further details on the aspects of measurement of g_1 and g_2 see Ref. [8,9]

The NA47 Experiment at CERN by the Spin Muon Collaboration

The SMC experiment measured spin structure functions with a polarized muon beam scattered off of a polarized solid state target. Table 1 shows the experimental conditions for the five years of running. Only a brief description of individual components of the experiment (the polarized target, the polarimeter(s) and the spectrometer) are given here, for details see [8].

The SMC polarized target system consisted of three subsystems: a) the magnet system, b) the dilution refrigerator and c) the twin targets with the NMR measurement setup. The magnet system provided a 2.5 T transverse field (superconducting coils) and a dipole field of 0.5 T (used for spin reversal). The $^3He/^4He$ dilution refrigerator was capable of 30 mK in frozen spin mode and about 100mk-300mk during the dynamic nuclear polarization (DNP). The target had two halves polarized in opposite direction. The polarization was measured using a NMR system. The muon beam polarization was measured using two independent methods: a) the decay method, and b) the $\mu - e$

Table 1. The experimental run conditions for SMC,SLAC and HERMES experiments. Symbol "⋆" indicates unpublished data at this time.

SMC Experimental Runs					
Year		Beam	Beam Pol.	Target	Tgt.Pol
1992		100 GeV	-0.80	d-Butanol	0.30
1993		190 GeV	-0.80	p-Butanol	0.85
1994		190 GeV	-0.80	d-Butanol	0.50
1995		190 GeV	-0.80	d-Butanol	0.50
1996		190 GeV	-0.80	p NH_3	0.90
$\delta P/P$			$\leq 3\%$		$\leq 3\%$
SLAC Experiments					
1992	E142	26 GeV	0.40	^{3}He	0.30–0.40
1993	E143	29 GeV	0.80	NH_3	0.65–0.80
				ND_3	0.25
1995	E154	48 GeV	0.82	^{3}He	0.40
1997⋆	E155	48 GeV	0.82	NH_3	0.85
				LiD	0.25
1999⋆	E155 (g_2)	48 GeV	–	NH_3	–
				LiD	–
$\delta P/P$			$\leq 2\%$		$\leq 3\%$
HERMES Experimental Runs					
1995		27.5 GeV	0.55	^{3}He	0.50
1996⋆		27.5 GeV	0.55	p	0.80
		27.5 GeV	0.55	d	
1997⋆		27.5 GeV	0.55	p	0.90
$\delta P/P$			$\leq 6\%$		$\leq 6\%$

scattering method. The decay method consists of determination of the shape of the Michel spectrum of the decay positron from the muon. The $\mu - e$ scattering method consisted of calibration of muon beam polarization using elastic scattering of polarized muons off of polarized electrons in a metalic target with known polarization. The two methods gave consistent values of the beam polarization. The SMC spectrometer had beam hodoscope for beam definition, a large 5 T spectrometer magnet and about 150 layers of wire chamber planes, muon detectors, electromagnetic and hadronic calorimeter.

The E142, E143, E154 and E155 Experiments at SLAC

The SLAC experiments measured the nucleon spin structure functions by scattering polarized electron beam from polarized gas and solid state targets.

Table 1 gives the details of running conditions for the series of SLAC experiments.

Operation of the SLAC polarized source involved laser photoemission from a (unstrained/strained) gallium arsenide (GaAs) photocathode. The polarized electrons produced this way were then accelerated to energies between 9 to 48 GeV. The beam polarization was measured using a Moller polarimeter. The polarized target setup consisted of a high powered ^{4}He evaporation refrigerator operating at 1 K and a 5 T superconducting split pair magnet. A single NMR coil measured the polarization of the target. Up to three large acceptance spectrometers situated in an angular range from 2.5^o to 9° were used for event reconstruction. Each spectrometer arm consisted of dipole magnets, threshold Čerenkov detectors, scintillation hodoscopes, and arrays of lead-glass. For more details see [9,10].

The HERMES Experiment at DESY

The experiment utilizes the self polarized electron beam in the HERA ring at 27.6 GeV and a stationary polarized target situated the path of the electrons to measure polarized structure functions. Table 1 shows the experimental conditions of the HERMES experiment. For experimental details see [11].

The HERMES target consists of a windowless storage cell. The windowless storage cell improves the over all dilution factor involved in the DIS, and also allows fast polarization reversals of the target's spin orientation. The electron in the HERA ring become transversely polarized by an effect called Sokolov-Turnov effect. Electron spin rotators before and after the experimental area are used to achieve longitudinal polarization in the experimental area. The polarization of the beam is measured using an inverse compton scattering polarimeter situated in the north area. A longitudinal polarimeter is now operational, and final results from the measurements of this polarimeter are expected soon. The experiment can detect DIS events and with RICH counters it will do good particle identification in future.

The explored $x - Q^2$ region

The three experimental efforts at CERN, SLAC and at DESY have resulted in measurements in an $x - Q^2$ region shown in Fig. 1. Only the region with $Q^2 > 1$ GeV^2 (DIS) relevant here is shown. Measurements made with μ and e beams are complementary in their $x - Q^2$ coverage, however, individual experiments cover a rather small Q^2 range at any particular value of x. This is especially true at lower values of x. Compared to the SLAC and HERMES experiments the SMC experiment measures g_1 at lower values of x at the same values of Q^2 or at higher values of Q^2 for any particular value of x. Because of the higher electron beam intensity (compared to the muon beam) the measurements from SLAC have higher statistical accuracy.

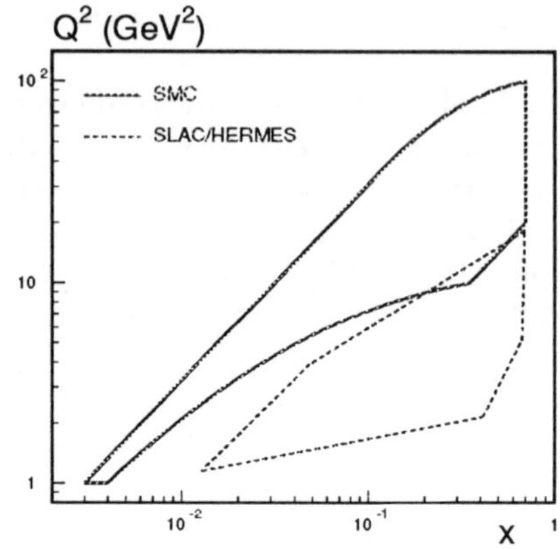

Figure 1. The $x - Q^2$ region explored by SMC, SLAC and HERMES experiment.

5 Data

The final data on virtual photon-proton and deuteron asymmetry, $A_1^{p/d}$ measured by the SMC experiments is shown in Fig. 2 along with the spin structure function measurements $g_1^{p/d}$[8]. SMC has made use of a new method (hadron tagging) of determining A_1 compared to their previous measurements[8] which reduces the dilution of event sample at low x due to radiative tails by as large as 40%. All events are analyzed in both ways, and the method that gives better result in each x bin is taken. For details see [8].

Experimental measurements from the SLAC experiment (E143 and E155) for proton and deuteron targets are shown in Fig.3. For comparison of statistical uncertainty, SMC's data along with almost final (but not quite the set shown before) data set. Most results on the sum rules given in this paper will have only the E143 data in them. E155 g_1 data is preliminary when this article goes to the press.

The E142 and E154 collaboration at SLAC and the HERMES collaboration at DESY have made direct measurements on the neutron spin structure function. Fig. 4 shows these measurements. All measurements are consistent with each other. E154 measurements are the latest to be made and published. The impressive improvement over its previous measurement (E142) was mainly due to significantly higher beam and target polarizations achieved. (See Table 1).

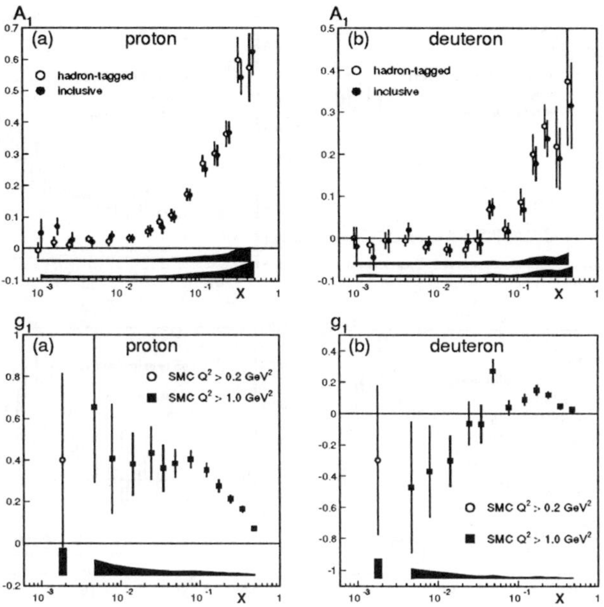

Figure 2. The $A_1^{p/d}$ and $g_1^{p/d}$ final results from SMC.

6 First Moments of Polarized Structure Functions

To evaluate the first moments, $\Gamma_1(Q_0^2)$, measurements of g_1 made at different $Q^2 = Q_m^2$ have to be evolved to a common $Q^2 = Q_0^2$. Before the advent of pQCD at next-to-leading order (NLO) for g_1, the evolution of g_1 was performed with the assumption that: $A_1(x)$ was independent of Q^2. Although the data are not precise enough to indicate clear Q^2 dependence, recently theoretical and experimental groups have relied on the pQCD techniques at NLO to extract the extent of Q^2 dependence and used that in the evaluation of $\Gamma_1^{p,n,d}$. Theoretical details of this procedure can be found in [12].

Polarized parton distribution functions (pdfs) are parametrized at an initial value of $Q^2 = Q_i^2$. DGLAP evolution equations[13] are used to evolve thm from one Q^2 to another using the coefficient[14] and splitting[15] functions evaluated at NLO. Singlet first moments for quarks and gluons are determined from the data, while the nonsinglet quark first moment (η_{ns}) is fixed using experimental results: $\eta_{ns} = \pm 0.75(g_A/g_V) + 0.25a_8$ where g_A/g_V has been determined from the β decay in the $n \to p$ transition to be 1.2601 ± 0.0025[6]

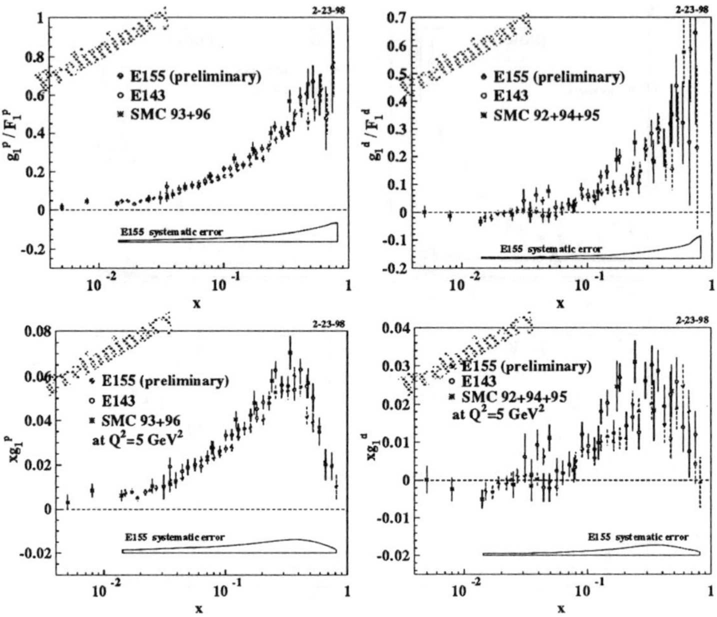

Figure 3. Proton and Deuteron spin structure function measurements from SLAC.

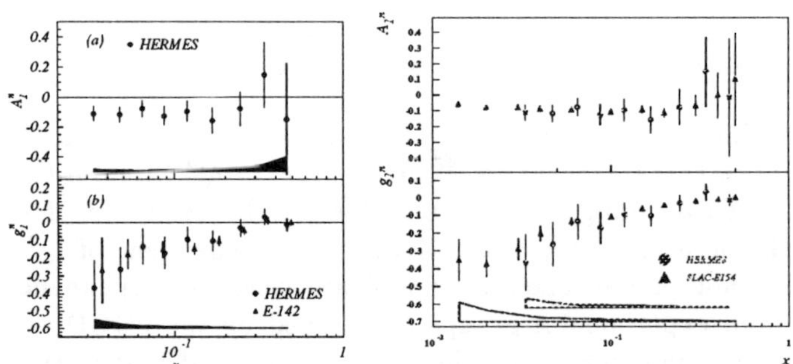

Figure 4. The A_1^n and g_1^n measurements from E142, HERMES and E154.

and a_8 has been determined from the hyperon weak decay in the baryon octet to be 0.575 ± 0.016[7]. By fixing the value of g_A/g_V from the experimental measurements the Björken sum rule is built into this analysis. Recently, by allowing the data to determine the value of g_A/g_V from pQCD fit, Börken

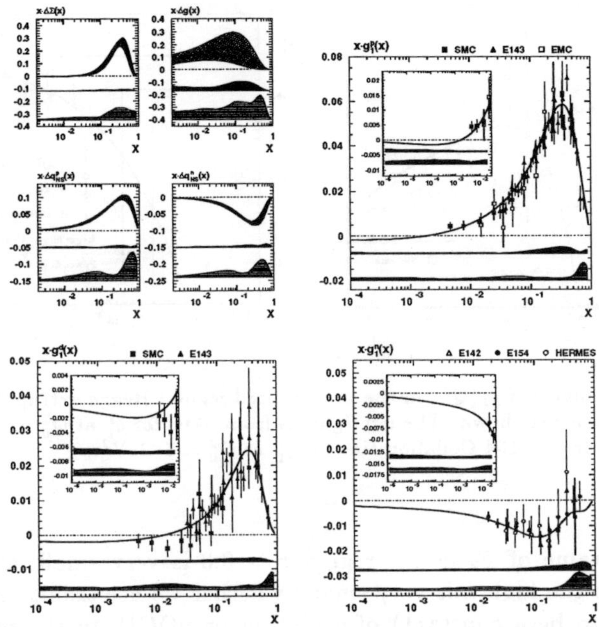

Figure 5. Parton distribution functions (four figures on top left) at NLO and their statistical, experimental systematic and theoretical systematic uncertainties. Top right and bottom left and right: $g_1^{p,d,n}$ world data and the pQCD best fit evaluated at $Q^2 = 5$ GeV^2. Figure from SMC analysis.

sum rule has been tested in a novel way[8,10,16]. Most analyses use the world sample of data with the condition $Q^2 > 1$ GeV^2 available to them at the time of the analysis. For a detailed description of the evaluation of systematic uncertainty please refer to [8,10].

Broad conclusions from all pQCD analyses are that out of the parton distributions the nonsinglet and singlet quark distributions are rather well determined, while the polarized gluon distribution has the largest uncertainty. Most significant source of uncertainty is theoretical in origin, but is related to the absence of data in the low x region ($x < 0.003$).

In the measured x range: $0.003 < x < 0.8$ integrating over the experimental data evolved to a fixed Q^2 gives a part of the first moment. Lets now discuss the regions outside of the measured x range, namely: a) $x > 0.8$ and b) $x < 0.003$.

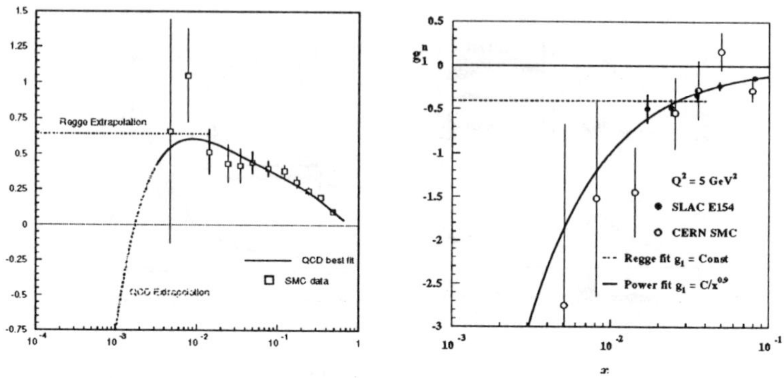

Figure 6. The behavior of $g_1(Q_0^2)$ in the unmeasured region: Regge extrapolation and the pQCD extrapolation are shown. The left figure is from SMC for g_1^p at $Q_0^2 = 10$ GeV^2 while the right figure is from E154 Collaboration for g_1^n at $Q_0^2 = 5$ GeV^2.

Since the value of F_2 in the region $x > 0.8$ is very small, the contribution to Γ_1 from the high x is very small[8] whether we use traditional methods (assuming A_1 to be a constant) of evolution or pQCD, furthermore the are consistent with each other. At $Q^2 = 5$ GeV^2 the results for proton are assuming $A_1 =$ constant are $\int_{0.8}^{1} g_1(x) = 0.0015 \pm 0.0006$ and using pQCD they are $\int_{0.8}^{1} g_1(x) = 0.0018 \pm 0.0003$.

In the low x region, however, the traditional method (based on Regge inspired models) and the pQCD analyses give significantly different results. The two low x extrapolations are shown in Fig. 6. At $Q^2 = 5$ GeV^2 the Regge extrapolation results in a contribution of 0.002 ± 0.002 to the first moment while the pQCD results in $-0.012^{+0.014}_{-0.025}$. Both approaches are questionable: a) The collaborations use $Q_0^2 > 3.0$ GeV^2 to evaluate their results. Validity of Regge behavior at $Q^2 > 1.5$ GeV^2 has been questioned by the HERA unpolarized DIS experiments. Also, the method to evaluate the uncertainty i.e. assigning an *adhoc* uncertainty of 100% of the Regge extrapolation integral is questionable[8,9]. b) On the other hand, although the uncertainty evaluation is on a better footing in the pQCD analysis, validity of the fits (pdfs) beyond the region of data available is questionable. Because pQCD uncertainty is rather conservative and covers the Regge value of low x extrapolation SMC has accepted pQCD method evaluate their results. E154 in their publication [10] have faced a similar problem with the evaluation of the first moment of the neutron structure function and chosen later to take the pQCD results[10].

Table 2. Recent results on Ellis-Jaffe and Börken sum rules from various collaborations. Note that although the detailed method of evaluating integrals and their uncertainties for different results is different, conclusions about sum rules are the same.

Ellis Jaffe Sum Rule				
Collaboration	Nucleon	Q^2 GeV^2	Theory	Exp. Result
SMC	Proton	10	0.170 ± 0.004	0.120 ± 0.016
	Deuteron	10	0.071 ± 0.004	0.019 ± 0.015
E143	Proton	3	0.160 ± 0.005	0.133 ± 0.010
	Deuteron	3	0.066 ± 0.004	0.047 ± 0.007
E154	Neutron	5	-0.018 ± 0.004	-0.058 ± 0.011
HERMES	Neutron	2.5		-0.037 ± 0.015

Börken Sum Rule			
Collaboration	Q^2 GeV^2	Theory	Exp. Result
SMC(a)	5	0.181 ± 0.003	$0.174^{+0.024}_{-0.012}$
E143	3	0.177 ± 0.004	0.164 ± 0.023
E154	5	0.181 ± 0.003	0.171 ± 0.013

7 Results: Spin Sum Rules

Results by experimental collaborations for Ellis Jaffe sum rule are summarized in Table 2 along with their theoretical expectations. Data show clear violation of the Ellis Jaffe sum rule. The violation indicates that the assumptions made in deriving the first moments of the nucleon spin structures i.e. SU(3) symmetry and $\Delta s = 0$ may not be correct. Using the values of first moments various collaborations have evaluated the resulting values of Δs and found them to be negative[8,9,10].

Table 2 summarizes the recent results on the verification of Börken sum rule from various experimental collaborations along with their theoretical expectations. The sum rules is clearly valid with about 10% relative accuracy on the experimental side. The theoretical calculation of this result is significantly better at this time.

8 Open Questions: Low x Behavior of g_1 and Gluon's Spin

Because of the large uncertainty in the low x behavior of g_1 and its non-negligible effect on the evaluation of Γ_1, as of now, this is the most important unknown quantity in evaluating Γ_1s. Different models for the low x behavior exist based on very different underlying physics. They predict very different behaviors. Only direct measurements can resolve the discrepancies. All recent papers on the subject from emphasize this point and have recommended a need for the measurement of g_1 in region $x < 0.003$ which is accessible with a polarized HERA collider at DESY[8,10,16,17]. Measuring g_1 in this low x region will be a valuable source of information to understand the parton dynamics in this kinematic region.

The only way one can access the polarized gluon distribution at this time is through the pQCD analysis of the g_1 structure functions in which the Δg appears at next-to-leading order. A recent result from SMC (a typical one, consistent with most other analyses) is: $\int_0^1 \Delta g(Q^2 = 1\ \mathrm{GeV}^2)dx = 1.0^{+1.2}_{-0.3}(stat)\ ^{+0.4}_{-0.2}(syst)\ ^{+1.4}_{-0.5}(theory)$. It has been argued[16,17] that pQCD analysis of g_1 measured over a larger x, Q^2 range would help reduce the uncertainty in the polarized gluon distribution significantly. In addition, a process which might involve polarized gluon distribution at leading-order (LO) (like in photon-gluon fusion) would be an independent and alternative way to access the polarized gluon distribution. Experimental investigations of these kinds of reactions, are presently being planned.

9 Experimental Outlook

To resolve the two out standing questions i.e. the low x behavior of g_1 structure function, and the determination of $\Delta g(x, Q^2)$ in a nucleon, three major experimental efforts are underway.

The spin program within COMPASS experiment at CERN will have a polarized muon beam scattering of a polarized solid target. Compared to SMC a major increase in the beam intensity, a much improved polarized target system and significantly improved detector system capable of particle identification (unlike SMC's) would allow access to the polarized gluon distribution through the photon gluon fusion process. The main goal is to measure Δg through at least two different approaches: a) photon-gluon fusion in open charm production, and b) pgf in high-p_T jets which also relate to the polarized gluon distribution. The engineering run is expected to be held later 1999, and the physics runs will start in year 2000. With three years of data taking starting 2000 the experiment is expected to give the first realistic glimpse of the polarized gluon distribution.

Relativistic Heavy Ion Collider (RHIC) is under construction at Brookhaven National Laboratory with a target date for completion of October 1999. The collider is designed to accelerate *polarized* proton beams (among other nuclei) and will have a capability to collide them at luminosities of $\mathcal{L} = 2 \times 10^{32}\ \mathrm{cm}^{-2}\ \mathrm{sec}^{-1}$ and variable $\sqrt{s}$ up to 500 GeV. A rich physics program with possibilities of measurement of the polarized gluon distribution using a) direct photons, b) inclusive jets and c) heavy quark production in photon-gluon fusion processes, will form the core of spin studies at RHIC. The measurements will be done with the STAR and PHENIX detector presently under construction at BNL. Because of the depolarizing resonances encountered while accelerating polarized protons to a high energy, feasibility of having *polarized* proton beams in the RHIC ring is one of the open question. Further, the polarimetry of high energy beams is yet another open question which is under investigation now[19].

Operation of the HERA collider with 820 GeV polarized protons and 27.6 GeV polarized electron/positron beams will allow a study of a wide variety of spin observables at values at $\sqrt{s}$ = 300 GeV. Measurements of spin observables such as $g_1(x, Q^2)$ and $\Delta g(x, Q^2)$ in the range $10^{-6} < x < 1$ and $0.2 < Q^2 < 10^5$ GeV^2 would be possible[17]. In addition, direct determination of the polarized gluon in the region $0.002 < x_g < 0.2$ from polarized di-jet rates and hadrons with high p_T will be possible[18]. Possibilities of determination of polarized quark distributions from weak structure functions and semi-inclusive asymmetries, parton distributions in polarized photons, information on the helicity structure of possible new physics at very high Q^2 could all be explored with the polarized HERA collider[17].

The possibilities of having polarized protons in the HERA ring is under consideration now. It is proposed that ZEUS and H1 detectors continue to be operational after the present unpolarized program ends. The decision of whether HERA would run with polarized proton beam is expected to occur around 2002/2003 after the technical feasibility of having polarized protons in the HERA ring is confirmed. If this does happen, this facility will address the outstanding problems of spin physics in the most profound way.

10 Summary

I have presented an overview of g_1 measurements from an experimentalist's point of view starting with the early beginnings of the field, going all the way into the future. Present experimental data indicate that the Börken sum rule is valid and that the Ellis Jaffe sum rule is violated. The valence quarks carry very little spin of the nucleon, while the sea quarks aligned themselves against the nucleon's spin. Data indicates that gluon play a significant role in the spin dynamics of the nucleon. To solve the outstanding questions of low x and Δg, three major experimental efforts are underway at CERN, BNL, and possibly in future at HERA.

Acknowledgments

I wish to thank Prof. B. Adeva for inviting me to this meeting in Spain. Without the organizational help from Ms. Mercedes Fatas the trip would have been impossible. Thanks are also due to A. Brüll and C. Young for information on the HERMES and SLAC experiments, respectively.

References

1. C.S.Wu *et al.* Phys. Rev. **105**, 1413 (1957).
2. V. Hughes, in *Symposium on Internal Spin Structure of the Nucleon*, World Scientific, Ed. V.W.Highes, C. Cavata, pp 1 (1994).
3. J.D.Björken, Phys. Rev. **148**, 1467 (1966);PR **D 1**, 1376 (1970).
4. J. Ellis *et al.*, Phys. Rev. **D 9**, 1444 (1974); **10**, 1669(E) (1974).
5. EMC, J.Ashman *et al.*, Nucl. Phys. **B328**, 1 (1989).
6. Particle Data Group, R. Barnett *et al.*, Phys. Rev. **D 54**, 1 (1996).
7. F.E.Close *et al.*, Phys. Lett. **B 316**, 165 (1993).
8. SMC, B. Adeva *et al.*, Phys. Rev. **D 58**, 112001 (1998); 112002 (1998); **D 56**, 5530 (1997).
9. E143, K. Abe *et al.*, Phys. Rev. **D 58**, 112003 (1998).
10. E154, K. Abe *et al.*, Phys. Rev. Lett. **79**, 26 (1997); Phys. Lett. **B 405**, 180 (1997).
11. HERMES, K. Ackeroff *et al.* Phys. Lett. **B404**, 383 (1997).
12. R. Ball *et al.*, Phys. Lett. **B 378**, 255 (1996); M. Glück *et al.*, Phys. Rev. **D 53**, 4775 (1996); T. Gehrmann *et al.*, Z. Phys. **C 65**, 461 (1995); Phys. Rev. **D 53**, 6100 (1996).
13. V.N. Gribov and L.N. Lipatov, Sov. Journ. Nucl. Phys. **15**, 438 (1972); **15**, 675 (1972); G. Altarelli and G. Parisi, Nucl. Phys. B **126**, 298 (1977); Yu. L. Dokshitzer, Sov. Phys. JETP **46**, 641 (1977).
14. J. Kodaira, Nucl. Phys. B **165**, 129 (1980) and references therein.
15. R. Mertig and W.L. van Neerven, Z. Phys. C **70**, 637 (1996); W. Vogelsang, Nucl. Phys. B **475**, 47 (1996).
16. Altarelli *et al.*, Nucl. Phys. **B 496**, 337 (1997).
17. A. Deshpande *et al.* and other articles in *Physics with Polarized Protons at HERA*, DESY-Proceedings-1998-01, February 1998.
18. A. de Roeck *et al.*, Eur. Phys. J. **C 6**, 121 (1999).
19. BNL Experiments E925/E950 Proton beam polarimetry at the AGS.

THE NUCLEON AND POMERON STRUCTURE FUNCTIONS

C. MERINO
Dpto. de Física de Partículas,
Universidade de Santiago de Compostela,
E-15706 Santiago de Compostela, Spain

The CKMT model for the nucleon structure function F_2 gives a good description of the recent HERA data at low and moderate Q^2. Also is discussed the fit to the same data obtained with a modified version of the model in which a logarithmic dependence on Q^2 has been included. We show how the parametrization of the CKMT model for the nucleon structure function F_2 explains the HERA experimental data presented in the Caldwell-plot as the x and Q^2 behavior of the logarithmic slope of F_2. Finally, it's shown that the HERA data on large rapidity gap events can be successfully described by the CKMT model in terms of the Pomeron structure function, where the parameters are determined from soft hadronic diffraction data.

1 The CKMT model

The CKMT model[1] for the parametrization of the nucleon structure function F_2 is a theoretical model based on Regge theory which provides a consistent formulation of this function in the region of low Q^2, and that can therefore be used as a safe and theoretically justified initial condition in the perturbative QCD evolution equation, to obtain the structure function at larger values of Q^2.

The CKMT model[1] proposes for the nucleon structure functions

$$F_2(x, Q^2) = F_S(x, Q^2) + F_{NS}(x, Q^2), \tag{1}$$

the following parametrization of its two terms in the region of small and moderate Q^2. For the singlet term, corresponding to the Pomeron contribution:

$$F_S(x, Q^2) = A \cdot x^{-\Delta(Q^2)} \cdot (1 - x)^{n(Q^2)+4} \cdot \left(\frac{Q^2}{Q^2 + a}\right)^{1+\Delta(Q^2)}, \tag{2}$$

which $x \rightarrow 0$ behavior is determined by an effective intercept of the Pomeron, Δ, which takes into account Pomeron cuts and, therefore (and this is one of the main points of the model) it depends on Q^2. This dependence was parametrized[1] as :

$$\Delta(Q^2) = \Delta_0 \cdot \left(1 + \frac{\Delta_1 \cdot Q^2}{Q^2 + \Delta_2}\right). \tag{3}$$

Thus, for low values of Q^2 (large cuts), Δ is close to the effective value found from analysis of hadronic total cross-sections ($\Delta \sim 0.08$), while for high values of Q^2 (small cuts), Δ takes the bare Pomeron value, $\Delta \sim 0.2$-0.25. The parametrization for the non-singlet term, which corresponds to the secondary reggeon (f, A_2) contribution, is:

$$F_{NS}(x,Q^2) = B \cdot x^{1-\alpha_R} \cdot (1-x)^{n(Q^2)} \cdot \left(\frac{Q^2}{Q^2+b}\right)^{\alpha_R} , \qquad (4)$$

where the $x \to 0$ behavior is determined by the secondary reggeon intercept α_R, which should be in the range α_R=0.4-0.5. The valence quark contribution can be separated into the contribution of the u, B_u, and d, B_d, valence quarks and the normalization condition for valence quarks fixes both contributions at one given Q_0^2 value of Q^2 (we use $Q_0^2 = 2.GeV^2$ in our calculations). For both the singlet and the non-singlet terms, the behavior when $x \to 1$ is controlled by $n(Q^2)$, with $n(Q^2)$ being

$$n(Q^2) = \frac{3}{2} \cdot \left(1 + \frac{Q^2}{Q^2+c}\right) , \qquad (5)$$

so that, for Q^2=0, valence quark distributions have the same power, given by Regge intercepts, as in Dual Parton Model [2], $n(0)=\alpha_R(0)-\alpha_N(0) \sim 3/2$, and the behavior of $n(Q^2)$ for large Q^2 is chosen to coincide with dimensional counting rules.

The total cross-section for real (Q^2=0) photons can be obtained from the structure function F_2 using the following relation:

$$\sigma_{\gamma p}^{tot}(\nu) = \left[\frac{4\pi^2 \alpha_{EM}}{Q^2} \cdot F_2(x,Q^2)\right]_{Q^2=0} . \qquad (6)$$

The proper $F_2(x,Q^2) \sim Q^2$ behavior at $Q^2 \to 0$, is fulfilled in the model due to the last factors in equations 2 and 4.

2 Description of the HERA data on F_2 at low Q^2

When the CKMT model was first used [1] to fit the available experimental data on F_2, the lack of experimental data at low and moderate Q^2 limited the accuracy in the determination of the values of the parameters in the model. Now, the publication of the new experimental data [3,4] on F_2 from HERA at low and moderate Q^2 gives us the opportunity to include in the fit of the parameters of the model experimental points from the kinematical region where the parametrization should give a good description without need of any perturbative QCD evolution.

Thus, we proceed [5] as one had done in the previous fit [1], but by adding the above mentioned experimental data on F_2 from H1 and Zeus at low and moderate Q^2, to those [6] from NMC and E665 collaborations, and to data [7] on cross-sections for real photoproduction, into a global fit which allows the test of the model in larger regions of x and Q^2. We take as initial condition for the values of the different parameters those obtained in the previous fit [1]. The result of the new common fit to $\sigma_{\gamma p}^{tot}$ and F_2 is presented in Figure1, and the final values of the parameters can be found in Table1(b). The quality of the description provided by the CKMT model of all the experimental data, and, in particular, of the the new experimental data from HERA is very good, with a value of $\chi^2/d.o.f.$ for the global fit, $\chi^2/d.o.f.$=106.95/167, where the statistical and systematic errors have been treated in quadrature, and where the relative normalization among all the experimental data sets has been taken equal to 1.

Also, since the small-x HERA experiments allow for the first time the experimental study of the question of the interplay between soft and hard physics, we have modified our model, with basically only power dependence on Q^2, to include a logarithmic dependence on Q^2 as the one predicted asymptotically by perturbative QCD.

To include the logarithmic dependence on Q^2 in our model, we take into account that the behavior of F_2 at small-x in QCD is given by the singularities of the moments of the structure functions [8], the rightmost singularity giving the leading behavior. Thus, we modify 1 in the following way [9]

$$F_2(x,Q^2) = \left(\frac{\alpha_s(Q_0^2)}{\alpha_s(Q^2)}\right)^{d_S(n_S)} \cdot F_S(x,Q^2) + \left(\frac{\alpha_s(Q_0^2)}{\alpha_s(Q^2)}\right)^{d_{NS}(n_{NS})} \cdot F_{NS}(x,Q^2). \tag{7}$$

The exponents of the new factors in 7, $d_S(n_S)$ and $d_{NS}(n_{NS})$, give us the singularities in n_i, i=S,NS, of the momenta, which, as we mentioned above, control the QCD small-x behavior of F_2. Therefore, in our model, these exponents have to be evaluated (see 2 and 4), at n_S=$1+\Delta(Q^2\to\infty)$= $1+\Delta_0(1+\Delta_1)$, and at n_{NS}=α_R, respectively.

Then, we use this modified version of the CKMT parametrization of F_2 to repeat the fit of the same experimental data, including the HERA data on F_2 at small and moderate Q^2. As starting point for the QCD evolution, we take the same value Q_0^2=$2.GeV^2$ we use to fix the normalization of the valence component. The result of this second fit is also presented in Figure1, and the final values of the parameters in the model are given in Table1(c).

As it can be seen in Figure 1, the quality of this second fit is also reasonable, although the value of $\chi^2/d.o.f.$ ($\chi^2/d.o.f.$=453.19/167), is now appreciably higher than in the fit obtained with the non-modified version of the CKMT model.

Table 1: Values of the parameters in the CKMT model obtained in former fits, (a), in the fit in which also the low Q^2 HERA data have been included, (b), and in the fit to the same data obtained with the modified version of the CKMT model in which a logarithmic dependence of F_2 on Q^2 has been taken into account, (c). All dimensional parameters are given in GeV^2. The valence counting rules provide the following values of B_u and B_d, for the proton case, when fixing their normalization at Q_0^2=2.GeV^2: (a) B_u=1.2064, B_d=0.1798; (b) B_u=1.1555, B_d=0.1722; (c) B_u=0.6862, B_d=0.09742. In previous fits, (a), the parameter Δ_1 had been fixed to a value Δ_1=2.

CKMT model	(a)	(b)	(c)
A	0.1502	0.1301	0.1188
a	0.2631	0.2628	0.07939
Δ_0	0.07684	0.09663	0.1019
Δ_1	2.0 (fixed)	1.9533	1.2527
Δ_2	1.1170	1.1606	0.1258
c	3.5489	3.5489 (fixed)	3.5489 (fixed)
b	0.6452	0.3840	0.3194
α_R	0.4150	0.4150 (fixed)	0.5872

3 The Caldwell-plot

The so-called Caldwell-plot shows [10] the logarithmic slope of the structure function F_2, $dF_2/dlnQ^2$, derived from the ZEUS data, as a function of x, by fitting $F_2 \sim a + blnQ^2$ in bins of fixed x, using only statistical errors. This plot allows the study of the QCD scaling violations of F_2, and, in particular, in the small-x domain now accessible at HERA, where $dF_2/dlnQ^2$ can be directly related to the gluon density, can be an useful tool to know down to which value of Q^2 the perturbative NLO DGLAP QCD predictions give a good description of the F_2 data. Thus, the logarithmic slope $dF_2/dlnQ^2$ can be used to investigate the fundamental question of the interplay between the hard and soft physics.

As it can be seen in reference (9), for values of x down to $3 \cdot 10^{-4}$, the slopes are increasing as x decreases, but at lower values of x and Q^2 the slopes decrease, what seems to indicate a deviation from the perturbative behavior of the hard regime.

The figure 2 shows how both Regge based parametrizations with a constant effective value for the Pomeron intercept, like the Donnachie-Landshoff Regge

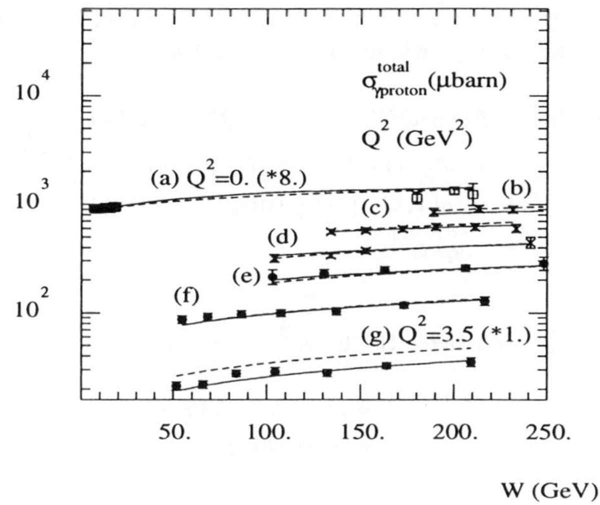

Figure 1: $\sigma^{total}_{\gamma p}$ and $\sigma^{total}_{\gamma^* p}$ in μbarns vs W in GeV for different values of Q^2. Theoretical fits have been obtained with the CKMT model (full line) and the modified version of the CKMT model (dashed line). Points at (a), Q^2=0.GeV^2 (*8.), (b), Q^2=0.15GeV^2 (*6.), (c), Q^2=0.25GeV^2 (*5.), (d), Q^2=0.5GeV^2 (*4.), (e), Q^2=0.8GeV^2 (*3.), (f), Q^2=1.5GeV^2 (*2.), and (g), Q^2=3.5GeV^2 (*1.). Experimental data on F_2 ($\sigma^{total}_{\gamma^* p}$) are from references (3), (4), and experimental data on $\sigma^{total}_{\gamma p}$ are from references (7).

fit or the ZEUSREGGE fit, and pure perturbative NLO QCD predictions, like GRV94 NLO QCD fit and the ZEUSQCD fit (see (9) and references therein for more details), fail in describing correctly the experimental data in the whole kinematical region. While the Donnachie-Landshoff and the ZEUSREGGE fits don't account for the data for values of x larger than $\sim 10^{-5}$, GRV and ZEUSQCD cannot describe the data when we go to values of x smaller than $\sim 6 \cdot 10^{-5}$.

On the contrary, the CKMT model described above, based on the Regge behavior, but with a Q^2-dependent Pomeron intercept, describes [11] the data in the region of low Q^2, and taken as a consistent initial condition for the QCD evolution equations, provides a complete description of the experimental results in the whole ranges of x and Q^2.

Thus, by using the formulae in section 1, we can write [11] for the CKMT model the slope $dF_2/dlnQ^2$ when x is kept fixed as:

$$\begin{aligned} \frac{dF_2(x,Q^2)}{dlnQ^2} = & F_S(x,Q^2)[\frac{\Delta_2}{Q^2+\Delta_2}\left(\Delta(Q^2)-\Delta_0\right) ln\frac{Q^2}{x(Q^2+a)} \\ & +\frac{c}{Q^2+c}\left(n(Q^2)-\frac{3}{2}\right) ln(1-x) + \frac{a\left(1+\Delta(Q^2)\right)}{Q^2+a}] \\ & + F_{NS}(x,Q^2)[\frac{c}{Q^2+c}\left(n(Q^2)-\frac{3}{2}\right) ln(1-x) \\ & +\frac{b\alpha_R(0)}{Q^2+b}], \end{aligned} \tag{8}$$

that in the limit $Q^2 \to 0$ takes the form

$$\begin{aligned} \frac{dF_2(x,Q^2)}{dlnQ^2} \sim & (1+\Delta_0)\, F_S(x,Q^2) \\ & + \alpha_R(0) F_{NS}(x,Q^2). \end{aligned} \tag{9}$$

If we consider the case when W is fixed we can take $x \sim cte \cdot Q^2$, and then, up to constant factors, we can write:

$$\begin{aligned} \frac{dF_2(x,Q^2)}{dlnQ^2} = & F_S(x,Q^2)[-\frac{\Delta_2}{Q^2+\Delta_2}\left(\Delta(Q^2)-\Delta_0\right) ln(Q^2+a) \\ & -\Delta(Q^2) + \frac{c}{Q^2+c}\left(n(Q^2)-\frac{3}{2}\right) ln(1-Q^2) \\ & -\frac{Q^2 n(Q^2)}{1-Q^2} + \frac{a(1+\Delta(Q^2))}{Q^2+a}] \\ & + F_{NS}(x,Q^2)[\frac{c}{Q^2+c}(n(Q^2)-\frac{3}{2}) ln(1-Q^2) \\ & +\frac{b\alpha_R(0)}{Q^2+b} + (1-\alpha_R(0)) - \frac{Q^2 n(Q^2)}{1-Q^2}], \end{aligned} \tag{10}$$

and when we make W fixed with $Q^2 \sim x \to 0$, we can easily see that this equation simply reduces to:

$$\frac{dF_2(x,Q^2)}{dlnQ^2} \sim F_2(x,Q^2). \tag{11}$$

One has to note that both equations 9 and 11 are valid for any well-behaved parametrization of F_2 in the sense of fulfilling the relation in equation 6.

In figure 2 is shown how the CKMT model accounts [11] for the slope $\frac{dF_2(x,Q^2)}{dlnQ^2}$ up to a value of $Q^2 \sim 2.GeV^2$, and how, from there, by using this parametrization as an initial condition for the perturbative QCD evolution equations, we obtain a good description of the data for any values of x and Q^2.

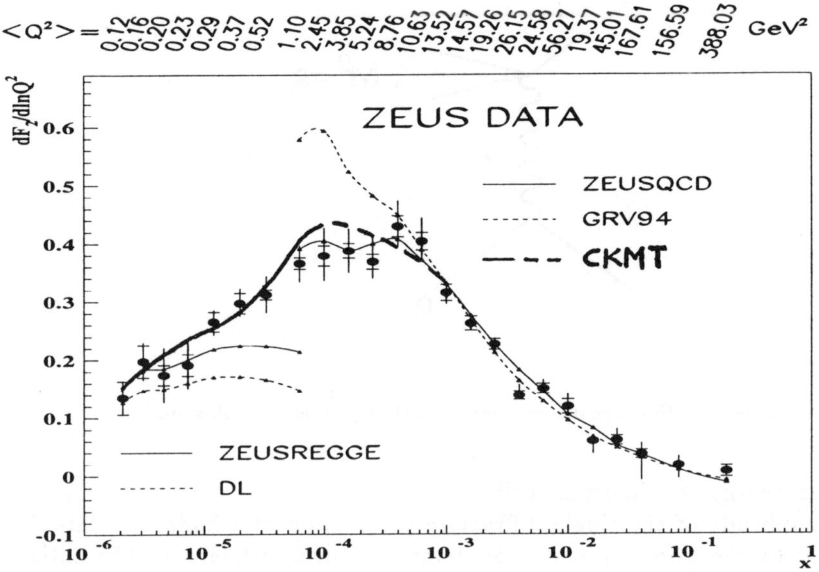

Figure 2: $dF_2/dlnQ^2$ as a function of x calculated by fitting ZEUS F_2 data in bins of x to the form $a + blnQ^2$ (see reference (10) and references therein for more details on the data, and on the ZEUSREGGE, ZEUSQCD, DL and GRV94 fits). For details on the CKMT fit, see reference (11).

4 Hard diffraction at HERA

Experiments at HERA provide new possibilities to study diffractive processes and the properties of the Pomeron. Recently, spectacular large rapidity gap events in deep inelastic scattering at very small-x have been observed by ZEUS [12] and H1 [13] collaborations at HERA. On the other hand, diffractive production has been studied for many years in high energy hadronic interactions and a large amount of information on these reactions has been obtained (see for example (14)). A new unexplored region of diffraction production by highly virtual photons is now accessible at HERA.

Here we shall consider the process of diffractive dissociation of highly virtual photons in the framework of Reggeon theory, which is the basis for the description of diffractive processes in hadronic interactions. We shall demonstrate that the characteristic features of the data on large rapidity gap events can be understood using the Regge based description of structure functions given by the CKMT model, together with our knowledge of diffractive produc-

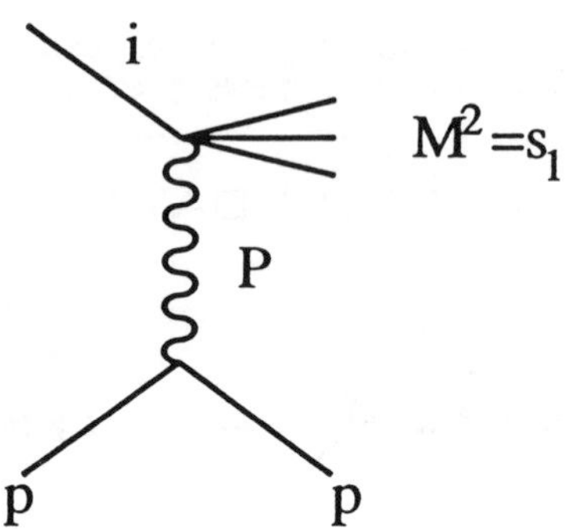

Figure 3: Diffractive dissociation of particle i in an i-p collision.

tion in high-energy soft hadronic collisions.

The amplitude of the single diffractive production of a hadronic state X is described by the Pomeron exchange diagram shown in figure 3. The initial particle i in this figure can be either a hadron or a photon (real or virtual). This diagram corresponds to single diffractive production without dissociation of the initial proton, while double diffraction corresponds to dissociation of both projectile and target. At high energies $s >> m^2$, when the total rapidity interval $\xi = ln\frac{s}{m^2}$ is large, these diagrams describe kinematical configurations of final particles in which large rapidity gaps (Δy) are present. Consider first the single diffraction dissociation of a real particle i (Figure 3). In the pole approximation for Pomeron exchange the cross-section of this process is given by

$$s_1 \frac{d\sigma_D}{dtds_1} = \frac{1}{16\pi}(g_{pp}^P(t))^2 exp\left[2\left(\alpha_P(t) - 1\right)\xi'\right]\sigma_{iP}^{tot}(s_1, t), \qquad (12)$$

where $s_1 \equiv M^2$ is the square of the invariant mass of the hadronic system X, $t = (p' - p)^2$ is the invariant momentum transfer to the final proton and $g_{pp}^P(t)$ is the vertex describing the Pomeron coupling to a proton. The Feynman-x of the Pomeron is defined by $x_P = s_1/s$, and $\xi' = ln(s/s_1) \approx \Delta y$. The last term in 12 is the total Pomeron-particle cross-section [15]. It is obtained by summation over all final states X and over their phase space. Note that this quantity, contrary to the cross-section of usual particles, is just defined by 12, but cannot be interpreted as a physical observable on its own. At large $s_1 >> m_i^2$ this cross-section is determined by the Pomeron and Regge exchanges and has the form (see Figure 4(a)):

$$\sigma_{iP}^{tot}(s_1,t) = \sum_k g_{ii}^k(0)\ r_{PP}^k(t) \left(\frac{s_1}{s_0}\right)^{\alpha_k(0)-1}, \tag{13}$$

with $s_0 = 1.GeV^2$. Here α_k denotes the trajectory of any reggeon k contributing to the i-P elastic amplitude, and $g_{ii}^k(0)$ and $r_{PP}^k(t)$ are the couplings of the reggeon r to particle i and to the Pomeron, respectively (see Figure 4(a)). The only reggeons k that contribute to 13 are the Pomeron itself and the f-trajectory. Note that in the first factor of 12 we have only considered the Pomeron contribution. The contribution of secondary reggeons such as the f-trajectory is important only for $x_P > 0.05$ to 0.1.

In the case when particle i is a virtual photon (with virtuality $q^2 = -Q^2$), $\sigma_{\gamma^* P}^{tot}$ and g_{ii}^k in 12 and 13 depend on Q^2. For large values of Q^2, it is convenient to introduce the structure function of the Pomeron [16], F_P, related to the total cross-section of a virtual photon $\sigma_{\gamma^* P}^{tot}$ in the same way as the total cross-section of a virtual photon on a proton is related to the proton structure function F_2 :

$$\sigma_{\gamma^* P}^{tot}(s_1, Q^2, t) = \frac{4\pi^2 \alpha_{EM}}{Q^2} F_P(s_1, Q^2, t). \tag{14}$$

Note that in the present case there is an extra variable t (the virtuality of the Pomeron). It is natural to assume[16] (and calculations of the simplest QCD diagrams confirm [17]) that at large Q^2 the function F_P obeys approximate Bjorken scaling (up to a logarithmic dependence on Q^2, given by QCD-evolution), i.e. depends essentially on the variable

$$\beta = \frac{Q^2}{2\,(p_P \cdot p_i)} \simeq \frac{Q^2}{s_1 + Q^2}. \tag{15}$$

It follows from 15 that the characteristic massses in the diffractive production by highly virtual photon are $s_1 \sim Q^2$. For fixed values of β the Q^2-dependence of the diffraction production process is the same $(1/Q^2)$ as for the total cross-section of the virtual photon on the proton, i.e. it is a main twist effect. On the other hand the diffractive production of a state with fixed mass (e.g. ρ, ω, ϕ, ...) decreases faster with Q^2.

In the case of virtual photons we also have the Bjorken variable

$$x = \frac{Q^2}{2\,(p_p \cdot p_i)} \simeq \frac{Q^2}{s + Q^2}. \tag{16}$$

The longitudinal momentum fraction taken by the Pomeron is $x_P = x/\beta$, and, consequently, the variable ξ' in equation 12 is now given by $\xi' = ln(\beta/x)$.

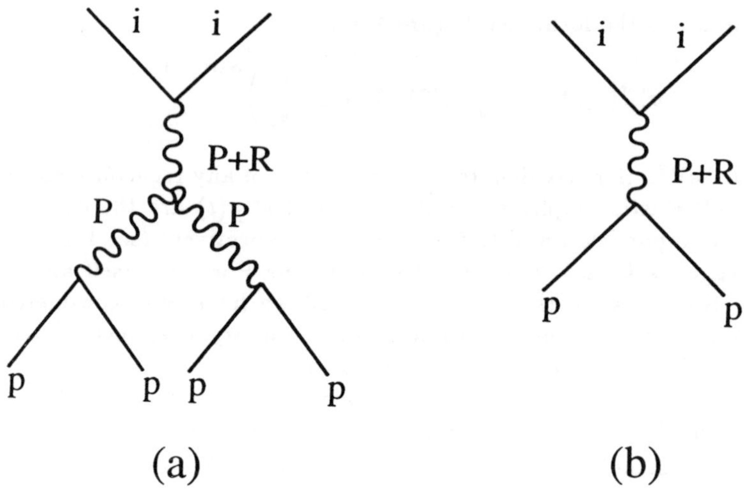

Figure 4: (a) Triple Reggeon diagram obtained by squaring (unitarity), the single diffraction dissociation diagram of Figure 3. (b) Reggeon diagram for elastic i-p scattering. When i is a virtual photon this diagram corresponds to the proton structure function.

The crucial point of our approach [19 20] is the following: using the reggeon factorization property in 13 it is possible to relate the Pomeron structure function F_P to the proton structure function F_2 (or more precisely to that of the deuteron F_2^d, since the isospin of the Pomeron is equal to zero). Furthermore, the parameters in F_P are entirely given in terms of those in F_2^d, plus a few reggeon couplings which can be obtained from soft hadronic diffraction in the framework of conventional Regge theory. By comparing the Pomeron structure function (Figure 4(a)) with that of the nucleon (Figure 4(b)), and using factorization, we see that the Pomeron structure function F_P is identical to F_2^d, given by 1, 2 and 4, except for the following changes in its parameters:

$$F_P(\beta, Q^2, t) = F_2^d\left(\beta, Q^2; A \to eA, B \to fB, n(Q^2) \to n(Q^2) - 2\right), \qquad (17)$$

and where

$$e = \frac{r_{PP}^P(t)}{g_{pp}^P(0)} \quad , \quad f = \frac{r_{PP}^f(t)}{g_{pp}^f(0)}. \qquad (18)$$

Note that the t-dependence of F_P is entirely due to that of the triple reggeon couplings $r(t)$. Comparison with experiment shows that the t-dependence of r_{PP}^P and r_{PP}^f is practically the same and very weak. We have incorporated it in the t-dependence of $g_{pp}^P(t) = g_{pp}^P(0)\ exp\ (Ct)$, with $C = 2.2$ GeV^{-2} [14]. In this way e, f and F_P become independent of t, and $B = B_u + B_d = 1.2$.

The parameters e and f in F_P are obtained from conventional triple reggeon fits to high mass single diffraction dissociation for soft hadronic processes. The most important point in relating soft and hard diffraction dissociation is the following. Absorptive (or shadowing) corrections due to rescattering are very small at large Q^2 but may be quite large at $Q^2 = 0$. This is particularly true for diffractive processes. Indeed, the absorption corrections to the proton diffractive cross-section are quite large: they reduce the value of the unabsorbed cross-section by a factor of 3 to 4 (in the case of pions and photons, vector mesons, the effect is somewhat smaller). Since absorptive corrections decrease very rapidly when Q^2 increases, it is clear that, in the above considerations, valid at moderate and large values of Q^2, we deal with unabsorbed cross-sections. Reasonable values for e and f are: $e \sim f \sim 0.07$. It should be stressed, however, that, while the value of e is rather well determined, there are uncertainties in the determination of f. In conventional triple reggeon fits [14] with the Pomeron intercept equal to 1, one obtains for the values of e and f in the absorbed diffractive cross-sections $e \simeq f$. With a Pomeron intercept above 1, as the one in the CKMT model, the value of e is practically unchanged while the value of f decreases by a factor of 2 to 3. Precise data from HERA should allow a good determination of the parameter f.

The Pomeron trajectory in the second factor of 12 is parametrized as $\alpha_P(t) = 1 + \bar{\Delta} + \alpha' t$ with $\alpha' = 0.25$ GeV^{-2}. Here we use the value of the intercept at $Q^2 \leq 1$ GeV^2. This is due to the fact that the high virtuality of the photon does not go down into the lower part of the diagram of Figure 4(a). We use [18] $\bar{\Delta} = 0.13$, which corresponds to the effective Pomeron intercept obtained without eikonal absorptive corrections (in view of the smallness of the triple reggeon coupling r, the eikonal absorptive corrections in the lower part of the diagram are expected to be small).

The only remaining parameter is the proton-Pomeron coupling $g_{pp}^P(0)$ for which we use the value $(g_{pp}^P(0))^2 = 23mb$ [18].

Apart from the change in the parameters resulting from Regge factorization, the Pomeron and proton (or deuteron) structure functions also differ in the $x \to 1$ behavior. The Dual Parton Model [2] arguments relevant for $Q^2 = 0$ lead to $n(0) = -1/2$ (as compared to 3/2 for the proton), and dimensional counting rules relevant for $Q^2 \neq 0$ lead [21] to $n = 1$ (as compared to 3 for the proton). This provides the justification for the change in $n(Q^2)$ introduced in 17.

As discussed above, the CKMT parametrization for the structure functions has to be used at moderate Q^2 (up to $Q^2 \sim 5GeV^2$). At larger values of

Q^2, we use 17 as an initial condition for QCD evolution. In doing so we determine F_P at all values of Q^2. Note that in order to perform the QCD evolution we have to know the gluon distribution function. In the proton case, the normalization of this distribution is obtained using the energy-momentum conservation sum rule. In the Pomeron case, $\sigma^{tot}_{\gamma^* P}$ is not an ordinary cross-section (see the discussion following 12), and therefore this sum rule cannot be applied. Fortunately, the normalization of the gluon structure function can be obtained in this case using Regge factorization, i.e. multiplying the gluon distribution in the proton, by the factor $e = 0.07$. However, one expects that the gluon distribution in the Pomeron to be harder than the one in the nucleon. In this case, the QCD evolution would be modified and this would produce some changes in our predictions (see below).

We perform the DGLAP Q^2-evolution using as initial condition the valence and sea quark distributions given in equations 2, 4 and 17, at $Q_0^2 = 5GeV^2$, while for the gluon distribution we use the same $\beta \rightarrow 0$ behavior as for the sea quark, but we leave the power n_g of $(1-\beta)$ as a free parameter, i.e. we put

$$\beta g_P^p(\beta) = eA_g g\beta^{-\Delta(Q_0^2)}(1-\beta)^{n_g(Q_0^2)}, \tag{19}$$

where $A_g = 1.71$ is the gluon normalization constant in a nucleon [1]. Note that the gluon normalization is obtained from Regge factorization and not energy-momentum sum ruled is used.

Numerical calculations have been performed using the QCD evolution program of reference (22). We present here some of the obtained results in the one loop approximation, but we have checked that practically the same results are obtained in two loops.

We find[19 20] that a good description of the Q^2-dependence is achieved with either $n_g = 0$ or $n_g = -0.5$ at $Q^2 = 5.GeV^2$. The 3-dimensional structure function $F_2^D(x, Q^2, x_P)$, integrated over t, is plotted versus x_p in Figure 5 and compared with H1 data. Figures 5 and 6 show that the HERA data confirm the predictions of the CKMT model for hard diffraction based on Regge theory, factorization, and the assumption that the Pomeron intercept and the triple-Reggeon couplings are the same in soft and hard diffraction. Also from a QCD analysis of the Q^2-dependence of the present data one gets indication of a hard gluon distribution with the gluons carrying substantially more momentum than the quarks. This can be reached without need of any momentum sum rule for the Pomeron, by using a normalization constant of the gluon distribution determined from Regge factorization.

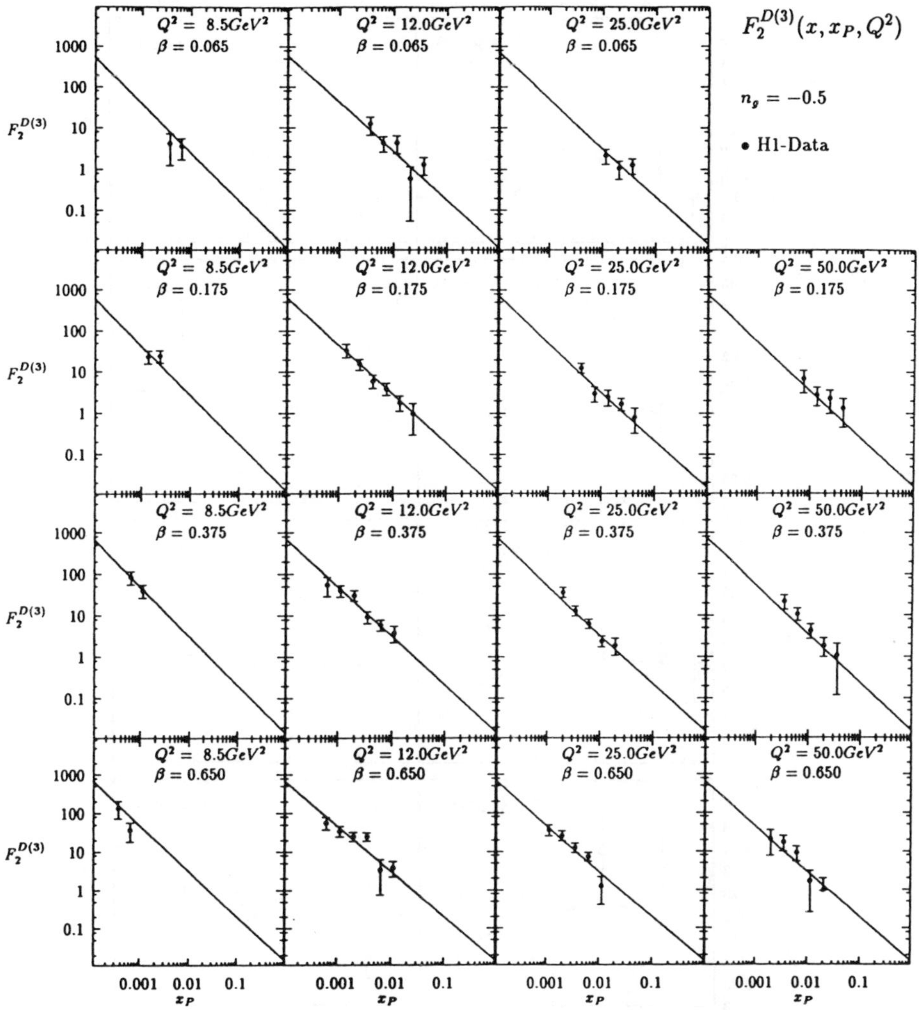

Figure 5: The 3-dimensional proton structure function fo r diffractive events $F_2^D(x, x_P, Q^2)$ is plotted versus x_P at va rious values of Q^2 and $\beta = x/x_P$ and compared with H1 data (13). The CKMT model results are for $n_g = -0.5$. Very similar results are ob tained with $n_g = 0$.

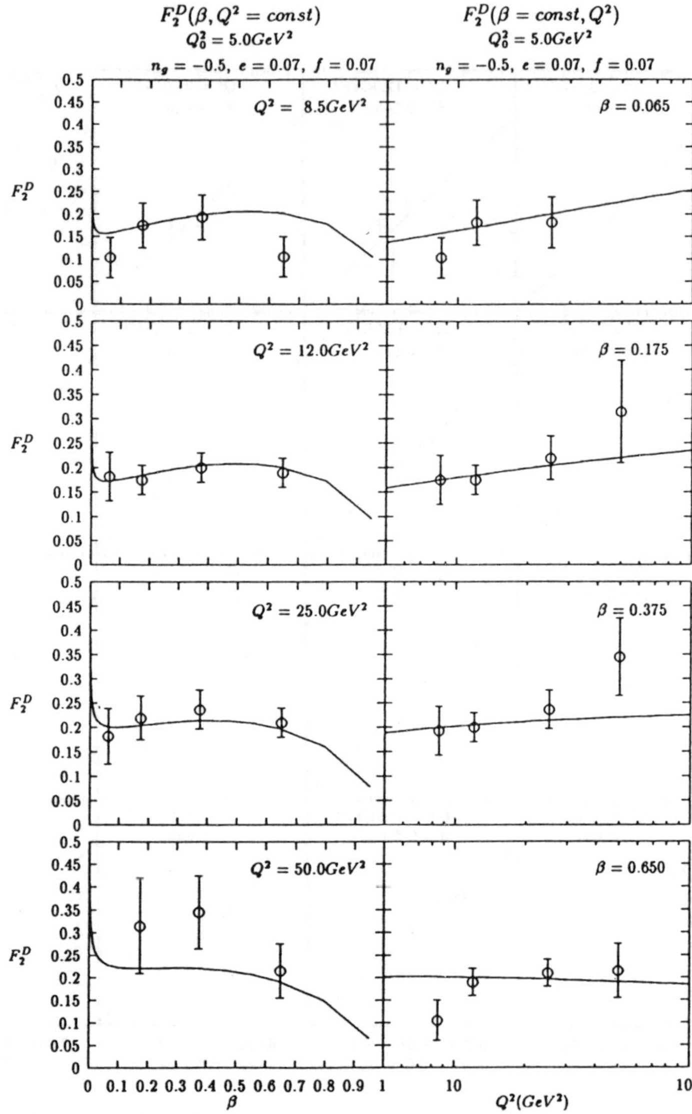

Figure 6: The 2-dimensional proton structure function fo r diffractive events $F_2^D(\beta, Q^2)$ is plotted versus Q^2 at fi xed β and versus β at fixed Q^2 and compared with H1 data (13). The CKMT model results are for $n_g = -0.5$. Very similar results are ob tained with $n_g = 0$. In order to take into account the presence of double diffraction in the data, the theoretical result has bee n multiplied by 1.3.

5 Conclusions

In conclusion, the CKMT model for the parametrization of the nucleon structure functions provides a very good description of all the available experimental data on $F_2(x, Q^2)$ at low and moderate Q^2, including the more recent small-x HERA points. Also the fit to the same data obtained with a modified version of the model in which a logarithmic dependence on Q^2 is included, has been presented. Even though the quality of this second description is also reasonable, its $\chi^2/d.o.f.$ is appreciably higher than that corresponding to the fit obtained with the non-modified version of the CKMT model.

Also, the CKMT model for F_2, used as an initial condition for QCD evolution describes the HERA experimental data presented in the so-called Caldwell-plot, where the x and/or Q^2-dependence of the logarithmic slope of the structure function, $dF_2/dlnQ^2$ is represented.

Finally, using the parametrization of the structure functions provided by the CKMT model, based on Regge theory and Reggeon factorization, we are able to describe the properties of the large rapidity gap events observed at HERA in terms of the Pomeron structure function with parameters determined from soft diffraction in hadronic processes.

Acknowledgments

It's a pleasure to thank Prof. B. Adeva and the members of the Grupo de Altas Enerxías de Santiago de Compostela (GAES) for the perfect organization of this meeting. I'm also grateful to Mercedes Fatas for her main rol in getting the nice atmosphere of the conference and for providing the speakers with very friendly material to the writing of the proceedings. Finally, also thanks are due to M. Gallas and M.J. Tobar for their assistance in improving the quality of the postscript version of this contribution.

References

1. A. Capella, A.B. Kaidalov, C. Merino and J. Tran Than Van, *Phys. Lett.* B **337**, 358 (1994).
2. A. Capella, U. Sukhatme, C.-I. Tan and J. Tran Than Van, *Phys. Rep.* **236**, 225 (1994).
3. C. Adloff *et al* (H1 Collaboration), *Nucl. Phys.* B **497**, 3 (1997).
4. J. Breitweg *et al* (ZEUS Collaboration), *Phys. Lett.* B **407**, 432 (1997).
5. A.B. Kaidalov and C. Merino, *hep-ph/9806367* and accepted for its publication in *Eur. Phys. J. C.*

6. P. Amaudruz *et al* (New Muon Collaboration), *Phys. Lett.* B **259**, 159 (1992). M.R. Adams *et al* (E665 Collaboration), FERMILAB-Pub 1995/396.
7. D.O. Caldwell *et al*, *Phys. Rev. Lett.* **40**, 1222 (1978). M. Derrick *et al* (ZEUS Collaboration), *Phys. Lett.* B **293**, 465 (1992), and *Z. Phys.* C **63**, 391 (1994). S. Aid *et al* (H1 Collaboration), *Z. Phys.* C **69**, 27 (1995).
8. A. de Rújula *et al*, *Phys. Rev.* D **10**, 1649 (1974).
9. K. Adel, F. Barreiro and F.J. Ynduráin, *Nucl. Phys.* B **495**, 221 (1997).
10. A. Caldwell, DESY Theory Workskop, DESY, Hamburg (Germany), October 1997. J. Breitweg *et al* (ZEUS Collaboration), *DESY-98-121* and *hep-ex/9809005.*
11. A.B. Kaidalov and C. Merino, in preparation.
12. M. Derrick *et al* (ZEUS Collaboration), *Phys. Lett.* B **332**, 228 (1994).
13. T. Ahmed *et al* (H1 Collaboration), *Nucl. Phys.* B **429**, 477 (1994), and *Phys. Lett.* B **348**, 681 (1995).
14. A.B. Kaidalov, *Phys. Rep.* **50**, 157 (1979).
15. A.B. Kaidalov and K.A. Ter-Martirosyan, *Nucl. Phys.* B **75**, 471 (1974).
16. G. Ingelman and P.E. Schlein, *Phys. Lett.* B **152**, 256 (1985).
17. N.N. Nikolaev and B.G. Zakharov, *Z. Phys.* C **53**, 331 (1992).
18. A.B. Kaidalov, L.A. Ponomarev and K.A. Ter-Martirosyan, *Sov. Jour. of Nucl. Phys.* 444681986.
19. A. Capella, A.B. Kaidalov, C. Merino and J. Tran Than Van, *Phys. Lett.* B **343**, 1995 (403).
20. A. Capella, A.B. Kaidalov, C. Merino, D. Pertermann and J. Tran Than Van, *Phys. Rev.* D **53**, 1996 (2309).
21. S.J. Brodsky and G. Farrar, *Phys. Rev. Lett.* **31**, 1153 (1973).
22. P. Aurenche, R. Baier, M. Fontannaz, M.N. Kienzle-Focacci and M. Werlen, *Phys. Lett.* B **233**, 1989 (517).

HIGH ENERGY NEUTRINOS: HORIZONTAL SHOWERS AND COHERENT RADIO PULSES

ENRIQUE ZAS
Departamento de Física de Partículas, Universidade E-15706 Santiago de Compostela, Spain
E-mail: zas@gaes.usc.es

I discuss the motivations for high energy neutrino detection and the possibilities offered by the radio technique and air shower arrays for their detection.

1 Introduction and Motivation

Neutrino detection constitutes a real experimental challenge because of their low interaction cross section. Precisely for this reason neutrinos travel through large amounts of matter bringing information from shielded regions of the Universe and opening a promising new window for Astronomy. Neutrinos also constitute a fundamental sector of the Standard Model which is least explored. Low energy neutrinos have so far been detected from the sun [1], from supernova 1987A [2] and from the interactions of cosmic rays with the atmosphere [3]. The potential of neutrino detection to do fundamental physics is illustrated by the few events from SN 1987A leading to the confirmation of the star collapse hypothesis and the establishment of the best bounds on neutrino mass [4]. Solar and atmospheric neutrino observations in disagreement with the theoretical expectations have motivated flavor oscillations strongly supported by recent data. If such oscillations are true neutrinos are in a privileged position to enlighten the physics beyond the Standard Model.

We will concentrate here on the high energy neutrinos which are also of extreme interest both from the astrophysical and particle physics points of view. The neutrino nucleon cross section rises linearly with energy until the scale of the W mass is reached ($\sim$ 100 TeV in the lab frame) and from then on it rises approximately as $E^{1/3}$ because of the low x behavior of the parton distributions. At the highest energies the cross section behavior is controlled by the unmeasured low x and high momentum transfer (Q^2) region of the parton distributions. Such rise on the one hand favours detection of higher energy neutrinos. This is of course compensated by the neutrino fluxes which are expected to drop with energy in agreement with the general behavior observed in cosmic rays and gamma rays over many decades of energy. This rise also makes the Earth opaque to neutrinos above about 100 TeV, and for this reason TeV neutrinos can be used in principle as probes for Earth tomography.

Conventionally high energy neutrino detectors have been designed to exploit the long range of the muons produced in the leptonic vertex of charged muon neutrino interactions [5]. The idea is to instrument large volumes of natural water or ice with photodetectors to see its Cherenkov light. Two of these experiments, AMANDA and Baikal, are already in operation and are currently seen as prototypes of the "km^3" initiative, to instrument 1 km^3 of water or ice with photodetectors, expected to be needed to address important physics issues in the light of calculated neutrino fluxes [6]. Other projects NESTOR and ANTARES are in the research and development phases. At the highest energies there are alternative techniques that have been proposed for detection whose potential has been known for decades but only more recently they have been seriously studied. They are based on detecting showers produced by neutrino interactions either in the air or in dense media such as water or ice too. Such showers which are produced in the hadronic vertex of all neutrino nucleon interactions and in the leptonic vertex of charged current electron neutrino interactions. In this sense they are complementary to the conventional muon detection technique and together with them they can provide useful information on neutrino flavor ratios. As high energy neutrino fluxes are not known with much certainty these techniques may turn out to be adequate for large detector designs and in any case should be further explored.

The establishment of cosmic rays well above the Greisen-Zatsepin-Kuz'min (GZK) cutoff ($\sim 5\ 10^{19}$ eV) provides a further strong motivation for high energy neutrino astronomy. This cutoff is due to the resonant production of Δ in the interactions of protons with the Cosmic Microwave Background (CMB) photons. In such interactions neutrinos are inevitably produced. Moreover most of the mechanisms proposed to explain their origin either through shock acceleration of protons or from fragmentation of supermassive particles, simultaneously produce high energy neutrinos at levels which are quite possibly observable in some future. There are also several detectors in project to study such high energy events. They provide further motivation for neutrino astronomy in a double way. On the one hand by addressing the physical origin of such events they may lead to more precise neutrino predictions once the mechanism is understood. On the other hand air shower detectors can be also used to detect neutrinos of the highest energies by looking for horizontal showers [7] as discussed below, since at high zenith angles the electromagnetic component of cosmic ray showers gets absorbed in the atmosphere.

The prospects for detecting both high energy cosmic rays and neutrinos in the near future has stirred much activity in this field. In this article we will first briefly comment on some of the existing predictions on high energy neutrinos from a variety of sources and then use them to discuss the potential of such alternative techniques with emphasis on later developments.

2 Fluxes

In most mechanisms neutrinos arise from the decay of charged pions (or kaons), produced in different type of high energy particle interactions. The pions can be produced in proton-proton or photon-proton interactions or alternatively from direct fragmentation of quarks, in the same way they are produced routinely in electron positron colliders. As a result in the absence of flavor oscillations the relative fluxes of muon and electron neutrinos are roughly independent as fixed by the pion decay chain:

$$\begin{aligned} \pi^+ &\rightarrow \nu_\mu + \mu^+ \\ &\qquad \mu^+ \rightarrow \overline{\nu}_\mu + \nu_e + e^+ \end{aligned}$$

and its charge conjugate reaction. For relativistic pions in flight it can be assumed that, on average, each of the four leptons produced in the reaction and subsequent muon decay carries one fourth of the parent energy [5].

The interaction of high energy cosmic rays with well understood targets leaves little doubt about neutrino production. Atmospheric neutrinos fall in this category and are known to within about 10% certainty at energies below 1 PeV [5]. These neutrinos constitute the background for observation of other neutrinos sources. Cosmic rays also interact with nucleons in the galaxy, such as those in dust, molecular clouds, the galactic disk matter or in compact objects like the sun.

The atmospheric neutrino flux is zenith angle dependent because of the competition between interaction and decay of the parent pions. At high enough energies the pion is much more likely to interact than to decay because of its Lorentz expanded lifetime. At some yet unknown energy in the 100 TeV-10 PeV interval "prompt" neutrinos from the decays of charmed particles (with considerably shorter lifetimes) dominate the atmospheric flux. The charged current charm decay produces electron and muon neutrinos on practically equal footings thus altering the naive 1:2 ratio from π decay above to a naive 1:1 for instance:

$$D^+ \rightarrow \nu_e + e^+ + X$$

or

$$\begin{aligned} D^+ &\rightarrow \nu_\mu + \mu^+ + X \\ &\qquad \mu^+ \rightarrow \overline{\nu}_\mu + \nu_e + e^+ \end{aligned}$$

The prompt component is very uncertain due to the poorly known charm production cross section. This problem is related to the observations of prompt

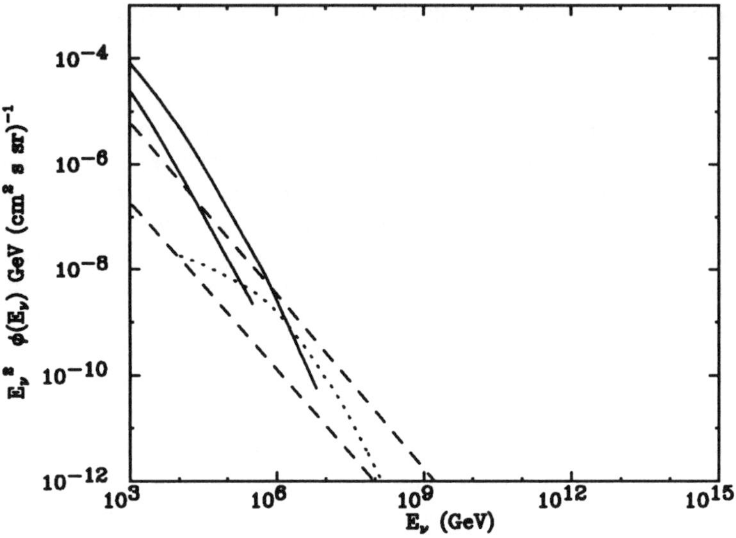

Figure 1: (from top to bottom where listed) Conventional Atmospheric Neutrino flux predictions from pion and kaon decays for 90^0 and 0^0 zenith angles (solid) and a prompt flux calculation from charmed particles (dotted) [9] compared to neutrino flux predictions from CR interactions with the galactic matter for 0^0 and 90^0 galactic latitude (dashes) [11].

atmospheric muons which are also produced in the charmed particle decays and it is of great interest as an alternative way to experimentally constrain open charm production in the Standard Model [9,10].

The vertical and horizontal atmospheric neutrino fluxes are shown in Fig 1 compared to a typical prompt neutrino prediction [9] and to a calculation of neutrino fluxes produced by cosmic ray interactions with galactic matter [11]. It is apparent that galactic matter neutrinos dominate the conventional atmospheric flux at the same energy at which prompt neutrinos are expected to dominate the atmospheric neutrino flux. This difficults their possible identification but it is nevertheless hoped that the prompt neutrinos can be also indirectly determined by measuring the atmospheric prompt muons which are also produced in the decays of charmed particles [9].

The interactions of cosmic rays with the CMB photons are also an unavoidable source of neutrinos assuming the higher energy cosmic rays are of extragalactic origin and hence universal as it is most generally believed. This is supported by dimensional arguments which limit proton acceleration in the

Galaxy to the required energies. I will refer to these as GZK neutrinos to stress their relation to the cosmic ray energy cutoff. The calculation of these neutrinos [5,12,13] implies extrapolating cosmic ray fluxes to energies above the maximum currently observed ($\sim 3\ 10^{20}$ eV) and making assumptions about the evolution of cosmic ray luminosity with cosmological time. The cosmic ray sources have to be integrated over time (or redshift) up some earlier epoch (z_{ult}) which is expected to be provided by the Galaxy formation era (z_{ult}=2-4).

These flux predictions are all fairly flat because the proton-photon interaction cross section has a threshold behavior at the resonant Δ production. Most neutrinos are produced with energies about a factor 20 (see next section) below the Greisen-Zatsepin-Kuz'min cutoff energy $\sim 10^{20}$ eV. Depending on (z_{ult}) the interactions of the highest energy neutrinos with the cosmic neutrino background can play a more important role altering their shapes in the highest energy region [14]. Fig. 2 includes some of these calculated fluxes indicating the levels of uncertainty. Note that for all flux figures the "y-axis" is multiplied by E^2 to enhance the features and also because detection is typically more efficient at higher energies, so that peaks in these kind of plots roughly corresponds to peaks of expected events.

All these mechanisms are quite certain, at least in the sense that if by some means the predicted neutrino fluxes were not found, the implications of such non-discovery would have a larger impact in physics and/or cosmology than their actual observation. There are other candidate sources in astrophysical objects that we know exist, where it is plausible that protons (or nuclei) are accelerated to then collide and produce pions. Some of them can be galactic such as accretion in binary systems, supernova remnants, but those reaching to highest energies are likely to be powerful objects such as Active Galactic Nuclei (AGN) or Gamma Ray Bursts (GRB). AGN may accelerate protons to high energies. The recent identification of the sources that emit the most powerful gamma rays (TeV) with *Blazars*, a subclass of AGN interpreted as those that have their jet axis pointing in our direction [15] suggests mechanisms in which protons are accelerated in the jets of these AGN which are boosted to high energies by high Lorentz factors that are believed to apply to the jet flow [16]. The γ-rays from the π^0 decays and their subsequent cascade give rise to the detected high energy photons. The neutrinos come from the charged pion decays and constitute a signature of proton acceleration which has also been suggested as the origin of the highest energy cosmic rays. Early flux predictions from AGN cores and jets [17,18] are show in Fig. 2. It has been recently remarked that neutrino fluxes for proton interactions in optically thin targets are bounded by the highest energy cosmic ray data [19].

Some other exotic candidate sources, not yet known to exist, are Primor-

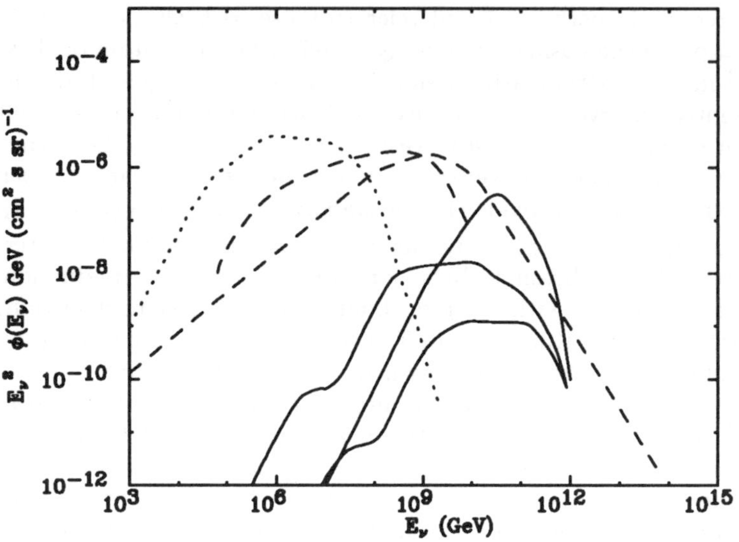

Figure 2: Fluxes of GZK neutrinos for $z_{ult} = 2.2$ [12], and $z_{ult} = 4$ and $z_{ult} = 2$ for ref. [13] (solid from top to bottom) compared to those from AGN jets in refs. [16] (dashed from left to right) and to earlier predictions from AGN cores [15] (dotted).

dial Black Holes (PBH), decays of topological defects (TD) or WIMP annihilation. TD models arise naturally in grand unified theories of particle interactions with spontaneous symmetry breakdown, when some field vacuum goes through a phase transition to a new degenerate vacuum as the Universe cools down in its expansion. Different regions of space go to different vacua and the net distribution may evolve later into a vacuum field with non-trivial topology, surrounding a point (monopole), a line (string) or a surface (domain wall). These cosmological objects accumulate energy and when they interact with themselves or with other objects of their own nature, they annihilate liberating large amounts of energy in the form of X particles, the Gauge bosons of the underlying grand unified theory. TD scenarios have been given a lot of attention recently because they avoid the difficulties inherent to the acceleration of cosmic rays to the highest energies.

TD models are very uncertain because they are significantly affected by a variety of unknown parameters besides the overall normalization. Several authors have normalized injection rates to cosmic or gamma ray measurements and bounds [20,21,22] obtaining neutrino flux predictions. The injection rates of

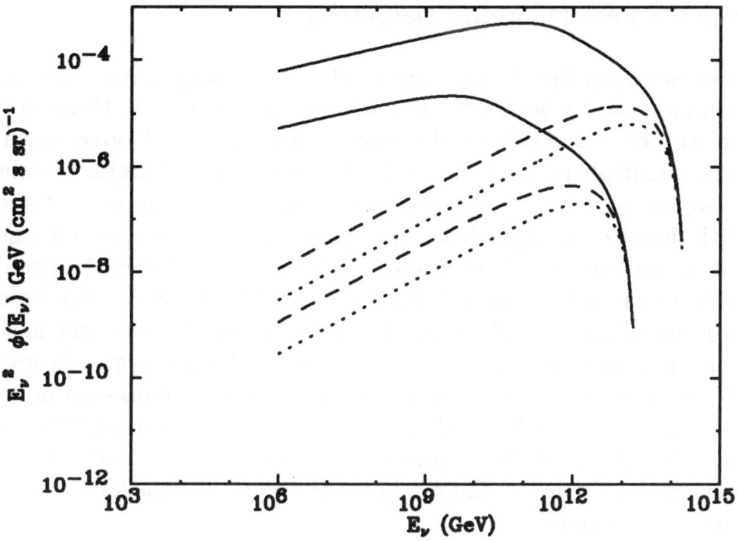

Figure 3: Neutrino flux predictions from models with decays of topological defect with evolution parameters $p = 0$ (full), 1 (dashed) and 2 (dots). The upper (lower) set corresponds to $M_X = 10^{15}$ GeV ($M_X = 10^{14}$ GeV).

X particles per comoving volume are usually parameterized as t^{p-4} with $p = 0$ for superconducting cosmic strings, $p = 1$ for monopoles and cosmic strings and $p = 2$ for constant injection. The flux predictions extend to very high energies dictated by the mass of the assumed X particles, expected to be of order $10^{14} - 10^{16}$ GeV. The spectral index of the neutrinos is governed by the combination of the parameter p and the fragmentation model used and quite generally they are pretty flat. Fig. 3 illustrates some predictions for different p, different M_X and using the injection rates compatible with observations [22].

It is presently impossible to predict the most interesting energy range for detecting high energy neutrinos as there is a large number of open possibilities. Chances for EeV neutrinos are however high because neutrinos from the GZK cutoff, from proton acceleration in AGN jets and from TD scenarios all extend up and beyond the EeV region. The pecularities of EeV neutrino detection put alternative detection techniques in a interesting complementary position.

3 Alternatives for EeV Neutrino Detection

To detect the low neutrino fluxes expected in the PeV energies and beyond large natural volumes are to be used. The idea is to instrument these volumes to be able to record the effects of neutrino interactions. Conventional detectors use photomultipliers in water or ice to detect the Cherenkov light emitted by the long range muons produced in charged current muon neutrino interactions. With directional capabilities upcoming events are exclusively due to interactions of neutrinos that have travelled through the Earth. Conventional detectors in water or ice may, in addition, detect Cherenkov light from the showers with the advantage of being also sensitive to the electron neutrino interactions. This may become most important at high energies (above ~PeV) when the showers will be very large and the Earth becomes opaque to neutrinos. They require directional and shower reconstruction capabilities to separate neutrino showers from those induced by atmospheric muons. These detectors are in good development and are likely to bring information on neutrino fluxes in very short time [24].

An alternative is provided by the radio technique, proposed almost 40 years ago [25] and aiming to instrument large volumes of ice (or sand) with radio antennas to capture the coherent Cherenkov emission from the showers produced by neutrino interactions in the radio band 100 MHz-1 GHz. As long as the wavelength of the emitted radiation is larger that the dimension of the shower the emission from each particle becomes coherent. As a result the radiated electric field becomes proportional to the excess charge and the emitted power scales as the square of the shower energy thus favoring detection of the highest energy events. This coherent enhancement for high energy showers is similarly advantageous in the suggestion to use existing hydrophone arrays to detect the sound pulse produced in the oceans by neutrino interactions [26].

At low energies the dominant interactions of electrons, positrons and photons, Möller and Bhabha scattering, electron positron annihilation and Compton scattering are responsible for an excess negative charge which amounts to about a 20%. This is because of the absence of positrons in the target shower positrons are annihliated while matter electrons are accelerated into the shower. As a result the energy in radio emission becomes proportional to the square of the shower energy, making the method most advantageous for the highest energies. Many experimental difficulties are however anticipated [27] and presently antennas are being tested deep under ice using the AMANDA boreholes in Antarctica (RICE) [28].

During the 60's the technique was largely explored for air showers and radio pulses in the 1-100 MHz band were routinely observed in coincidence with

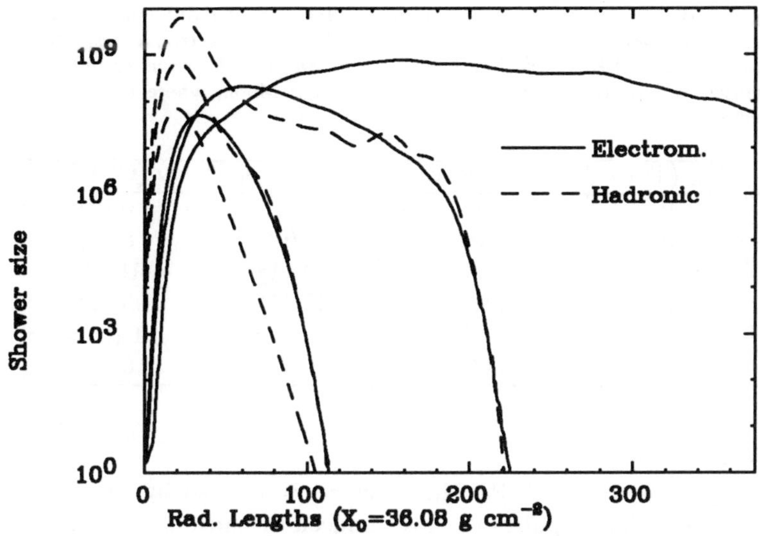

Figure 4: Depth development of 100 PeV, 1 EeV and 10 EeV electromagnetic and hadronic showers compared.

particle detector arrays[29]. For air showers electric dipole radiation and charge separation both in the earth magnetic field and in the atmospheric electric field play comparable roles to excess charge in generating radio signals. Unfortunately uncontrolled effects such as atmospheric conditions at large distance from the detection region constituted an inherent difficulty that prevented systematic shower measurements.

The first attempt at numerically calculating the coherent pulse from a shower was done for electromagnetic showers in ice up to 10 PeV that confirmed that the technique would become competitive for higher energies[30]. The simulation becomes increasingly problematic from the computational point of view for energies roughly also above 10 PeV as particles have to be tracked to energies in the 100 keV range. Alternative ways of calculation are crucial if this possibility is ever to be seriously considered. The simulation of EeV showers has only been recently approached both for electromagnetic and hadronic type showers with a combination of simulation, parameterizations and in the one dimensional approximation. This is also the first time that the Cherenkov light in a dense medium is studied in this energy range. The electric field spectrum of the radio pulse displays a complex diffraction like pattern which depends on

Table 1: Yearly neutrino event rates expected in a radiotelescope for different operational frequencies. The shower energy threshold is estimated for the case of electromagnetic showers assuming a noise level of 250 K, a 10% bandwidth and a 200 m dish. Number in brackets indicates the neutrinos expected in the central region of the Moon, where no cosmic ray signals are expected.

ν_{op} (MHz)	E_{th} (TeV)	Cosmic Rays	ν (AGN)Ref. [18]	ν (GZK) [12]
300	$2\ 10^8$	2,000	0 (0)	0 (0)
500	$4\ 10^7$	14,000	1 (0)	8 (0)
10^3	$9\ 10^6$	80,000	24 (1)	19 (1)
$3\ 10^3$	$3\ 10^6$	60,000	17 (5)	3 (1)
$5\ 10^3$	$3\ 10^6$	20,000	6 (2.5)	1 (0.5)
10^4	$4\ 10^6$	3,000	1 (0)	0 (0)

the frequency considered.

For energies below about 20 PeV in ice showers induced by electrons or photons produce very similar pulses to those induced by hadrons. At very high energies however showers can behave very differently in dense media because of the Landau-Pomerančuck- Migdal (LPM) effect. This manifests as a drastic reduction in the cross section for pair production and bremsstrahlung, the main interactions that govern shower development at high energies. As a result the showers can become very long (hundreds of radiation lengths) and consequently the radio emission becomes narrower [31]. In Fig. 4 the simulation results for the developments of hadronic and electromagnetic showers in ice are compared ice. Hadronic showers show much smaller elongation because of the LPM effect while electromagnetic showers are clearly distorted for energies above 20 PeV [31]. This effect is incidentally also important for all other techniques relying on detecting showers in dense media and in principle allows the separation of showers induced by electrons in charged current electron neutrino interactions from the other showers which are produced by the nucleon fragments [32].

It has been suggested that the radio technique can be used to detect radio pulses from showers just under the Moon surface [33]. It may be possible that radio signals from very high energy showers produced close to the surface can be detected from the Earth using radiotelescopes. The sensitive volume exceeds that achievable in the conceivable extensions of the other alternatives. The pulses produced by cosmic rays in fact constitute an important background and it may be a better technique for detecting ultra high energy cosmic ray showers.

Cosmic rays must hit the surface of the Moon tangentially on the rim as seen by us so that they can be detected from the Earth. This is a geometric effect obtained by demanding that the Earth lies in the Cherenkov cone, because of the refraction index of the Moon surface. Allowing some deviation of the observing direction from the Cherenkov angle opens a small rim around the Moon from which coherent radio signals from sufficiently energetic showers could reach us. The rim size is frequency dependent since the the radio pulse angular spread rises as the antenna frequency decreases. The inner region of the Moon, excluding this cosmic ray rim, can be used for searching the neutrino showers but high cosmic ray rejection factors are anticipated. Rates are very sensitive to the telescope parameters and the noise which has to be determined experimentally. As a guide calculations are tabulated in Table 1 for different frequencies both for cosmic rays and for neutrinos from a number of predictions [34].

Another alternative is provided by Horizontal Air Showers (HAS) measurements known for long to be sensitive to deeply penetrating particles such as muons and neutrinos [35]. In this case atmospheric air constitutes the large volume in which horizontal showers that resemble vertical ones are searched. Comic ray induced showers in the horizontal direction have to travel close to 100 times more atmospheric depth to reach the surface than vertical ones. As a result such showers hardly reach the surface and when they do it is only through their penetrating component, the muons, and they practically, do not have any electromagnetic component, that is photons and electrons, as compared to vertical showers. As a result ordinary looking showers in the horizontal direction have to be induced by deeply penetrating particles such as muons or neutrinos. Any air shower detector that is capable of recognizing a deep shower from those induced by cosmic rays, such as an array with capability to distinguish the electromagnetic part from the muons in the shower, is thus in principle sensitive to neutrinos. Fluorescence detectors such as Fly's Eye, and other proposals to look for horizontal and upcoming showers from satellites are also sensitive to neutrinos for the same reason. The horizontal cosmic ray showers have a muon component, largely unstudied, which constitutes the background to HAS. This is a topic of interest in its own but beyond the scope of this article.

Existing air shower array data have already been used to constrain models for diffuse neutrino fluxes. Some experimental bounds on neutrino fluxes from horizontal showers, AKENO [36] and EAS-TOP and Fly's Eye [37] (looking for upcoming showers) and from and underground muon detector, Frèjus [38], are shown in Fig. 5 together with some of the neutrino fluxes used for reference. Some of the fluxes are clearly in conflict with experiment. Some of the earlier models for diffuse neutrino fluxes from AGN and from the annihilations of Superconducting Cosmic Strings are ruled out by underground detectors and

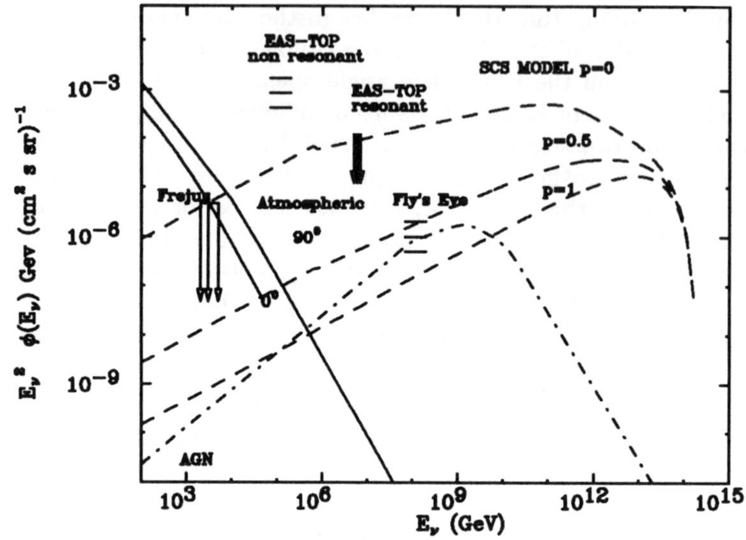

Figure 5: Neutrino flux predictions compared to existing bounds from experiment. Parallel lines indicate the uncertainty associated to the spectral index assumed for the conversion to a flux bound.

by the muon poor horizontal shower bound from AKENO [23]. It is generally accepted that intermediate-size horizontal showers (energies ~1 PeV) will be effective in establishing the prompt component of the atmospheric muon flux [9], what constitutes a background for underground muon detectors. Muons interact mainly through bremsstrahlung to produce muon poor horizontal showers up to the highest measured shower sizes of about 10^5 particles. As the shower size (energy) rises the rate of horizontal showers induced by atmospheric muons is expected to drop in a similar fashion to the cosmic ray rate becoming too low to be observed.

Table 2: Range of yearly neutrino event rates expected in one of the 3000 km^2 arrays in the Pierre Auger Project for several fluxes discussed above. The spread in the value is indicative of array acceptance and neutrino cross section uncertainties.

	AGN-cores [17]	AGN-jets [18]	GZK $z_{ult} = 4$ [13]	TD $p = 1.5$ [20]
Rate	0.2-1.5	2-7	0.1-0.4	2-10

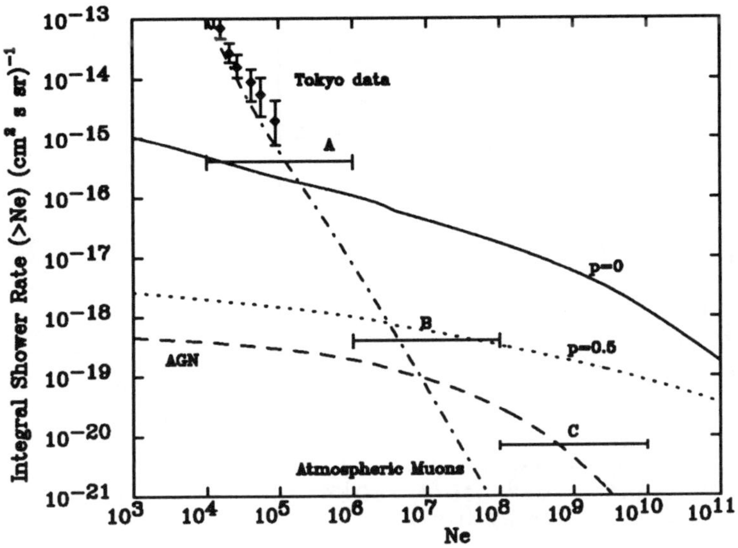

Figure 6: Integral horizontal shower rate as a function of shower size (Ne) for different models. TD neutrinos with $p = 0$ (solid line), $p = 0.5$ (dotted), AGN neutrinos (dashed) and atmospheric muons (dot-dashed). Diamonds represent the HAS data from ref 24. Horizontal lines marked as A, B and C represent the expected limit of non observation in a year made with ideal detectors of areas 10^5 m^2 (A), 10^8 m^2 (B), and 10^{10} m^2 (C). See text.

At energies well above the PeV horizontal showers are only expected to be induced by neutrinos if the fluxes are high enough. Fig. 6 illustrates the horizontal shower rates expected for some of the discussed neutrino fluxes compared to the old HAS data from the university of Tokyo [36] and to 90% confidence upper bounds that could be established on the basis of non observation of horizontal events in a year [23] for three different array sizes. For this reason it is expected that future projects for detecting the highest energy showers such as the Pierre Auger Observatory, the Telescope Array and HiRes will play an important role also as neutrino telescopes. The Pierre Auger Observatories, the largest of such installations now commissioned (3000 km^2 each), will be particularly well suited for neutrino detection because they include an array of water Cherenkov tanks which are effective for detecting particles equally from vertical and horizontal directions.

A large effort is being made to understand HAS well to establish the capabilities of these observatories as neutrino telescopes. Fig. 7 displays its expected

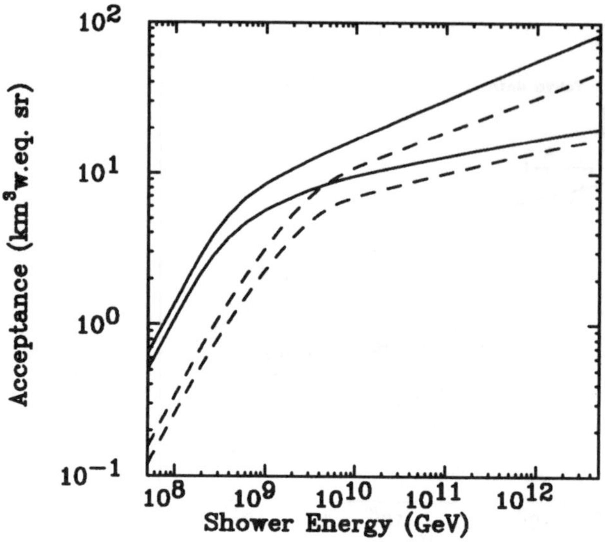

Figure 7: Expected acceptance of the Pierre Auger observatory as a function of shower energy to near horizontal neutrinos ($\theta_{zenith} \geq 75$ degrees) for electromagnetic showers (dashed) and hadronic (solid). Lower set of curves corresponds to only showers with axis falling in the array. Upper set considers all showers. Volume units are km^3 of water equivalent.

acceptance for horizontal showers above $\theta_{th} = 75^o$ [7,39]. In table 2 we give some expected neutrino event rates for a number of neutrino flux predictions from different sources [7]. Part of this effort is geared to understanding horizontal events in Haverah Park, an old array for cosmic rays that stopped operation in 1987, which consisted on similar water Cherenkov tanks to the ones that will be used in the Pierre Auger Observatories. In Fig. 8 we illustrate the muon distribution on the ground produced by a cosmic ray of 80° which is quite outstanding because of the magnetic field of the Earth. This background showers are easy to distinguish from deep showers not only because they lack electromagnetic component but also because of their geometry and their time distributions.

I have stressed the importance that in my opinion the highest energy neutrinos have for the future of this field, particularly in the light of recent theoretical developments on the alternative techniques: HAS and the radio technique. One can conceive large detectors combining them in the future to detect very high energy neutrinos with shower reconstruction and thus some flavor identi-

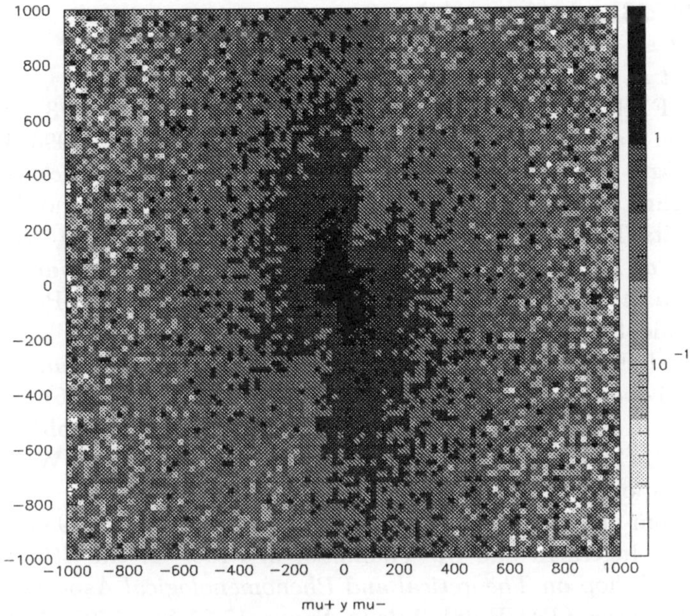

Figure 8: Muon density (m^{-2}) on the ground (scale in m) for a simulated proton air shower of 10^{19}eV incident at 80°. The two ellipses correspond μ^+ and μ^- separation.

fication capabilities.

Acknowledgments

I am indebted to Jaime Alvarez Muñiz, Jose Juan Blanco-Pillado, Máximo Ave, Gonzalo Parente and Ricardo Vázquez with whom I have done much of the work reported here and I thank CICYT (AEN96-1773) and Xunta de Galicia (XUGA-20602B98) for supporting my work.

References

1. J. Bahcall, to appear in *Proc. of the 18th Texas Symposium on Relativistic Astrophysics* eds. A. Olinto, J. Frieman and D.Schramm (World Scientific, Singapore, 1997). e-Print Archive: astro-ph/9702057.
2. See for instance D.N. Schramm, *Comments Nucl. Part. Phys.* **17** 5

(1987) pp 239-278 and references therein.
3. K.S. Hirata *et al.*, *Phys. Lett.* **B280** (1992) 146-152.
4. L.F. Abbott *et al. Nucl. Phys.* **B299** (1988) 734-756.
5. T.K. Gaisser, F. Halzen and T. Stanev, Phys. Rep. 238 (1995) 173.
6. F. Halzen, The Case for a Kilometer-Scale Neutrino Detector: 1996, in *Proc. of the Sixth International Symposium on Neutrino Telescopes*, ed. M. Baldo-Ceolin, Venice (1996); F. Halzen, High Energy Neutrino Astronomy and its Telescopes: the Case for a Kilometer-Scale Detector, in *Proceedings of 1994 Snowmass Summer Study: Particle and Nuclear Astrophysics in the Next Millenium*, ed. by E.W. Kolb and R.D. Peccei (World Scientific, Singapore, 1996) p. 256.
7. G. Parente and E. Zas, in *Proceedings of the 7th Int.Symposium on Neutrino Telescopes*. p. 345, ed. by M. Baldo Ceolin, Venice (1996).
8. R.J. Protheroe *in "Towards the Millennium in Astrophysics: Problems and Prospects"*, Erice 1996, eds. M.M. Shapiro and J.P. Wefel (World Scientific, Singapore); e-print Archive: astro-ph/9612213.
9. E. Zas, F. Halzen and R.A. Vázquez, *Astropart. Phys.* **1**, 297 (1993).
10. P. Gondolo, G. Ingelman and M. Thunman, in *Proc. of the 4th International Workshop on Theoretical and Phenomenological Aspects of Underground Physics (TAUP 95)*, Toledo, Spain, 17-21 Sep 1995. Published in Nucl.Phys.Proc.Suppl.48:472-474,1996 e-Print Archive: hep-ph/9602402.
11. G. Domokos *et al.*, *J. Phys. G: Nucl. Part. Phys.* **19** (1993) 899.
12. F.W. Stecker, C. Done, M.H. Salamon and P. Sommers, *Phys. Rev. Lett.* **66** (1991) 2697; **69**, 2738(E) (1992).
13. S. Yoshida and M. Teshima, Prog. of Theo. Phys. 89 (1993) 833.
14. T.J. Weiler, *Ap. J.* **285** (1984) 495.
15. C.M. Urry *ct al.*, *Pub. Astr. Soc. Pacif.* **107**, 803–845 (1995).
16. M.J. Rees, Nature **211**, 468 (1966).
17. F.W. Stecker and M.H. Salamon, "TeV Gamma-ray Astrophys" (1994) 341-355 (QCD953:I48:1994) e-Print Archive: astro-ph/9501064.
18. K. Mannheim, Astrop. Phys. 3 (1995) 295-302; R.J. Protheroe, in Proc. IAU Colloq. 163, Accretion Phenomena and Related Outflows, ed. D. Wickramasinghe et al., in press (1996).
19. E. Waxman and J. Bahcall, *Phys. Rev.* D59 (1999) 3002; K. Mannheim et. al, preprint archive astro-ph/9812398.
20. P. Battacharjee *et al.*, Phys. Rev. Lett. **69** (1992) 567.
21. G. Sigl, S. Lee and D.N. Schramm, *Phys. Lett.* **B392** (1997) 129–134.
22. R.J. Protheroe and T. Stanev, *Phys. Rev. Lett.*, **77**, 3708 (1996), and *Erratum* **78**, 3420 (1997).

23. J.J. Blanco-Pillado *et al.*, *Phys. Rev. Lett.*, **78**, 3614 (1997).
24. F. Halzen; *Lectures on Neutrino Astronomy: Theory and Experiment* TASI School, July 1998; e-print Archive astro-ph/9810368.
25. G.A. Askar'yan, *Soviet Physics* JETP **14,2** 441–443 (1962); **48** 988–990 (1965).
26. L.G. Dedenko *et al.*, *Proc. XXVth ICRC*, Durban, South Africa, Ed. M.S. Potgieter *et al.* (1997), Vol. 7, pp. 89.
27. J.V. Jelley. *Astropart. Phys.* **5** (1996)255-261.
28. C. Allen *et al.*, *Proc. XXVth International Cosmic Ray Conference*, Durban, South Africa, Ed. M.S. Potgieter *et al.* (1997), Vol. 7, pp. 85–88.
29. H.R. Allan, *Progress in Elementary Particles and Cosmic Ray Physics* **10** (1971) pp. 171–302 (North Holland Publ. Co.) and references therein.
30. F. Halzen, E. Zas, E. and T. Stanev, *Phys. Lett.* **B257** 432 (1991).
31. J. Alvarez-Muñiz and E. Zas, *Phys. Lett.* **B411** (1997) 218-224; *Phys. Lett.* **B434** (1998) 396-406.
32. J. Alvarez-Muñiz, R.A. Vazquez and E. Zas, in preparation (1999) .
33. R.D. Dagkesamansky and I.M. Zheleznykh, *Pis'ma Zh. Eksp. Teor. Fiz. (JETP Lett.* **50** (1989) 233; *Proc. of Astrophysical aspects of the most energetic cosmic rays* Kofu (1990), Ed. M. Nagano and F. Takahara, World Scientific (1991) p. 373.
34. J. Alvarez-Muñiz and E. Zas, *Proc. XXVth ICRC*, Durban, South Africa, Ed. M.S. Potgieter *et al.* (1997), Vol. 7, pp. 309–312.
35. V.S. Berezinsky and G.T. Zatsepin *Phys. Lett.* **B28** (1969) 423; P. Kiraly *et al.*, *J. Phys. A:Gen. Phys.* **4** (1971) 367. V.S. Berezinsky and A. Yu. Smirnov, *Astrophys. Space Science* **32** (1975) 461.
36. M. Nagano *et al.*, *J. Phys. Soc. Japan.* **30** (1971) 33; S. Mikamo *et al.*, *Lett. al Nuov. Cim.* **34** (1982) 273.
37. R.M. Baltrusaitis *et al.*, *Phys. Rev* **D31** (1985) 2192.
38. W. Rhode, K. Daum, *et al.* Proc. 24th ICRC, Rome 1995. vol. 1, p. 781. Also, W. Rhode, *et al.* *Astropart. Phys.* **4** (1996) 217.
39. J. Capelle, J.W. Cronin, G. Parente and E. Zas; *Astropart. Phys.* **8** (1998) 321.

PROSPECTS FOR SOLAR AXIONS SEARCHES WITH CRYSTALS VIA BRAGG SCATTERING

I. G. IRASTORZA, S. SCOPEL, S. CEBRIÁN, E. GARCÍA, D. GONZÁLEZ, A. MORALES, J. MORALES, A. ORTIZ de SOLÓRZANO, J. PUIMEDÓN, A. SALINAS, M. L. SARSA, J. A. VILLAR
Laboratorio de Física Nuclear. Universidad de Zaragoza
50009, Zaragoza, SPAIN

Abstract

A calculation of the expected signal due to Primakov coherent conversion of solar axions into photons via Bragg scattering in several solid–state detectors is presented and compared with present and future experimental sensitivities. The axion window $m_a \gtrsim 0.03$ eV (not accessible at present by other techniques) could be explored in the foreseeable future with crystal detectors to constrain the axion–photon coupling constant $g_{a\gamma\gamma}$ below the latest bounds coming from helioseismology. On the contrary a positive signal in the sensitivity region of such devices would imply revisiting other more stringent astrophysical limits derived for the same range of the axion mass.

1 Introduction

Introduced twenty years ago as the Nambu–Goldstone boson of the Peccei–Quinn symmetry to explain in an elegant way CP conservation in QCD[1], the axion is remarkably also one of the best candidates to provide at least a fraction of the Non Barionic Dark Matter of the Universe.

Axion phenomenology is determined by its mass m_a which in turn is fixed by the scale f_a of the Peccei–Quinn symmetry breaking, $m_a \simeq 0.62$ eV $(10^7\ \mathrm{GeV}/f_a)$. No hint is provided by theory about where the f_a scale should be. A combination of astrophysical and nuclear physics constraints, and the requirement that the axion relic abundance does not overclose the Universe, restricts the allowed range of viable axion masses into a relatively narrow window[2,3,4]:

$$\begin{aligned} 10^{-6}\mathrm{eV} \lesssim \quad & m_a \quad \lesssim 10^{-2}\mathrm{eV} \\ 3\ \mathrm{eV} \lesssim \quad & m_a \quad \lesssim 20\ \mathrm{eV}. \end{aligned} \tag{1}$$

The physical process used in axion search experiments is the Primakov effect. It makes use of the coupling between the axion field ψ_a and the electromagnetic tensor:

$$\mathcal{L} = -\frac{1}{4} g_{a\gamma\gamma} \psi_a \epsilon_{\mu\nu\alpha\beta} F^{\mu\nu} F^{\alpha\beta} = -g_{a\gamma\gamma} \psi_a \vec{B} \cdot \vec{E} \tag{2}$$

and allows for the conversion of the axion into a photon.

Like all the other axion couplings, $g_{a\gamma\gamma}$ is proportional to m_a[2,5]:

$$g_{a\gamma\gamma} \simeq 0.19 \frac{m_a}{\text{eV}} \left(\frac{E}{N} - \frac{2(4+z)}{3(1+z)} \right) 10^{-9} \text{GeV}^{-1} \tag{3}$$

where E/N is the PQ symmetry anomaly and the second term in parenthesis is the chiral symmetry breaking correction. The anomaly E/N depends on the particular axion model, while the symmetry breaking correction is a function of the parameter $z \equiv m_u/m_d \simeq 0.56$. Two popular models are the GUT–DFSZ axion[6] (E/N=8/3) and the KSVZ axion[7] (E/N=0). However, it is possible to build viable axion models with different values of E/N[8] and the determination of the parameter z is subject to some theoretical uncertainties[9]. This implies that a very small or even vanishing $g_{a\gamma\gamma}$ cannot be in principle excluded.

Solid state detectors provide a simple mechanism for axion detection [10,11]. Axions can pass in the proximity of the atomic nuclei of the crystal where the intense electric field can trigger their conversion into photons. In the process the energy of the outgoing photon is equal to that of the incoming axion.

Axions can be efficiently produced in the interior of the Sun by Primakov conversion of the blackbody photons in the fluctuating electric field of the plasma. The resulting flux has an outgoing average axion energy E_a of about 4 keV (corresponding to the temperature in the core of the Sun, $T \sim 10^7 K$) that can produce detectable x–rays in a crystal detector. Depending on the direction of the incoming axion flux with respect to the planes of the crystal lattice, a coherent effect can be produced when the Bragg condition is fulfilled, leading so to a strong enhancement of the signal. A correlation of the expected count–rate with the position of the Sun in the sky is a distinctive signature of the axion which can be used, at the least, to improve the signal/background ratio.

The process described above is independent on m_a and so are the achievable bounds for the axion–photon coupling $g_{a\gamma\gamma}$. This fact is particularly appealing, since other experimental techniques are limited to a more restricted mass range: "haloscopes"[12], that use electromagnetic cavities to look for the resonant conversion into microwaves of non relativistic cosmological dark halo axions, do not extend their search beyond $m_a \simeq 50\ \mu$eV, while the dipole magnets used in "helioscope"[13] experiments are not sensitive to solar axions heavier than $m_a \simeq 0.03$ eV.

A pilot experiment carried out by the SOLAX Collaboration[14] has already searched for axion Primakov conversion in a germanium crystal of 1 kg obtaining the limit $g_{a\gamma\gamma} \lesssim 2.7 \times 10^{-9}$ GeV^{-1}. This is the (mass independent but solar model dependent) most stringent laboratory bound for the axion–photon coupling obtained so far, although less restrictive than the globular cluster bound[15] $g_{a\gamma\gamma} \lesssim 0.6 \times 10^{-10}$ GeV^{-1}. Notice however that the experimental accuracy of solar observations is orders of magnitude better than for any other star.

Nevertheless the solar model itself already requires[16] $g_{a\gamma\gamma} \lesssim 10^{-9}$ GeV^{-1}, whereas the above Ge crystal bound has not yet reached such sensitivity. The 10^{-9}GeV^{-1} limit sets a minimal goal for the sensitivity of future experiments, prompting the need for a systematic discussion of present efforts and future prospects for axion searches with crystals. In the following we give the result of such an analysis, focusing on Germanium, TeO_2 and NaI crystals which are currently operating as dark matter detectors [21,22,23,24,25].

2 Primakov conversion in crystals

We will make use of the calculation of the flux of solar axions of Ref. [17] with the modifications introduced in Ref. [11] to include helium and metal diffusion in the solar model. A useful parametrization of the flux is the following:

$$\frac{d\Phi}{dE_a} = \sqrt{\lambda}\frac{\Phi_0}{E_0}\frac{(E_a/E_0)^3}{e^{E_a/E_0}-1} \tag{4}$$

where λ=$(g_{a\gamma\gamma} \times 10^8/\text{GeV}^{-1})^4$ is an dimensionless coupling introduced for later convenience, Φ_0=5.95 $\times 10^{14}$ cm^{-2} sec^{-1} and E_0=1.103 keV.

In the general case of a multi–target crystal, we calculate the expected axion–to–photon conversion count rate in a solid–state detector, integrated in the (electron–equivalent) energy window $E_1 < E_{ee} < E_2$, which is given by[18]:

$$\begin{aligned} R(E_1,E_2) &= (2\pi)^3 2\hbar c \frac{V}{v_a^2}\sum_G \frac{d\Phi}{dE_a}\frac{1}{|\vec{G}|^2}\frac{g^2_{a\gamma\gamma}}{16\pi^2} \times |\sum_j F^0_{a,j}(\vec{G})S_j(\vec{G})|^2 \cdot \\ & \sin^2(2\theta)\frac{1}{2}\left[\text{erf}\left(\frac{E_a-E_1}{\sqrt{2}\sigma}\right) - \text{erf}\left(\frac{E_a-E_2}{\sqrt{2}\sigma}\right)\right] \end{aligned} \tag{5}$$

where we have used the cross–section of the conversion process calculated in Ref.[19]. The first sum is over the vectors $\vec{G}$ of the reciprocal lattice, defined by the property $\exp i\vec{G}_i \cdot \vec{x}_i \equiv 1$, where x_i indicate the positions in space of the target nuclei. V is the volume of the detector, v_a that of the elementary cell, 2θ the scattering angle, σ the resolution of the detector, FWHM=2.35 σ, while:

$$S_j(\vec{G}) = \sum_i e^{i\vec{a}^j_i\vec{G}} \tag{6}$$

is the structure function of the crystal and

$$F^0_{a,j}(\vec{q}) = \frac{Z_j e k^2}{r_j^{-2}+q^2}. \tag{7}$$

$k \equiv |\vec{k}| \simeq E_a$ is the axion momentum. The crystal is described by a Bragg lattice with a basis whose sites are occupied by atoms of different types. The $\vec{a}_i^j$ indicate the i'th basis vector occupied by the j'th target–nucleus type, Z_j is the atomic number of the j–th type target nucleus while $r_j \simeq 1$ Å is the screening length of the corresponding atomic electric field parametrized with a Yukawa–type potential.

The energy distribution of Eq.(4) implies that the transferred momentum $q \equiv |\vec{q}| = 2k \sin\theta$ corresponds to a wavelength of a few Å, which is of the order of the distances between atoms in a typical crystal lattice. This is the reason why a Bragg–reflection pattern arises in the calculation and in Eq.(5) the integral over the transferred momentum has been replaced by a sum over the vectors of the reciprocal lattice, i.e. over the peaks that are produced when the Primakov conversion verifies the Bragg condition $\vec{q} = \vec{G}$ and the crystal interacts in a coherent way. The Bragg condition implies that in Eq. (5) $E_a = \hbar c |\vec{G}|^2 / 2 \ \hat{u} \cdot \vec{G}$ where the unitary vector $\hat{u}$ points toward the Sun. This term induces a time dependence in the expected signal as the detector moves daily around the Sun.

3 Time correlation and background rejection

In the expected signal the dependence on λ can be factorized: $R \equiv \lambda \bar{R}$. An example of the function $\bar{R}$ for several materials is shown in Fig. 1 as a function of time during one day. In all the plots the main axis of the lattice have been taken parallel to the South, West and upward direction. For a summary of the crystallographic information needed to perform these calculations, see ref.[18].

The signal is peaked around the maximum of the flux of Eq.(4) and presents a strong sub–diary dependence on time, due to the motion of the Sun in the sky. The time duration of the peaks decreases with growing energies, from tens of minutes in the lowest part of the axion energy window, down to a minimum of about one minute in the higher one, and is related to the energy resolution of the detector.

In order to extract the signal from the background for each energy interval $E_k < E < E_k + \Delta E$ we introduce, following Ref.[14], the quantity:

$$\chi = \sum_{i=1}^{n} \left[\bar{R}(t_i) - < \bar{R} > \right] \cdot n_i \equiv \sum_{i=1}^{n} W_i \cdot n_i \tag{8}$$

where the n_i indicate the number of measured events in the time bin $t_i, t_i + \Delta t$ and the sum is over the total period T of data taking. The brackets indicate time average.

By definition the quantity χ is expected to be compatible with zero in absence of a signal, while it weights positively the events recorded in coincidence with the expected peaks.

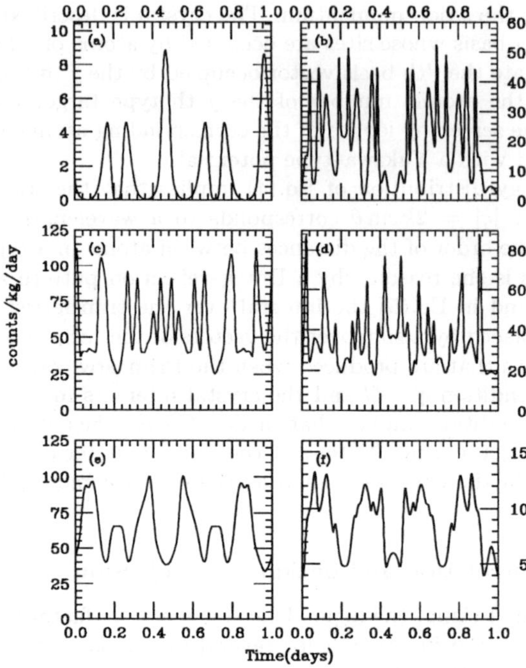

Figure 1: Expected axion signals for Primakov conversion in various crystals as a function of time for $\lambda = 1$. In the calculation the representative day of 1 april 1998 and the coordinates of the LNGS laboratory have been assumed. From top–left to bottom–right: a) Ge, 2 keV$\leq E_{ee} \leq$2.5 keV; b) Ge, 4 keV$\leq E_{ee} \leq$4.5 keV; c) TeO_2, 5 keV$\leq E_{ee} \leq$7 keV; d) TeO_2, 7 keV$\leq E_{ee} \leq$9 keV; e) NaI, 2 keV$\leq E_{ee} \leq$4 keV; f) NaI, 4 keV$\leq E_{ee} \leq$6 keV.

The time distribution of n_i is supposed to be Poissonian, with mean:

$$< n_i >= \left[\lambda \bar{R}(t_i) + b\right] \Delta t. \tag{9}$$

Assuming that the background b dominates over the signal the expected average and variance of χ are given by:

$$\begin{aligned} < \chi > &= \lambda \cdot A & (10) \\ \sigma^2(\chi) &\simeq b/A & (11) \end{aligned}$$

with $A \equiv \sum_i W_i^2 \Delta t$. Each energy bin $E_k, E_k + \Delta E$ with background b_k provides an independent estimate $\lambda_k = \chi_k / A_k$ so that one can get the most probable combined value of λ:

$$\lambda = \sum_k \chi_k / \sum_l A_l$$

$$\sigma(\lambda) = \left(\sum_k A_k / b_k \right)^{-\frac{1}{2}} . \quad (12)$$

The sensitivity of an axion experimental search can be expressed as the upper bound of $g_{a\gamma\gamma}$ which such experiment would provide from the non–appearance of the axion signal, for a given crystal, background and exposure. If λ is compatible to zero, then at the 95% C.L. $\lambda \lesssim 2\times\ 1.64\times\sigma(\lambda)$. It is easy to verify that the ensuing limit on the axion–photon coupling $g_{a\gamma\gamma}^{lim}$ scales with the background and exposure in the following way:

$$g_{a\gamma\gamma} \le g_{a\gamma\gamma}^{lim} \simeq K \left(\frac{\mathrm{b}}{\mathrm{cpd/kg/keV}} \times \frac{\mathrm{kg}}{\mathrm{M}} \times \frac{\mathrm{years}}{\mathrm{T}} \right)^{\frac{1}{8}} \times 10^{-9}\ \mathrm{GeV}^{-1} \quad (13)$$

where M is the total mass and b is the average background. The factor K depends on the parameters of the crystal, as well as on the experimental threshold and resolution. The application of the statistical analysis described above results in a background rejection of about two orders of magnitude.

4 Discussion and conclusions

In order to perform a systematic analysis of the axion–detection capability of crystal detectors, we have applied the technique described in the previous section to several materials. The result is summarized in Table 1, where the limit given by the experiment of Ref.[14] is compared to those attainable with running[21], being installed[22,23] and planned[24,25] crystal detector experiments.

In Table 1 a Pb detector is also included, to give an indication of the best improvement that one would expect by selecting heavy materials to take advantage of the proportionality to Z^2 of the cross section (crystals of $PbWO_4$ have ben considered in the setup of CUORE[22]). These results can be easily extended to other elements with lower atomic numbers than those considered in Table 1 (for instance in Al_2O_3 or LiF crystals, also in operation or planned dark matter experiments) although they are expected to yield less stringent limits.

In Fig.2 the result of our analysis is compared to the present astrophysical and experimental bounds in the plane m_a–$g_{a\gamma\gamma}$. The horizontal thick line represents the constraint $g_{a\gamma\gamma} \lesssim 3\times10^{-10}\ \mathrm{GeV}^{-1}$ taken from

Table 1: Axion search sensitivities for running (COSME[20],DAMA[21]), being installed (CUORICINO[22], CANFRANC[23]) and planned (CUORE[24], GENIUS[25]) experiments are compared to the result of SOLAX[14]. To estimate the best improvement that would be possible by selecting high-Z materials, a calculation is also shown for a Pb lattice, assuming the same conditions of background, threshold and resolution of the GENIUS experiment. The coefficient K is defined in Eq.(13).

	K	M (kg)	b (cpd/ /kg/keV)	E_{th} (keV)	FWHM (keV)	$g_{a\gamma\gamma}^{lim}$(2 years) (GeV^{-1})
Ge[14]	2.5	1	3	4	1	2.7×10^{-9}
Ge[20]	2.3	0.234	0.7	3	0.4	2.4×10^{-9}
Ge[25]	2.5	1000	1×10^{-4}	4	1	3×10^{-10}
TeO_2[22]	3	42	0.1	5	2	1.3×10^{-9}
TeO_2[24]	2.8	765	1×10^{-2}	3	2	6.3×10^{-10}
NaI[21]	2.7	87	1	2	2	1.4×10^{-9}
NaI[23]	2.8	107	2	4	2	1.6×10^{-9}
Pb	2.1	1000	1×10^{-4}	4	1	2.5×10^{-10}

Table 1. The mass intervals $m_a \lesssim 10^{-6}$ eV, 10^{-2} eV $\lesssim m_a \lesssim 3$ eV, $m_a \gtrsim 20$ eV are excluded respectively by cosmological overclosure[2], SN1987A[3] and oxygen excitation in water Cherenkov detectors[4], the edges of all the excluded regions being subject to many astrophysical and theoretical uncertainties. The horizontal lines represent the limits from SOLAX[14], helioseismology[16], and globular clusters[15]. The theoretical predictions for $g_{a\gamma\gamma}$ in the case of the DFSZ and the KSFZ axion models are also given.

As shown in the expression of the $g_{a\gamma\gamma}$ bound of Eq.(13) the improvement in background and accumulation of statistics is washed out by the 1/8 power dependence of $g_{a\gamma\gamma}$ on such parameters. It is evident, then, that crystals have no realistic chances to challenge the globular cluster limit. A discovery of the axion by this technique would presumably imply either a systematic error in the stellar–count observations in globular clusters or a substantial change in the theoretical models that describe the late–stage evolution of low–metallicity stars.

On the other hand, the sensitivity required for crystal–detectors in order to explore a range of $g_{a\gamma\gamma}$ compatible with the solar limit of Ref.[16], appears to be within reach, provided that large improvements of background as well as substancial increase of statistics be guaranteed. Collecting a statistics of the order of a few tons×year could be not so difficult to achieve by adding properly the results of various experiments. In such a case the exploration of a particular axion window, not accessible to de-

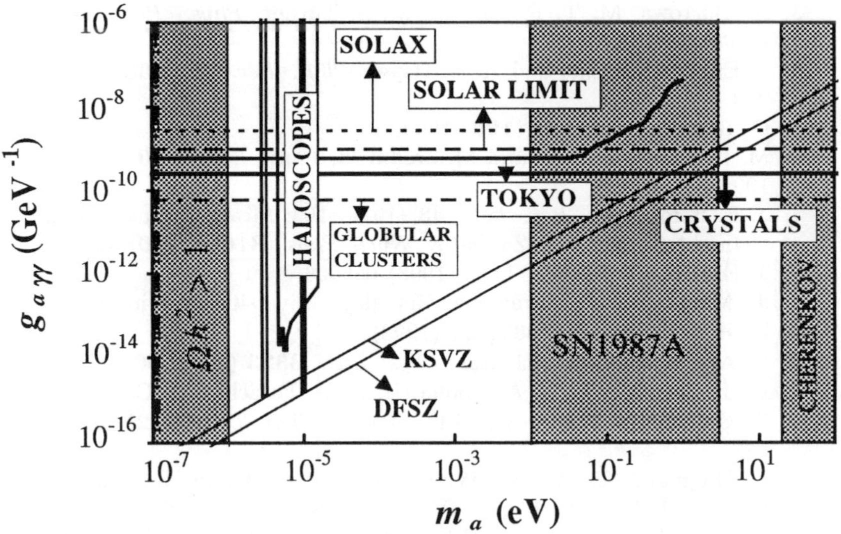

Figure 2: The solar axion limit attainable with crystal detectors (horizontal thick line) is compared to the present astrophysical and experimental bounds and to the DFSZ and KSVZ axion theoretical predictions. The shaded mass intervals $m_a \lesssim 10^{-6}$ eV, 10^{-2} eV $\lesssim m_a \lesssim 3$ eV, $m_a \gtrsim 20$ eV are excluded respectively by cosmological overclosure[2], SN1987A [3] and oxygen excitation in water Cherenkov detectors[4]. The horizontal lines represent the upper limits for $g_{a\gamma\gamma}$ from helioseismology[16] (dashes), SOLAX[14] (dots) and globular clusters[15] (dot–dashes). The regions excluded by haloscope searches[12] and the upper limit on $g_{a\gamma\gamma}$ from the Tokio Helioscope experiment[13] are also shown.

tectors of other types, could be only a question of time, as a bonus from current and future dark matter searches.

Acknowledgements

This search has been partially supported by the Spanish Agency of Science and Technology (CICYT) under the grant AEN98–0627. S.S. is supported by a fellowship of the INFN (Italy). The authors wish to thank Frank Avignone for useful discussions.

References

1. R. D. Peccei and H. R. Quinn, *Phys. Rev. Lett.* **38** (1977) 1440.
2. E. W. Kolb and M. S. Turner, *The Early Universe*, Addison Wesley Publishing, (1990).

3. A. Burrows, M. T. Ressel and M. S. Turner, *Phys. Rev.* **D42** (1990) 3297.
4. J. Engel, D. Seckel and A. C. Hayes, *Phys. Rev. Lett.* **65** (1990) 960.
5. G. Raffelt, *Phys. Rep.* **198** (1990) 1.
6. M. Dine, W. Fischler and M. Srednicki, *Phys. Lett.* **B104** (1981) 199.
7. J. E. Kim, *Phys. Rev. Lett.* **43** (1979) 103; M. A. Shifman, A. I. Vainschtein and V. I. Zakharov, *Nucl. Phys.* **B166** (1980) 493.
8. J. E. Kim, *Phys. Rev.* **D58** (1998) 055006.
9. T. Moroi and H. Murayama, Berkeley University preprint UCB-PTH-98/20, hep-ph/9804291 (1998).
10. E. A. Pascos and K. Zioutas, *Phys. Lett.* **B323** (1994) 367.
11. R. J. Creswick, F. T. Avignone III, H. A. Farach, J. I. Collar, A. O. Gattone, S. Nussinov and K. Zioutas, *Phys. Lett.* **B427** (1998) 235.
12. C. Hagmann, D. Kinion, W. Stoeffl, K. van Bibber, E. Daw, H. Peng, L.J. Rosenberg, J. LaVeigne, P. Sikivie, N.S. Sullivan, D. Tanner, F. Nezrick, Michael S. Turner, D.M. Moltz, J. Powell and N.A. Golubev, *Phys. Rev. Lett.* **80** (1998) 2043; S. Matsuki, talk given at the II International Workshop on the Identification of Dark Matter (IDM98), Buxton, England, 7–11 September 1998.
13. S. Moriyama, M. Minowa, T. Namba, Y. Inoue, Y. Takasu and A. Yamamoto, *Phys. Lett.* **B434** (1998) 147.
14. A. O. Gattone, D. Abriola, F. T. Avignone, R. L. Brodzinski, J. I. Collar, R. J. Creswick, D. E. Di Gregorio, H. A. Farach, C. K. Guérard, F. Hasenbalg, H. Huck, H. S. Miley, A. Morales, J. Morales, S. Nussinov, A. Ortiz de Solórzano, J. H. Reeves, J. A. Villar and K. Zioutas (The SOLAX Collaboration), *Nucl. Phys.* **B**(Proc. Suppl.)**70** (1999) 59.
15. G. Raffelt, *Stars as Laboratories for Fundamental Physics* (University of Chicago Press, Chicago, 1996); G. Raffelt and D. Dearborn, *Phys. Rev.* **D36** (1987) 2211.
16. H. Schlattl, A. Weiss and G. Raffelt, submitted to *Astrop. Phys.*, hep-ph/9807476.
17. K. van Bibber, P.M. McIntyre, D.E. Morris and G. Raffelt, *Phys. Rev.* **D39** (1989) 2089.
18. S. Cebrián, E. García, D. González, I. G. Irastorza, A. Morales, J. Morales, A. Ortiz de Solórzano, J. Puimedón, A. Salinas, M. L. Sarsa, S. Scopel, J. A. Villar, submitted to *Astrop. Phys.*, hep-ph/9811359.
19. W. Buchmüller and F. Hoogeveen, *Phys. Lett.* **B237** (1990) 278.
20. COSME Collaboration (University of Zaragoza, University of South Carolina, Pacific National Northwest Lab.)

21. R. Bernabei, P. Belli, F. Montecchia, W. Di Nicolantonio, A. Incicchitti, D. Prosperi, C. Bacci, C.J. Dai, L.K. Ding, H.H. Kuang and J.M. Ma, *Phys. Lett.* **B424** (1998) 195.
22. E. Fiorini, *private communication*; CUORICINO, *Letter of intent*, submitted to INFN and LNGS.
23. A. Morales, "Dark Matter and its Detection", Summary Talk given at the NUPECC Workshop on Present and Future of Neutrino Physics, Frascati, *NUPECC Report in Highlights and Opportunities in Nuclear Physics*, Ed. by J. Vervier et al., December 1997, To be published in *Physics Reports*.
24. O. Cremonesi, Future of Cryogenic Detectors, talk given at NEUTRINO 98, Takayama, Japan, 4–9 June 1998, to be published in *Nucl. Phys.*B(Proc. Suppl.)1999.
25. B. Majorovits, talk given at the II International Workshop on the Identification of Dark Matter (IDM98), Buxton, England, 7–11 September 1998.

Tests of the Electroweak Symmetry Breaking Sector at the Tevatron

J.F. DE TROCONIZ
Dpto. de Física Teórica
Universidad Autónoma de Madrid,
E28049 Cantoblanco, Spain

The phenomenology of the electroweak symmetry breaking sector is expected to involve resonances decaying preferentially into heavy fermions. Three recent CDF analyses are reviewed and used to constrain Higgs and Technicolor models.

1 Introduction

For many years the standard model has been remarkably successful at explaining and predicting experimental data. However the details of the electroweak symmetry breaking (EWSB) sector of the theory remain largely unknown, and one of the primary goals of present and future colliders is to uncover the mechanism responsible for the symmetry breaking.

In the standard model and many extensions to it, the electroweak symmetry is spontaneously broken by introducing fundamental scalar particles into the theory. These are eventually identified with W_L, Z_L, and one or more physical Higgs bosons [1]. In fact the standard model incorporates the simplest implementation, with just one scalar field doublet. This leaves a single observable scalar particle, the Higgs boson, with unknown mass but fixed couplings to other particles. It should be noted that this simple model explains several striking facts, and any complication (even if possible) tends to weaken it [2].

Alternatively, the electroweak symmetry may be broken dynamically. This is the hallmark of technicolor (TC) theories [3] in which a new strong gauge force (technicolor) and new fermions (technifermions) are introduced. The technicolor force is modelled after QCD, scaled up to the TeV scale, with the technifermions being the analogs of ordinary quarks. Technicolor acts between the technifermions to form bound states (technihadrons). In particular, the technipions include the longitudinal weak bosons, W_L and Z_L, as well as the pseudo-Goldstone bosons of dynamical symmetry breaking. Thus the dynamics of the technifermions assume the role of the scalar Higgs fields in theories with spontaneous symmetry breaking.

Both models coincide in that they can be extended to generate all masses in the theory, notably the fermion masses. In the first case, explicit Yukawa couplings are used. In the second, the first TC model was augmented by

Extended Technicolor (ETC) [4]. In fact, it is attractive to assume that a connection between the physics of the EWSB and the physics of flavour exists even more generally than the two previous examples. The important point is that both models now predict that the scalars (fundamental or composite) will have couplings to ordinary fermions that are proportional to mass, i.e., these scalars will decay predominantly in channels involving heavy quarks and heavy leptons. The experimental consequence of this is to provide a general frame to look for the particles involved in the EWSB, namely, to look for resonances decaying into heavy quarks and/or heavy leptons produced in pairs or in association with a W or Z.

In the following we describe three experimental searches recently published by the CDF Collaboration using $\sim$ 100 pb^{-1} of $p\bar{p}$ collisions at the Tevatron [5,6,7]:

(1) Search for $X \to b\bar{b}$ in association with $W \to \ell\nu$.

(2) Search for $X \to b\bar{b}$ in association with $V \to q\bar{q}$, where $V = W, Z$.

(3) Search for events including a $\tau^+\tau^-$ pair and two extra jets.

No signal beyond standard model expectations has been observed in the current data sample. So in the rest of the article we will investigate the implications of this null result on the models cited at the beginning of the section, emphasizing the first-hand information on the performance of the different experimental techniques and direct evaluation of involved backgrounds. In particular we use searches (1) and (2) to constrain the production of a light standard model Higgs, and search (3) to investigate technicolor models containing a technifamily.

There are other analyses not discussed here for lack of space that can be used to find information about the EWSB sector: top physics, charged Higgs searches, etc. The reader will find information in references [8].

2 Light Standard Model Higgs

The possible range for the mass of the standard model Higgs extends from a lower bound of about 88 GeV/c^2 from the LEP experiments [9] to $\mathcal{O}(1)$ TeV. The present analysis has to be restricted to light masses for two reasons:

- The production cross sections at the Tevatron are small and decrease rapidly as a function of M_{H^0}. In $p\bar{p}$ collisions at $\sqrt{s} = 1.8$ TeV, the Higgs production mechanism with the most promising detection possibilities is $p\bar{p} \to V + H^0$, with $V = W, Z$. In the framework of the standard model, the production cross section in this channel is 1.3 to 0.11 pb for Higgs masses between 70 to 140 GeV/c^2 [10].

- The technique used is to look for resonances in the $b\bar{b}$ channel. $H^0 \to b\bar{b}$ is the dominant decay mode of the Higgs boson only up to $M_{H^0} \sim 130$ GeV/c^2.

However there are two important motivations to emphasize the region of light Higgs masses:

- Precision electroweak experiments suggest that the Higgs boson mass may lie at the lower end of the open range[11].

- A similar symmetry breaking mechanism occurs in the minimal supersymmetric extension of the standard model, where several observable scalar states are predicted, the lightest of which is expected to decay predominantly into $b\bar{b}$ states and to have a mass below 135 GeV/c^2 [12].

The cross sections expected in the standard model are out of the scope of the present analysis, using ~ 100 pb^{-1} of $p\bar{p}$ collisions. We therefore report on a general search for a Higgs scalar produced in association with a vector boson with unknown cross section σ_{VH^0}. We look for H^0 decays to a $b\bar{b}$ pair with unknown branching ratio β, and for two possible decays of the vector boson: (i) $W \to \ell\nu$ with $\ell = e, \mu$ and (ii) $V \to q\bar{q}$. Finally, the limits obtained in both channels are combined.

2.1 Leptonic Analysis

The experimental signature considered is WH^0 with $W \to e\nu$ or $\mu\nu$, and $H^0 \to b\bar{b}$, giving final states with one high-p_T lepton, large missing transverse energy ($\not{E}_T$) due to the undetected neutrino, and two b jets [5]. The ability to tag b jets with high efficiency and a low mistag rate is vital for searching for the decay of $H^0 \to b\bar{b}$. We use the secondary vertex (SECVTX) and soft-lepton (SLT) b-tagging algorithms developed for the top quark discovery [13].

A three-level trigger selects events that contain an electron or muon for this analysis. The event selection starts with the requirement of a primary lepton, either an isolated electron with $E_T > 20$ GeV [14] or an isolated muon with $p_T > 20$ GeV/c, in the central region $|\eta| < 1.0$. A W boson sample is selected by requiring $\not{E}_T > 20$ GeV. Events which contain a second, same flavour lepton with $p_T > 10$ GeV/c are removed as possible Z boson candidates if the reconstructed ee or $\mu\mu$ invariant mass is between 75 and 105 GeV/c^2. The events must also not be accepted by the CDF top dilepton analysis [13]. To

Table 1: The expected number of tagged $W + n$ jet events from QCD backgrounds and top production compared the observed number of events.

	$W + 1$ jet	$W + 2$ jet	$W + 3$ jet	$W + \geq 4$ jet
		Events with single tag		
Expected	67 ± 13	30 ± 5	16 ± 2	13 ± 3
Observed	66	36	11	12
		Events with double tag		
Expected	0	3.0 ± 0.6	3.6 ± 0.6	5.2 ± 1.1
Observed	0	6	6	2

further reduce the dilepton backgrounds, we reject events with an additional high-p_T isolated track with opposite charge to that of the primary lepton. The remaining events are classified according to jet multiplicity. Jets are defined as localized energy depositions in the calorimeters and are reconstructed using an iterative clustering algorithm with a fixed cone of radius $\Delta R = \sqrt{\Delta\eta^2 + \Delta\phi^2} = 0.4$ in $\eta - \phi$ space [15]. Jet energies are then corrected for energy losses in uninstrumented detector regions, energy falling outside the clustering cone, contributions from underlying event and multiple interactions, and calorimeter nonlinearities.

The $W + 2$ jet bin is expected to contain most of the signal, while the other bins are used to check the background calculation. In order to enhance the signal in the $W + 2$ jet bin, we require that one or both of the jets be identified ('tagged') as coming from a b hadron. We require at least one jet to be tagged by the SECVTX algorithm, which has a higher signal-to-noise ratio than the SLT algorithm. For the single-tag analysis, the other jet must not be tagged, while for the double-tag analysis the second jet must be tagged by either SECVTX or SLT.

The SECVTX tagging algorithm begins by searching for secondary vertices that contain three or more displaced tracks. If none are found, the algorithm searches for two-track vertices using more stringent track criteria. A jet is tagged if the secondary vertex transverse displacement from the primary one exceeds three times its uncertainty.

The SLT tagging algorithm identifies electrons and muon from semileptonic b decays by matching tracks with $p_T > 2$ GeV/c with clusters of electromagnetic energy in the calorimeters or tracks in the muon chambers. To gain additional background rejection, we require the track to lie within a cone $\Delta R < 0.4$ of the axis of a jet and to be displaced in the transverse plane from the primary vertex by at least two standard deviations in the jet direction. This latter requirement reduces lepton misidentifications by a factor of five

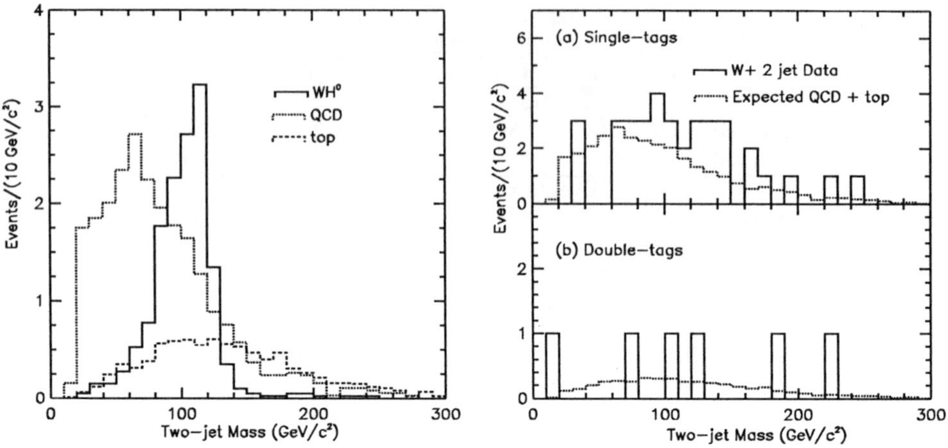

Figure 1: (Left) The expected tagged two-jet mass distribution from QCD (dotted), top (dashed), and WH^0 production (solid) with $M_{H^0} = 110$ GeV/c^2. The WH^0 distribution has been scaled up by a factor of 50. (Right) The measured two-jet mass distribution in the data along with background expectations from (a) single-tagged events and (b) double-tagged events.

while retaining 65% of the efficiency.

The total acceptance is calculated as a product of the kinematic and geometric acceptance, trigger, lepton identification, and b-tagging efficiencies, and the W leptonic branching ratios. The acceptance for identifying WH^0 is calculated from data and a standard model simulation of Higgs production, where the Higgs is forced to decay into $b\bar{b}$ with 100% branching ratio. The acceptance increases monotonically from $0.53 \pm 0.13\%$ ($0.17 \pm 0.04\%$) to $1.1 \pm 0.3\%$ ($0.42 \pm 0.11\%$) for single (double) tagging as M_{H^0} increases from 70 to 120 GeV/c^2. A 25% systematic in the acceptance comes from uncertainties in the modeling of initial and final state radiation, jet energy, and b-tag, trigger, and lepton identification efficiencies.

Background events come predominantly from the direct production of W bosons in association with heavy quarks (estimated using the HERWIG Monte Carlo program), mistags (from generic jet data), and $t\bar{t}$ production (normalized to the CDF measured cross section $\sigma_{t\bar{t}} = 7.6^{+1.8}_{-1.5}$ pb [16]). Other small backgrounds are estimated from a combination of Monte Carlo simulations and data.

The numbers of observed single-tagged and double-tagged events and the corresponding background estimates are shown in Table 1. By construction, data and expectations are in reasonably good agreement in the $W+\geq 3$ jet bins, which, along with other $t\bar{t}$ decay channels, were used to measure the $t\bar{t}$ production cross section. The number of b-tags in the $W+2$ jet bin can be compared to the background calculation. This bin shows a small excess of events corresponding to one standard deviation.

To increase the sensitivity of the search we look for a resonant mass peak in the reconstructed two-jet invariant mass distribution using the 4-momenta of the jets as measured by the calorimeter. The expected two-jet invariant mass shape for WH^0 production is shown in Figure 1 (left) for $M_{H^0}=110$ GeV/c^2. The distributions for the data are shown in Figure 1 (right), along with the background expectation.

We set an upper limit on the production cross section times branching ratio of $p\bar{p}\rightarrow WH^0$ as a function of M_{H^0}, by fitting the number of events in the $W+2$ jet samples and the shape of the two-jet mass distributions. The fit yields $\sigma_{WH^0}\cdot\mathcal{B}(H^0\rightarrow b\bar{b})$ in the range from $0.2^{+4.7}_{-0.0}$ to $5.7^{+4.2}_{-3.0}$ pb for a new particle of mass between 70 and 120 GeV/c^2, statistically compatible with no signal. From the 95% C.L. limits on $\sigma(p\bar{p}\rightarrow WH^0)$, the corresponding limits on VH^0 production were calculated. We used the program PYTHIA to compute the standard model prediction for the ratio $\sigma(ZH^0)/\sigma(WH^0)$. The leptonic analysis efficiency for ZH^0 events relative to that for WH^0 events was estimated to be $(10\pm 2)\%$. The limits are summarized in Table 2.

2.2 Hadronic Analysis

The experimental signature considered is four jets in the final state, with two of them identified as b jets [6]. The hadronic channel described here has the advantage of a larger branching ratio, and is sensitive to both WH^0 and ZH^0 production ($\sigma(ZH^0)/\sigma(WH^0)\sim 0.6$), but suffers from a larger QCD background.

The data sample was recorded with a trigger which requires four or more clusters of contiguous calorimeter towers, each with transverse energy $E_T\geq 15$ GeV, and a total transverse energy $\sum E_T\geq 125$ GeV. Offline, events are required to have four or more jets with uncorrected $E_T>15$ GeV and $|\eta|<2.1$. After this initial selection the sample contains 207,604 events. In addition, we require that at least two among the four highest-E_T jets in the event tagged by the SECVTX algorithm.

There are 764 events with four or more jets and two or more b-tags. In these events, only the four highest-E_T jets are considered for the mass reconstruction: the two highest-E_T b-tagged jets are assigned to the Higgs boson, and the other

Table 2: Summary of the hadronic analysis fit results, standard model predictions for $\beta\sigma$, and 95% C.L. limits from the hadronic, leptonic, and combined analyses.

M_{H^0} (GeV/c^2)	$\beta\sigma$ (pb) fit	$\beta\sigma$ (pb) SM	$\beta\sigma$ (pb) had. limit	$\beta\sigma$ (pb) lep. limit	$\beta\sigma$ (pb) comb. limit
70	44 ± 42	1.13	117.3	21.9	23.1
80	0^{+19}_{-0}	0.76	53.2	28.2	23.8
90	$0.0^{+9.7}_{-0.0}$	0.55	28.9	29.0	18.0
100	$0.0^{+7.6}_{-0.0}$	0.41	22.8	27.2	16.8
110	$0.0^{+6.3}_{-0.0}$	0.30	18.7	30.1	17.1
120	$0.0^{+5.9}_{-0.0}$	0.20	17.6	25.0	16.0
130	$0.0^{+5.5}_{-0.0}$	0.12	16.7	38.5	19.7
140	$0.0^{+5.1}_{-0.0}$	0.06	15.3	34.5	17.2

two to the vector boson. The $b\bar{b}$ invariant mass distribution in signal events contains a Gaussian core with a sigma of $\sim 0.14 \times M_{H^0}$. The tails of the distribution are dominated by the cases (25-30%) where the jet assignment in the mass reconstruction is incorrect. In most of these cases, one of the jets assigned to the Higgs is a heavy quark jet from the decay of the V boson.

The challenge of this analysis is to understand the sample composition. The main source of background events is QCD heavy flavor production. The heavy flavor content of QCD hard processes has been modelled with PYTHIA. We generated all QCD jet production channels and retained the events that contained a heavy quark produced either in the hard scattering or in the associated radiation process. Events with a heavy quark are conventionally classified in three groups: direct production, gluon splitting, and flavor excitation. Direct production events are characterized by a high value of the invariant mass, $M_{b\bar{b}}$, and a low value of the transverse momentum of the $b\bar{b}$ system, $p_T(b\bar{b})$. The same is true for flavor excitation events. The kinematics of final state gluon splitting events favor a relatively smaller invariant mass value and a large $p_T(b\bar{b})$, since both jets tend to be emitted along the same direction. In this plane, the Higgs signal shows a greater tendency to large $M_{b\bar{b}}$ and $p_T(b\bar{b})$ values. A cut on $p_T(b\bar{b}) \geq 50$ GeV/c is $\sim$ 80% efficient for the signal and strongly discriminates against direct production and flavor excitation of heavy quarks. After the $p_T(b\bar{b})$ requirement is applied to the data 589 events remain.

Other backgrounds are $t\bar{t}$ production, Z + jets events with $Z \to b\bar{b}/c\bar{c}$ and fake double-tags. The first two are estimated from Monte Carlo and the last one from data. Using the CDF measured $t\bar{t}$ production cross section and a

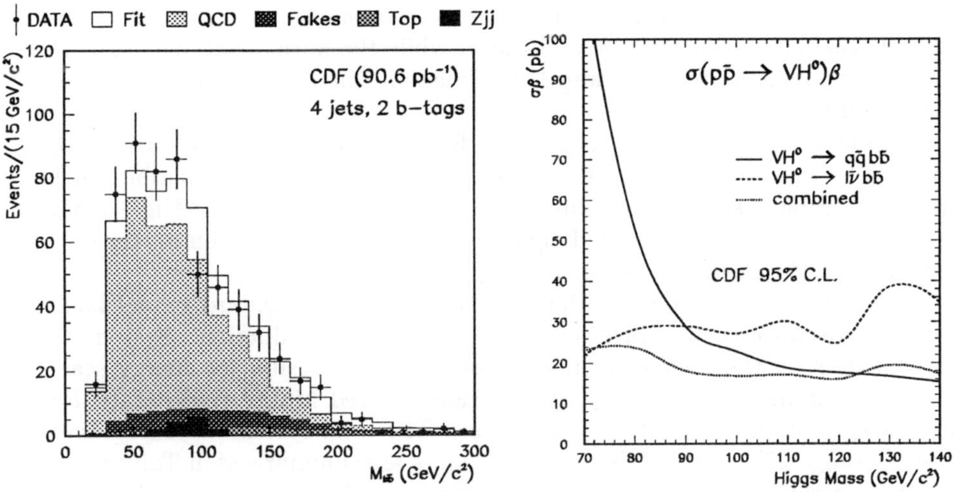

Figure 2: (Left) The measured invariant mass distribution from double-tagged events (points) compared to the fit prediction. The solid line is the sum of the QCD, fakes, $t\bar{t}$, and Z + jets components. (Right) The CDF 95% C.L. upper limits on $\sigma(p\bar{p} \to VH^0) \cdot \beta$ where $\beta = \mathcal{B}(H^0 \to b\bar{b})$.

top quark mass of $M_t = 175$ GeV/c^2, HERWIG predicts 26 ± 7 $t\bar{t}$ events in the data, after trigger, kinematic and b-tag requirements. The same generator predicts 17 ± 4 Z + jets background events. Fake double-tags are defined as events in which at least one of the two tagged jets contains a false secondary vertex in a light quark or gluon jet. Fake tag probabilities are parameterized by measuring in several inclusive jet data samples the proportion of jets in which a secondary vertex is reconstructed on the wrong side of the primary vertex with respect to the jet direction [13]. The current data set is estimated to contain 89 ± 11 fake double-tag events. Finally, other minor sources of background account for less than 1% of the total number of events, have a broad invariant mass distribution, and are neglected in the final fit.

The total signal detection efficiency is defined as the product of the trigger efficiency, the kinematical and geometrical acceptances, the double b-tagging efficiency, the $p_T(b\bar{b})$ cut efficiency, and the V hadronic branching fractions. The total efficiency increases linearly from $0.6 \pm 0.1\%$ to $2.2 \pm 0.6\%$ for Higgs masses ranging from 70 GeV/c^2 to 140 GeV/c^2.

The shape of the observed b-tagged dijet invariant mass distribution is fit, using a binned maximum-likelihood method, to a combination of signal, fake double-tag events, and QCD, $t\bar{t}$ and Z + jets backgrounds. The QCD and signal normalizations are left free in the fit while the normalizations of the $t\bar{t}$, Z + jets and fakes are constrained by Gaussian functions to their expected values and uncertainties.

The fit yields $\sigma_{VH^0} \cdot \beta = 44 \pm 42$ pb for $M_{H^0} = 70$ GeV/c^2, statistically compatible with zero signal. For larger masses, zero signal contribution is preferred. Table 2 shows the result of the fits as a function of the Higgs mass. Figure 2 (left) shows the b-tagged dijet invariant mass distribution for the data compared to the results of the fit for $M_{H^0} \geq 80$ GeV/c^2.

Since the observed distribution is consistent with standard model background expectations, we place limits on $p\bar{p} \rightarrow VH^0$ production. Systematic uncertainties on the 95% C.L. limits arise from luminosity, jet energy scale, double b-tagging efficiencies, QCD radiation, limited Monte Carlo statistics, and background normalizations and shapes. The total systematic uncertainty is in the range 26% − 30%. The 95% C.L. limits are summarized in Table 2 and Figure 2 (right). The resulting bounds fall rapidly from 117 pb at $M_{H^0} = 70$ GeV/c^2 to values between 15 and 20 pb for $M_{H^0} > 105$ GeV/c^2.

2.3 Combined results

To combine the two results presented above, the data from both channels were then fitted simultaneously [6]. Correlations between systematic uncertainties due to luminosity, QCD radiation, and b-tagging efficiency were taken into account. All other systematic uncertainties were considered uncorrelated. The 95% C.L. limits range from 16 to 24 pb and are shown in Table 2 and Figure 2 (right).

3 Technicolor

Now we turn to the second mechanism of EWSB discussed in the Introduction[7]. Particularly interesting from the present experimental point of view [17,18] are TC models containing a technifamily, i.e. a set of technifermions with the same structure and quantum numbers of a complete standard model generation of quarks and leptons, and carrying an additional TC quantum number. By convention, technifermions which are color-triplets of ordinary QCD are called techniquarks, and color-singlet technifermions are called technileptons. The particle spectrum of these models includes color-singlet, -triplet and -octet technipions. The technipions (π_T) decay via ETC interactions. Since these are also responsible for the fermion masses, technipions are expected to have Higgs-boson-like couplings to ordinary fermions, i.e. to decay preferentially to third-generation quarks and leptons. In particular, the color-triplet technipions

Table 3: The 95% confidence level upper limits on the leptoquark (color-triplet technipion) production cross section times branching ratio squared as a function of $M(\pi_{LQ})$ and $M(\rho_T)$, for $\Delta M = 50$ GeV/c^2. Numbers are given in pb.

$M(\pi_{LQ})$	$M(\rho_T)$ (GeV/c^2)										
(GeV/c^2)	200	250	300	350	400	450	500	550	600	650	700
100	12.7	9.8	8.2	7.4	7.2	7.7	8.5	9.4	9.8	10.0	10.2
125		6.4	5.3	4.6	4.1	3.9	3.9	4.1	4.5	4.8	5.0
150			4.7	4.1	3.6	3.3	3.1	3.0	3.1	3.2	3.5
175				3.7	3.3	3.1	2.9	2.7	2.6	2.6	2.7
200					3.4	3.0	2.8	2.5	2.3	2.2	2.1
225						2.9	2.7	2.5	2.3	2.2	2.1
250							2.8	2.5	2.3	2.2	2.1
275								2.5	2.3	2.2	2.0
300									2.3	2.2	2.0

are an example of scalar third-generation leptoquarks (π_{LQ}). In this section, we use the results of a search for third-generation leptoquarks by CDF, in order to explore TC models containing a technifamily. Other experimental constraints on these models come from precision electroweak measurements at LEP [19,20], and from measurements of the $b \rightarrow s\gamma$ decay rate [21].

Let's consider technicolor models containing a family of color-singlet technileptons and color-triplet techniquarks. In these models, there is a color-octet vector resonance, called technirho (ρ_T), with the quantum numbers of the gluon. Leptoquarks are assumed to be pair produced via gluon-gluon fusion and $q\bar{q}$ annihilation. In $q\bar{q}$ and gg collisions, the ρ_T couples to the gluon propagator enhancing s-channel reactions (Fig. 3), analogously to the vector-meson-dominance description of the process $e^+e^- \rightarrow \pi^+\pi^-$ [22]. Two decay modes may exist for the technirho [18]: $\rho_T \rightarrow q\bar{q}, gg$ and $\rho_T \rightarrow \pi_T\bar{\pi}_T$. If the ρ_T mass is less than twice the π_T mass, only the $q\bar{q}, gg$ decay mode is possible, resulting in resonant dijet production. A search result for the dijet signal of ρ_T has already been reported by CDF. The CDF-measured dijet mass spectrum excludes ρ_T with masses in the range $260 < M(\rho_T) < 480$ GeV/c^2 at the 95% C.L. [23]. If the ρ_T mass is larger than twice the π_T mass, the ρ_T decays preferentially into π_T pairs.

The technipion spectrum of the technifamily model was estimated in [24,18]. It contains color-singlet, -triplet and -octet (π_8) technipions. The octets are heavier than the triplets, and these are heavier than the singlets. We make the simplifying assumption that there is no mass splitting among the different octet and triplet technipions. We consider the class of color-triplet technipions

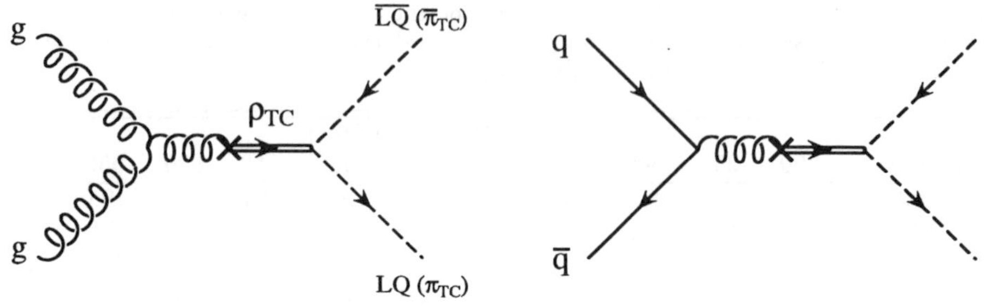

Figure 3: The resonant production of technipion pairs. The technirho couples directly to the gluon via vector-meson-dominance enhancing s-channel production.

decaying via $\pi_{LQ} \to \bar{b}\tau^-$ ($\bar{\pi}_{LQ} \to b\tau^+$) with branching fraction β.

The leading-order leptoquark pair production cross section depends only on the technirho mass ($M(\rho_T)$), the leptoquark mass ($M(\pi_{LQ})$), and the technirho width ($\Gamma(\rho_T)$). $M(\pi_{LQ})$ and $M(\rho_T)$ are treated as independent free parameters. $\Gamma(\rho_T)$ can be calculated as a function of four more basic quantities, $\Gamma(\rho_T) = \Gamma(M(\rho_T), M(\pi_{LQ}), \Delta M, N_{TC})$, where $\Delta M = M(\pi_8) - M(\pi_{LQ})$, and N_{TC} is the number of technicolors. We consider $M(\rho_T)$, $M(\pi_{LQ})$, ΔM, and N_{TC} as the four continuous parameters of the theory. We set limits in the $M(\pi_{LQ}) - M(\rho_T)$ plane. We probe the dependence of the production cross section on $\Gamma(\rho_T)$ by fixing $N_{TC} = 4$, while allowing ΔM to take one expected and two limiting values. ETC and QCD corrections to $M(\pi_8)$ and $M(\pi_{LQ})$ are responsible for ΔM, analogously to the QED corrections to $M(\pi^o)$ and $M(\pi^\pm)$. ΔM is expected to be around 50 GeV/c^2 [18]. We take $\Delta M = 0$ and $\Delta M = \infty$ as two extreme values. The resulting variation in $\Gamma(\rho_T)$ could also have been obtained changing N_{TC} by a factor of 4, for a fixed $\Delta M = 50$ GeV/c^2.

The experimental signature considered is $\tau^+\tau^-$ plus two jets in the final state, in the case where one τ decays leptonically and the other decays hadronically. The analysis selects a 110 pb^{-1} data set containing an isolated electron or muon in the region $|\eta| < 1$ with $p_T > 20$ GeV/c [14], and an isolated, highly-collimated hadronic jet consistent with a hadronic tau decay. Hadronic τ candidates (τ-jets) are selected from jets that have an uncorrected total transverse energy of $E_T > 15$ GeV in the region $|\eta| < 1$. The associated charged particles with $p_T > 1$ GeV/c in a 10° cone around the jet direction must satisfy the following requirements: (*i*) the τ-jet must have one or three charged particles; (*ii*) if there are three, the scalar sum p_T must exceed 20 GeV/c and the invariant mass must be smaller than 2 GeV/c^2; and (*iii*) the leading charged

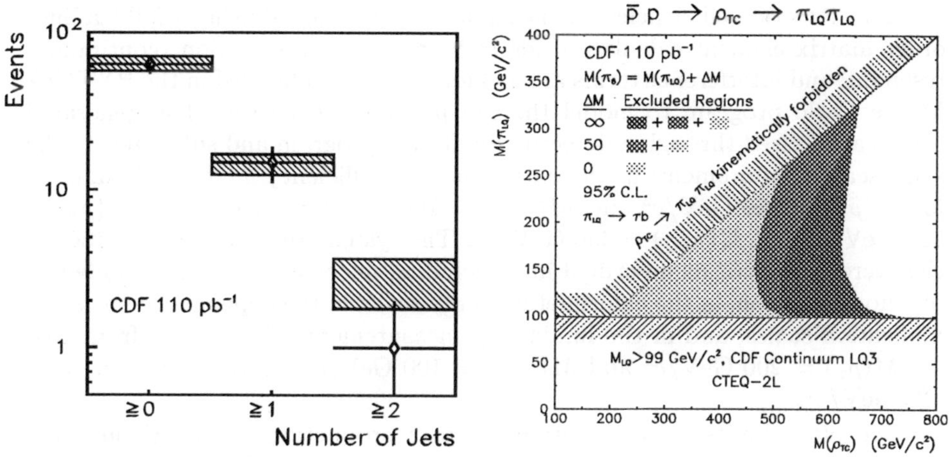

Figure 4: (Left) The inclusive jet multiplicity in $\tau^+\tau^-$ candidate events (diamonds) compared to standard model expectations. (Right) The 95% C.L. exclusion regions in the $M(\pi_{LQ}) - M(\rho_T)$ plane. The three shaded areas correspond (from left to right) to technipion mass splitting values of 0, 50 GeV/c^2 and ∞, respectively.

particle must have $p_T > 10$ GeV/c and must point to an instrumented region of the calorimeter. The efficiency of the τ-jet identification criteria grows from 32% for τ-jets in the range $15 < E_T < 20$ GeV to a plateau value of 59% for $E_T > 40$ GeV. Isolated τ-jets must have no charged particles with $p_T > 1$ GeV/c in the annulus between 10° and 30° around the jet axis. Events where the high-p_T lepton is consistent with originating from a $Z \to ee$ or $Z \to \mu\mu$ decay are removed. In addition, the analysis uses the missing transverse energy characteristic of neutrinos from tau decays. The requirement $\Delta\Phi < 50°$, where $\Delta\Phi$ is the azimuthal separation between the directions of the missing transverse energy $\not{E}_T$ and the lepton, distinguishes $\tau^+\tau^-$ events from backgrounds such as $W +$ jets. Figure 4 (left) shows the jet multiplicity in $\tau^+\tau^-$ candidate events. The agreement with the standard model background prediction is excellent. Finally, two or more jets with $E_T > 10$ GeV and $|\eta| < 4.2$, assumed to originate from b quark hadronization, are required. One leptoquark pair candidate event survives these selection criteria. The observed yield is consistent with the $2.4^{+1.2}_{-0.6}$ expected background events from standard model processes, dominated by $Z \to \tau\tau +$ jets production (2.1 ± 0.6) with the remainder from diboson and $t\bar{t}$ production [7].

The detection efficiencies for the signal are determined using a full leading-order matrix element calculation for technipion pair production (continuum, resonant, and interference terms are included)[18] and embedded in the PYTHIA Monte Carlo program to model the full $p\bar{p}$ event structure. The generated events are passed through a detector simulation program and subjected to the same search requirements as the data. The total efficiency increases from 0.3% for $M(\rho_T) = 200$ GeV/c^2 and $M(\pi_{LQ}) = 100$ GeV/c^2, to 1.8% for $M(\rho_T) = 700$ GeV/c^2 and $M(\pi_{LQ}) = 300$ GeV/c^2. The systematic errors in the efficiencies were estimated as described in [7], including uncertainties in the modelling of gluon radiation, in the calorimeter energy scale, in the dependence on renormalization scales, and in the luminosity measurement. They range from 15% for $M(\rho_T) = 200$ GeV/c^2 and $M(\pi_{LQ}) = 100$ GeV/c^2, to 10% for $M(\pi_{LQ}) \geq 125$ GeV/c^2.

We place limits on the leptoquark pair production cross section times branching ratio squared within the framework of the technicolor model described above. Table 3 lists the leptoquark 95% confidence level upper limits on the production cross section times branching ratio squared as a function of $M(\pi_{LQ})$ and $M(\rho_T)$, for $\Delta M = 50$ GeV/c^2. These numbers differ by at most 1 pb from the corresponding limits for $\Delta M = 0$ and $\Delta M = \infty$ when $M(\pi_{LQ}) < 175$ GeV/c^2. For larger values of $M(\pi_{LQ})$ the differences are negligible. Comparing to the theoretical expectations for $\sigma(p\bar{p} \to \pi_{LQ}\bar{\pi}_{LQ}) \cdot \beta^2$, we place bounds in the $M(\pi_{LQ}) - M(\rho_T)$ plane. Figure 4 (right) shows the 95% C.L. mass exclusion regions. The upper part of the plot corresponds to the kinematically forbidden region where $M(\rho_T) < 2M(\pi_{LQ})$. The bottom region is the exclusion area from the continuum leptoquark analysis, $M(\pi_{LQ}) \geq 99$ GeV/c^2 [7]. The three shaded areas from left to right correspond to technipion mass splitting values of $\Delta M = 0$, 50 GeV/c^2 and ∞, respectively. Although more information is presented in Figure 4, it is useful to summarize our technirho excluded region using a single number. For $\Delta M = 0$ (50, ∞) and $M(\pi_{LQ}) < M(\rho_T)/2$, we exclude color octet technirhos with mass less than 465 (513, 622) GeV/c^2 at 95% confidence level.

4 Discussion and Prospects

We have described several attempts to explore the phenomenology of the EWSB sector at the Tevatron. A first kind of analyses tried to reconstruct $b\bar{b}$ resonances. From the experimental point of view, this involved the study and development of a number of techniques of broad interest for the hadron collider experiments of the next decade. These include jet spectroscopy, b-tagging algorithms, and quantitative evaluation of challenging new backgrounds as QCD production of heavy quarks in a multijet environment.

No signal has been found yet, and the results have been used to constrain the production of a light standard model Higgs. The Higgs searches have not provided yet any lower Higgs mass bound. The sensitivity of the present search is limited by statistics to a cross section approximately two orders of magnitude larger than the predicted cross section for standard model Higgs production [10]. It also should be noted that, because these limits were derived from a shape fit, they only apply to a very restricted region of parameter space in the minimal supersymmetric extension of the standard model.

For the next Tevatron run CDF hopes for an approximately twenty-fold increase in the total integrated luminosity and a factor of two improvement in the double b-tagging efficiency. D0 will be as good as CDF in this respect. However, this is still insufficient to reach say a 120 GeV/c^2 Higgs mass, unless the total detection efficiency be improved by one order of magnitude. The viability of this improvement is at present under study [25]. Plans include the installation of dedicated Higgs triggers, systematic inclusion of all relevant channels (notably ZH^0 production, with $Z \to \ell\ell$ and $Z \to \nu\nu$), finer mass resolution, and additional, more efficient selection algorithms (e.g., neural networks).

Another analysis described a virtually background-free search of resonances decaying into τ leptons. The four objects appearing in the selected events can be arranged in two ways. Looking for τ-jet leptoquark resonances the analysis can be used to directly constrain TC models including one technifamily. The expected factor $\times 20$ in luminosity will help to push the technirho mass limits closer to the TeV region. Interpreted as containing a $\tau^+\tau^-$ resonance, the same events can be used to test effectively the large $\tan(\beta)$ region of the SUSY extension of the standard model [26].

In conclusion, we are just at the very beginning of this kind of physics at hadron colliders. Most importantly we need data to look at, and for that reason the CDF and D0 Collaborations are looking forward to Run 2 with great anticipation.

Acknowledgments

I would like to thank Prof. B. Adeva and collaborators for the stimulating atmosphere of the meeting and warm hospitality.

References

1. P.W. Higgs, Phys. Lett. **12**, 132 (1964), Phys. Rev. Lett. **13**, 508 (1964), Phys. Rev. **145**, 1156 (1966); F. Englert and R. Brout, Phys. Rev. Lett. **13**, 321 (1964).

2. M. Veltman, Report No. CERN-97-05.
3. S. Weinberg, Phys. Rev. D **19**, 1277 (1979); L. Susskind, Phys. Rev. D **20**, 2619 (1979).
4. S. Dimopoulos and L. Susskind, Nucl. Phys. **B155**, 237, (1979); E. Eichten and K. Lane, Phys. Lett. B **90**, 125 (1980).
5. F. Abe *et al.*, Phys. Rev. Lett. **79**, 3819 (1997).
6. F. Abe *et al.*, Phys. Rev. Lett. **81**, 5748 (1998).
7. F. Abe *et al.*, Phys. Rev. Lett. **78**, 2906 (1997); Report No. Fermilab-Pub-98-352-E.
8. S. Abachi *et al.*. Report No. Fermilab-Conf-96-258-E, 1996; F. Abe *et al.*, Phys. Rev. Lett. **79**, 357 (1997); B. Bevensee, Report No. Fermilab-Conf-98-155-E, 1998; K. Tollefson, Report No. Fermilab-Conf-98-389-E, 1998.
9. M. Acciarri *et al.*, Phys. Lett. **B 431** 437 (1998); R. Barate *et al.*, Phys. Lett. **B 440** 403 (1998); G. Abbiendi *et al.*, Report No. CERN-EP-98-173.
10. S.L. Glashow, D.V. Nanopoulos and A. Yildiz, Phys. Rev. D **18**, 1724 (1978); A. Stange, W. Marciano and S. Willenbrock, Phys. Rev. D **49**, 1354 (1994); Phys. Rev. D **50**, 4491 (1994).
11. F. Teubert, Invited Talk at the IVth International Symposium on Radiative Corrections, Barcelona, 1998.
12. M. Carena *et al.*, Nucl. Phys. **B 461** 407 (1996); H. E. Haber *et al.*, Z. Phys. **C75** 539 (1997).
13. F. Abe *et al.*, Phys. Rev. Lett. **74**, 2626 (1995).
14. In the CDF coordinate system, ϕ and θ are the azimuthal and polar angles with respect to the proton beam direction. The pseudorapidity η is defined as $-\ln[\tan(\theta/2)]$. The transverse momentum of a particle is $p_T = p\sin\theta$. The analogous quantity using calorimeter energies is called the transverse energy E_T. The difference between the vector sum of all the transverse energies and zero is the missing transverse energy $\not{E}_T$.
15. F. Abe *et al.*, Phys. Rev. D **45**, 1448 (1992).
16. F. Abe *et al.*, Phys. Rev. Lett. **80**, 2773 (1998).
17. E. Eichten, I. Hinchliffe, K. Lane, and C. Quigg, Rev. Mod. Phys. **56**, 579 (1984); Phys. Rev. D **34**, 1547 (1986).
18. K. Lane and M.V. Ramana, Phys. Rev. D **44**, 2678 (1991).
19. M. Martinez *et al.*, Report No. CERN-EP-98-27, 1998, submitted to Rep. Mod. Phys..
20. R.S. Chivukula, S. B. Selipsky, and E.H. Simmons, Phys. Rev. Lett. **69**, 575 (1992); R.S. Chivukula, E.H. Simmons, and J. Terning, Phys. Lett. B **331**, 383 (1994).

21. B. Balaji, Phys. Rev. D **53**, 1699 (1996); C. Lu and Z. Xaio, Phys. Rev. D **53**, 2529 (1996).
22. J.J. Sakurai, *Currents and Mesons* (Chicago University Press, Chicago, IL, 1969).
23. F. Abe *et al.*, Phys. Rev. D **55**, 5263 (1997); Report No. Fermilab-Pub-98-290-E.
24. M.E. Peskin, Nucl. Phys. **B175**, 197 (1980); J. Preskill, Nucl. Phys. **B177**, 21 (1981).
25. J. Conway, Talk at the Tevatron Run 2 SUSY/Higgs Workshop, Fermilab, 1998.
26. M. Drees *et al.*, Phys. Rev. Lett. **80**, 2047 (1998); Erratum ibid. **81**, 2394 (1998).

CP - VIOLATION AND B - PHYSICS

J. BERNABEU
Departament de Física Teórica,
Universitat de València
46100 Burjassot (Valencia), Spain

The following topics in CP Violation studies are discussed:

I Status of $K^o - \bar{K}^o$ problem.

II The future in K-physics.

III The rare decay $K_L \longrightarrow \pi^o \nu \bar{\nu}$.

IV B-physics.

V Principle of CP violating observables.

VI Rate Asymmetries.

VII Summary.

1 Introduction

In 1964, Christenson, Cronin, Fitch and Turlay[1] discovered that the long-lived neutral kaon K_L also decays to $\pi^+\pi^-$ with a branching ratio of $\sim 2\times 10^{-3}$.This discovery established CP violation and the fact that K_L is not identical to the CP-eigenstate with CP-eigenvalue equal to -1. Similarly, the short-lived neutral kaon K_S is not identical to the CP-eigenstate with CP-eigenvalue equal to +1.

CP-violation was confirmed later by the decay[2] $K_L \longrightarrow \pi^o\pi^o$ and by the charge asymmetry[3] in the K_{l3} decays $K_L \longrightarrow \pi^\pm l^\mp \nu_l$. This semileptonic asymmetry measures whether CP-violation is present in the physical eigenstates of the $K^o - \bar{K}^o$ Mass Matrix, referred to as Indirect CP Violation. The present value of the world average[4] of the charge asymmetry gives a CP-violation in the Mixing Re $(\epsilon_K) = (1.63\pm0.06)\times 10^{-3}$, where ϵ_K is the CP-mixing in $K_{L,S}$.

For the non-leptonic $K_{S,L} \longrightarrow 2\pi$ decays, the angular momentum of the pions vanishes. The spatial part of the (2π)- state is therefore symmetric, and since pions are bosons, the isospin state must be symmetric too. The two symmetric combinations of two I=1 states have I=0 and I=2, and the four transition amplitudes are thus $< 0|K_S >, < 2|K_S >, < 0|K_L >, < 2|K_L >$.

By normalizing to the dominant CP- conserving $\Delta I = 1/2$ transition amplitude $< 0|K_S >$, we define three complex numbers

$$\epsilon_0 = < 0|K_L > / < 0|K_S >$$

$$\epsilon_2 = \frac{1}{\sqrt{2}} < 2|K_L> / < 0|K_S> \qquad (1)$$
$$\omega = < 2|K_S> / < 0|K_S>$$

Non-vanishing values of $\epsilon_{0,2}$ demonstrate CP-violation, whereas ω parametrizes the relative contribution of the $\Delta I = 3/2$ amplitude.

The experimentally observable quantities are

$$\eta_{+-} = < \pi^+\pi^-|K_L> / < \pi^+\pi^-|K_S> \qquad (2)$$
$$\eta_{oo} = < \pi^o\pi^o|K_L> / < \pi^o\pi^o|K_S>$$

Relating the isospin states to the physical (2π)-states, one obtains

$$\eta_{+-} = \frac{\epsilon_0 + \epsilon_2}{1 + \frac{\omega}{\sqrt{2}}}, \quad \eta_{oo} = \frac{\epsilon_0 - 2\epsilon_2}{1 - \sqrt{2}\omega} \qquad (3)$$

The validity of the $\Delta I = 1/2$ rule for CP-conserving weak non-leptonic decays leads to $|\omega| << 1$. Its actual value is $|\omega| \simeq 1/22$.

A possible choice for the phase of the $K^o \longrightarrow 2\pi$ (I=0) amplitude is obtained by putting all its complexity as coming from final-state-interactions leading to a phase-shft δ_o.In this Wu-Yang phase convention,

$$\left.\begin{array}{r} < 0|K^o>= iA_oe^{i\delta o} \\ < 0|\bar{K}^o>= -iA_oe^{i\delta o} \end{array}\right\} A_o\text{real, Im }(A_o)=0 \qquad (4)$$

whereas

$$\left.\begin{array}{r} < 2|K^o>= iA_2e^{i\delta 2} \\ < 2|\bar{K}^o>= -iA_2^*e^{i\delta 2} \end{array}\right\} \qquad (5)$$

Then one gets

$$\epsilon_0 = \epsilon_K, \quad , \epsilon_2 = i\frac{1}{\sqrt{2}}\frac{\text{Im}A_2}{A_o}e^{i(\delta_2-\delta_o)} \equiv \epsilon' \qquad (6)$$

as well as the semileptonic charge asymmetry

$$\delta = \frac{\Gamma(K_L \to \pi^- l^+ \nu_e) - \Gamma(K_L \to \pi^+ l^- \bar{\nu}_e)}{\Gamma(K_L \to \pi^- l^+ \nu_e) + \Gamma(K_L \to \pi^+ l^- \bar{\nu}_e)} = 2 \text{ Re } (\epsilon_K) \qquad (7)$$

We observe:

(i) With $\epsilon_K$defined by CP-mixing of the physical $K_{L,S}$ state, Eq. (7) is independent of the phase convention and measures Indirect CP Violation.

(ii) Eq. (6) says, however, that the cellebrated separation of Indirect (ϵ_K) plus Direct (ϵ') CP-violation in the experimental quantities η_{+-}, η_{oo} [see Eq.(3)] for 2π-decays is a phase-convention-dependent language.

(iii) With $\eta_{+-} = \epsilon_K + \epsilon'$, $\eta_{oo} = \epsilon_K - 2\epsilon'$, the observable quantities satisfy the Wu-Yang triangle relation. The experimental aim is the separation of ϵ_K and ϵ'.

Eq. (6) says that the phase of ϵ' is

$$\text{Arg } (\epsilon') = \frac{\pi}{2} + (\delta_2 - \delta_o) = (48^o \pm 4^o) \tag{8}$$

where the numerical value is based on $\pi - \pi$ scattering [5].

The phase of ϵ_K can be identified from the diagonalization of the mass matrix $M - \frac{i}{2}\Gamma$ which, in the flavour basis $K^o - \bar{K}^o$, reads

$$\epsilon_K \simeq e^{i\emptyset_{SW}} \frac{\text{Im}M_{12} - \frac{i}{2}\text{Im}\Gamma_{12}}{\sqrt{\Delta M_K^2 + \frac{1}{4}\Delta\Gamma_K^2}} \tag{9}$$

where $\emptyset_{SW}$ is the so-called superweak phase

$$\emptyset_{SW} \equiv \arg\tan\left(\frac{-2\Delta M_K}{\Delta\Gamma_K}\right) \tag{10}$$

and $\Delta M_K - \frac{i}{2}\Delta\Gamma_K \equiv (M_S - M_L) - \frac{i}{2}(\Gamma_S - \Gamma_L)$.

Taking into account the approximate empirical relation $\Delta\Gamma_K \simeq -2\Delta M_K$, we find that $\emptyset_{SW}$ is again near to $\pi/4$. More precisely, $\emptyset_{SW} = 43.49^o \pm 0.08^o$. The phase introduced by the second factor of ϵ_K in Eq. (9) is very small, so one concludes that the phases of ϵ_K and ϵ' are nearly equal. The equality of the phases $\emptyset_{+-}$ and $\emptyset_{oo}$ of the observable amplitudes η_{+-} and η_{oo}, respectively, is a test of CPT- invariance in Nature.

The experimental results citeb4

$$|\eta_{+-}| = (2.285 \pm 0.019) \times 10^{-3}, \quad |\eta_{oo}| = (2.275 \pm 0.019) \times 10^{-3} \tag{11}$$

show their equality within errors, indicating that $|\epsilon'| << |\epsilon|$. This is in part a consequence of the approximate $\Delta I = 1/2$ rule.

In order to extract the corresponding phases $\emptyset_{+-}$, $\emptyset_{oo}$, one needs a mechanism of $K_L - K_S$ interference. This is possible from either (i) Coherent Regeneration of K_S from K_L, or (ii) Time-Distribution of $K \longrightarrow \pi\pi$ events as obtained from a pure flavour (strangeness) state prepared at t = 0. The first method (i) needs the preparation of K_L beams. The second method (ii) makes use of flavour tag: either K^oor $\bar{K}^o$. The results for $\pi^+\pi^-$ decays obtaimed[6] by the CP-LEAR Collaboration at CERN are shown in Fig. 1, where the first observation of a difference in partial decay rates between particles and antiparticles is reported.

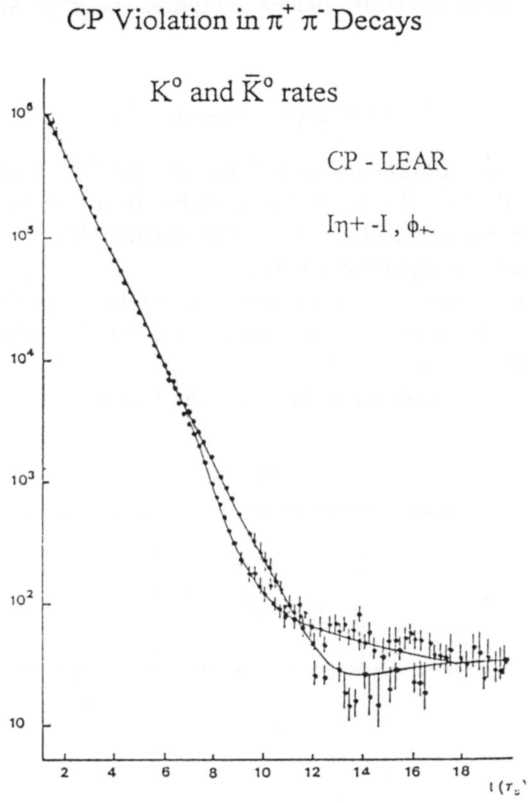

First direct observation of difference of decay rates between particles and antiparticles.

Figure 1: Time Evolution of K^o and $\bar{K}^o$ rates at CP-LEAR

The world averages give

$$\emptyset_{+-} = 43.5^o \pm 0.6^o; \quad \emptyset_{oo} = 43.4 \pm 1.0^o \tag{12}$$

compatible with CPT and in agreement with the superweak phase.

2 The future in K-physics

The main objective in $K \longrightarrow 2\pi$ decays is the separation of ϵ and ϵ' with a sensitivity better than 10^{-4}. The method followed by the CERN and FermiLab experiments is the "ratio of ratios", which compares the $\pi^o\pi^o$ and $\pi^+\pi^-$ decay channels

$$\mathrm{Re}(\epsilon'/\epsilon) = \frac{1}{6}\left\{1 - |\eta_{oo}/\eta_{+-}|^2\right\} \tag{13}$$

Present results give $(23.0\pm6.5)\times10^{-4}$ for the NA31 - Collaboration [7] at CERN and $(7.4\pm5.9)\times10^{-4}$ for the E731- Collaboration [8] at FermiLab. These conflicting results are statistically compatible within 8%. New experiments will push the precision to significant values.

The standard mechanism [9] to incorporate CP Violation in the ElectroWeak Theory is the CKM Complex Mixing Matrix for 3 families of quarks. It generates CP- Violating effects in both:

(i) the ΔS=2 $K^o\bar{K}^o$ transition, through the box diagram

Box-
-Diagram

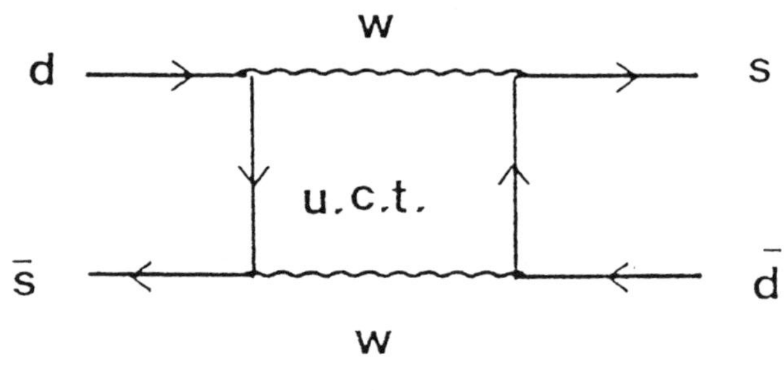

Figure 2: Quark diagram for $\Delta S = 2$ $K^o \rightleftharpoons \bar{K}^o$

The real part of the $K^o - \bar{K}^o$ mixing, which contributes to ΔM, is dominated by intermediate charm quarks running in the loop. Its main uncertainty comes from the hadronic matrix element of the effective four-quark $\Delta S = 2$ operator, when the higher degrees of freedom are integrated out. Its value is parametrized by B_K, measuring it with respect to the vacuum-contribution in the intermediate state.

The imaginary part of the box diagram provides the explanation of indirect CP violation ϵ_K in the Standard Model and it is dominated by intermediate top quarks running in the loop. Its experimental value can be analyzed in terms of the V_{td} matrix-element of the CKM Mixing Matrix, using the Wolfenstein parametrization [10].

(ii) the $\Delta S = 1$ decay amplitude, through the penguin diagram

Penguin-
-Diagram

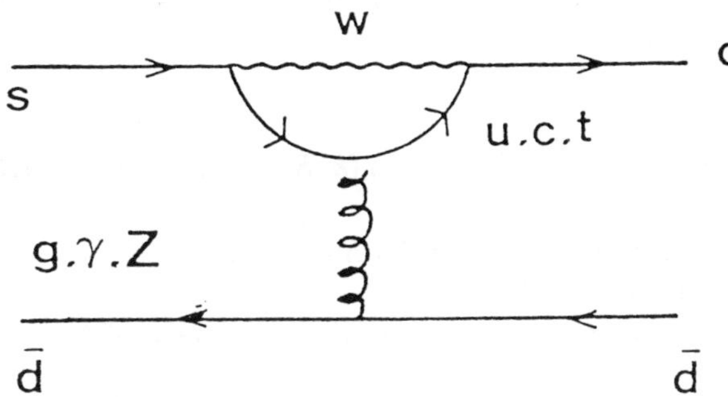

Figure 3: Quark diagram for $\Delta S = 1$ K^o Decays

A theoretical estimate of the direct CP-violation ϵ'/ϵ is very involved. The quark diagrams induce up to 10 four-quark operators once the higher degrees of freedom are integrated out. A control of the long-distance effects becomes both fundamental and difficult. Although the gluon-mediated penguin was initially assumed to be the main ingredient, the Z-penguin contains nondecoupling m_t^2 terms in the amplitude. The final result is very sensitive to m_t. The present theory provides values for ϵ'/ϵ ranging from -3×10^{-4} to $+10^{-3}$, depending on the hadronic physics. For a recent review, see [11].

There is a second method, envisaged [12] by the dedicated $\emptyset$-Factory at Frascati, to extract ϵ'/ϵ: that of Coherent Decay of $(K^o \bar{K}^o)$. The two-particle $K^o - \bar{K}^o$ system, such as is being produced by $\emptyset$- decay, has a total angular momentum J=L and it is an eigenstate of charge-conjugation C with eigenvalue C=$(-1)^L$.

Bose statistics requires that the physical state has to be symmetric under the product of C× Ş, where C is charge-conjugation and Ş is permutation of space coordinates. In other words, the physical state $|\psi>$ has to be an eigenstate of CŞ with eigenvalue +1.

Thus, as a consequence of Quantum Mechanics indistinguishibility, we can write

$$|\psi> = \frac{1}{\sqrt{2}}(1 + C\c{S})|LM; K^o\bar{K}^o> \tag{14}$$

$$= \frac{1}{\sqrt{2}} \left\{ |K^o(\vec{k})\bar{K}^o(-\vec{k}) > +(-1)^L |\bar{K}^o(\vec{k})K^o(-\vec{k}) > \right\}$$

If we impose the conditions of the e^+e^- machine at the ∅-peak, like DA∅NE at Frascati, we have L=1 and C=-. The produced state of $|K^o\bar{K}^o>$, when written in terms of the physical neutral kaon states with definite mass and lifetime, is

$$|K^o\bar{K}^o(C=-)>= \frac{1+|\alpha|^2}{2\sqrt{2}\alpha} \left\{ |K_L(\vec{k})K_S(-\vec{k}) > -|K_S(\vec{k})K_L(-\vec{k}) > \right\} \quad (15)$$

with no component of K_SK_S or K_LK_L states.

One concludes [13]:

(i) Bose stadistics for L=1 says that the state of two identical bosons is forbidden.

(ii) The result (i) applies to any time in the evolution of the system, and thus we have simultaneously NO K^oK^o, NO $\bar{K}^o\bar{K}^o$.

(iii) The preparation of $(K^o\bar{K}^o)$ from a ∅-factory leads to correlated (simultaneous) beams of K_L and K_S.

The parameter α in the normalization of the state (15) is the ratio of amplitudes

$$\alpha \equiv \frac{p}{q} = \frac{1+\epsilon_K}{1-\epsilon_K} \quad (16)$$

The study of the decay amplitude from this correlated system leads to the question for the appearance of X_1 at time t_1 and X_2 at time t_2, as illustrated in Fig.4

The result is

$$A(X_1,X_2) = \frac{1+|\alpha|^2}{2\sqrt{2}\alpha} < X_1|K_S >< X_2|K_S > e^{-i(\lambda_L+\lambda_S)t/2} \times \left\{ \eta_1 e^{-i\Delta\lambda\Delta t/2} - \eta_2 e^{i\Delta\lambda\Delta t/2} \right\} \quad (17)$$

where

$$\begin{aligned} \eta_j &\equiv < X_j|K_L > / < X_j|K_S > ; \ ; j=1,2 \\ \lambda_{S,L} &= M_{S,L} - \frac{i}{2}\Gamma_{S,L}; \ \ \Delta\lambda = \lambda_S - \lambda_L \\ t &= t_1 + t_2; \quad \Delta t = t_2 - t_1 \end{aligned} \quad (18)$$

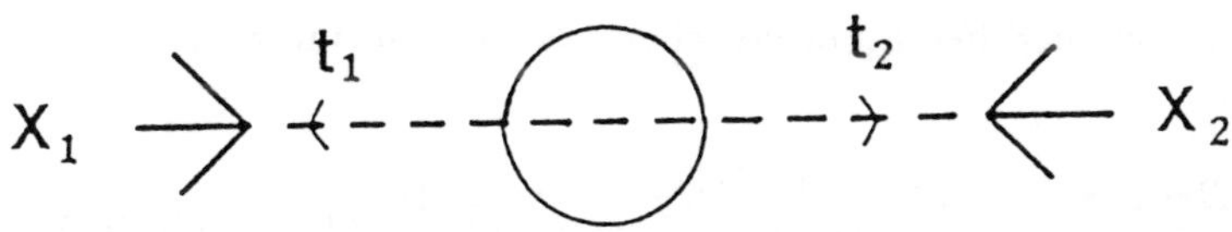

Figure 4: The Decay of the Correlated $(K^o\bar{K}^o)$

Suppose that we detect the two decay products X_1 and X_2 at equal times $t_1 = t_2$. For $\Delta t = 0$, the decay amplitude gives

$$A(X_1, X_2; \;\; \Delta t = 0) \propto (\eta_1 - \eta_2)$$

In particular, for $X_1 = X_2$ one finds

$$A[X_1(t_1), \;\; X_1(t_1)] = 0 \tag{19}$$

which is a quantum mechanical correlation [14] of the Einstein-Podolsky-Rosen type implied by Bose statistics.

If we select the two different 2π decay channels of the neutral kaon, i.e., $X_1 = \pi^+\pi^-$ and $X_2 = \pi^o\pi^o$, we find at equal times

$$A[\pi^+\pi^-(t_1), \, \pi^o\pi^o(t_1)] \propto (\eta_{+-} - \eta_{oo}) = 3\epsilon'$$

This result has been discussed [15] in order to maximize the ϵ' effect in the context of $\emptyset$-factories. The intensity asymmetry, for a given Δt, under the change $\Delta t \longrightarrow -\Delta t$, is sensitive to Im (ϵ'/ϵ) when Δt is of the order of ζ_S, the K_S lifetime, whereas it is very sensitive to Re(ϵ'/ϵ) for large Δt in these units.

Instead of comparing the rates at different time intervals, for a given decay channel $\pi^+\pi^-\pi^o\pi^o$ of the $K^o\bar{K}^o$ system, one can adopt a different strategy:

the comparison of different decay channels in time integrated rates [16]. In particular,

$$\frac{Br(\pi^+\pi^-,\pi^+\pi^-)}{Br(\pi^o\pi^o,\pi^o\pi^o)} = \left[\frac{Br(K_S \longrightarrow \pi^+\pi^-)}{Br(K_S \longrightarrow \pi^o\pi^o)}\right]^2 \{1 + 6\mathrm{Re}(\epsilon'/\epsilon)\} \tag{20}$$

3 The rare decay $K_L \longrightarrow \pi^o\nu\,\bar{\nu}$

This decay is a flavour-changing-neutral current (FCNC) process which, in the Standard Model, proceeds through a CP violating amplitude. CP-Violation is here induced by the interference between the two flavour amplitudes $K^o \longrightarrow \pi^o\nu\,\bar{\nu}$ and $K^o \longrightarrow \bar{K}^o \longrightarrow \pi^o\nu\,\bar{\nu}$. Similarly, from $\bar{K}^o$ one has the two paths $\bar{K}^o \longrightarrow \pi^o\nu\,\bar{\nu}$ and $\bar{K}^o \longrightarrow K^o \longrightarrow \pi^o\nu\,\bar{\nu}$.

One needs therefore a mechanism of interplay between mixing, written for K_L as $|K_L >= p|K^o > -q|\bar{K}^o >$, and decay, with amplitudes A(K^o) and $\bar{A}(\bar{K}^o)$, respectively. As discussed in Section II, the mixing is dominated by charm quarks running in the loop, so that

$$\left(\frac{q}{p}\right)_K = \frac{Vcs^*Vcd}{VcsVcd^*} \tag{21}$$

The FCNC decay amplitudes, able to generate CP-Violation, originate in the SM from box and weak penguin diagrams with top quarks running in the loop, i.e.,

The ratio of decay amplitudes is thus

$$\frac{\bar{A}}{A} = \frac{VtsVtd^*}{Vts^*Vtd} \tag{22}$$

due to the fact that both diagrams in Fig.5 contribute to a unique weak amplitude.

It is of interest to normalize the rate $\Gamma(K_L \longrightarrow \pi^o\nu\,\bar{\nu})$ to $\Gamma(K^+ \longrightarrow \pi^+\nu\,\bar{\nu})$, the last being induced by the same FCNC decay amplitudes but no mixing. In the approximation of top dominance again for this decay, one has

$$\frac{\Gamma(K_L \longrightarrow \pi^o\nu\bar{\nu})}{\Gamma(K^+ \longrightarrow \pi^+\nu\bar{\nu})} = \frac{1 - \mathrm{Re}\left[\frac{q}{p}\frac{\bar{A}}{A}\right]}{2} \equiv a_{\pi\nu\bar{\nu}} \tag{23}$$

The quantity $a_{\pi\nu\bar{\nu}}$ represents the probability for a CP-violating decay. In the SM, the use of (21) and (22) leads immediately cto

$$a_{\pi\nu\bar{\nu}} = \sin^2\beta \tag{24}$$

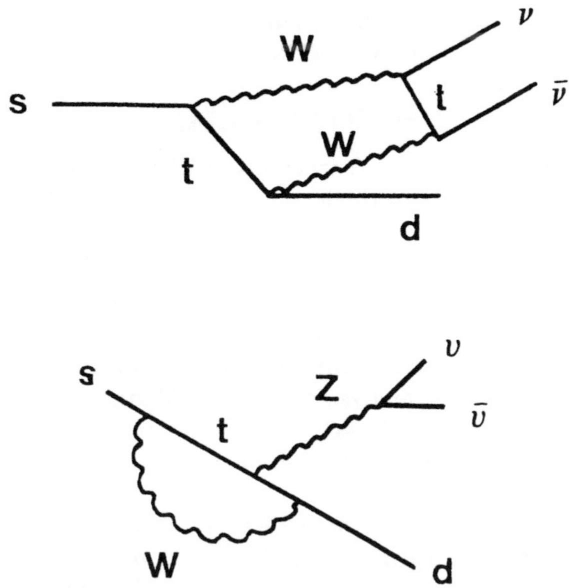

Figure 5: Box and weak penguin diagrams for FCNC $s \longrightarrow d\nu\,\bar{\nu}$

where β is the same CP-violating phase which also appears in the unitarity triangle associated with (b d)-physics [see next Section]. Note the result given by the square of a CP-violating angle β. In the approximations discussed, the result (24) is independent of hadronic uncertainties in matrix elements. The connection between the observable (23) and the CP-asymmetries for Bd-physics is most impressive.

Buchalla and Buras have scrutinized [17] the approximations involved and they have included the internal charm contribution in Fig. 5 to the decay amplitude for $K^+ \longrightarrow \pi^+\nu\,\bar{\nu}$. One still concludes that the hadronic uncertainties are reasonably small. The SM predictions are

$$\begin{aligned} Br(K^+ \longrightarrow \pi^+\nu\bar{\nu}) &= (9.1 \pm 3.8) \times 10^{-11} \\ Br(K_L \longrightarrow \pi^o\nu\bar{\nu}) &= (2.8 \pm 1.7) \times 10^{-11} \end{aligned} \tag{25}$$

where the error comes mainly from the poor determination of CKM-parameters. Ref. [18] analyzes the impact of new physics on this problem.

On the experimental front, a first event in $K^+ \longrightarrow \pi^+\nu\,\bar{\nu}$ has been reported by the BNL787-Collaboration, with a branching ratio

$$Br(K^+ \to \pi^+\nu\bar{\nu}) = (4.2^{+9.7}_{-3.5}) \times 10^{-10} \tag{26}$$

Limits for $Br(K_L \longrightarrow \pi^o\nu\,\bar{\nu})$ from FNALE799 are at the level of 10^{-6}, but a proposal at BNL aims to reach a sensitivity of 10^{-12}!!

4 B-physics

The main question is whether the origin of CP-violation can be explained within the Standard Model or it needs physics beyond the Standard Model. Inside the Standard Model, the analysis for Flavour Mixing starts from a Charged Current Lagrangian, with the Quark Mixing Matrix described by 4 independent parameters. In the Wolfenstein parametrization, for example, these are [λ, A, ρ, η].

It has to be emphasized that these are fundamental couplings of the Standard Model Lagrangian, so their determination is of primordial importance. In order to answer the question on the possible need of beyond-the-standard-model physics, one has to look for ways to overdetermine thase parameters and show the internal consistency of the model.

From the Unitary Quark Mixing Matrix

$$V_{CKM} = \begin{pmatrix} V_{ud} & V_{us} & V_{ub} \\ V_{cd} & V_{cs} & V_{cb} \\ V_{td} & V_{ts} & V_{tb} \end{pmatrix} = \begin{bmatrix} 1-\frac{\lambda^2}{2} & \lambda & A\lambda^3(\rho - i\eta) \\ -\lambda & 1-\frac{\lambda^2}{2} & A\lambda^2 \\ A\lambda^3(1-\rho-i\eta) & -A\lambda^2 & 1 \end{bmatrix} \tag{27}$$

one can establish 6 off-diagonal Unitarity Relations which, represented in the complex plane, can be visualized (for 3 families) by 6 Unitarity Triangles. One of these, that built from the d- and b- columns, is written as

$$(b,d) \longrightarrow V_{ub}^*V_{ud} + V_{cb}^*V_{cd} + V_{tb}^*V_{td} = 0 \tag{28}$$

and has the peculiarity that the three sides are of the same order magnitude: $O(\lambda^3)$. The Wolfenstein parametrization violates unitarity only to higher orders in λ.

In the Standard Model, a single phase parameter (η) is the only possible source of CP violation, so that the predictions for CP-violating phenomena are quite constrained. Moreover, the CKM mechanism requires several necessary conditions in order to generate an observable CP-violating effect: (i) All three families are required to play an active role. With only two fermion families, the Quark Mixing mechanism cannot give rise to CP-violation. In the kaon system, for instance, CP-violation effects can only appear at the one-loop level, where the top quark is present. In B-physics, however, the decays generated by tree level diagrams can induce CP-violation effects; (ii) The quarks of a given charge must be non-degenerate in mass. With degeneracy, the physical quark states could be redefined in order to vanish the quark mixing matrix element; (iii) all CKM-matrix elements must be non-zero. If any of these conditions were not satisfied, the CKM-phase could be rotated away by a redefinition of the quark fields. CP-violation effects are then necessarily proportional to the product of all CKM angles.

All these necessary conditions are summarized in a single requirement [20] on the original quark-mass matrices M_u and M_d:

$$\text{CP-violation} \Longleftrightarrow \text{Im}\{\det[M_u M_u^+ ,\, M_d M_d^+]\} \neq 0 \tag{29}$$

In the case of 3 families, there is a unique combination of angles and phases to generate CP-violation for all flavours that can be considered:

$$J \simeq A^2 \lambda^6 \eta \leq 10^{-4} \tag{30}$$

Any CP-violation observable involves the product J and, thus, violations of the CP symmetry are necessarily small.

One can make some general statements:

(i) In order to generate sizeable CP-violating asymmetries $[(\Gamma-\bar{\Gamma})/(\Gamma+\bar{\Gamma})]$ one should look for very suppressed decays, where the decay widths already involve small CKM matrix elements.

(ii) In the Standard Model, CP violation is a low-energy phenomenon, in the sense that any effect should disappear with $\frac{m_c^2 - m_u^2}{s} << 1$.

(iii) B decays are the optimal place for CP-violation signals to show up. They involve small CKM matrix elements in the decays and are the lowest mass processes where the three quark families play a tree-level role.

The (bd) Unitarity Triangle is shown in Fig. 6 where it has been scaled by dividing its sides by $|V_{cb}^* V_{cd}|$. In the Wolfenstein parametrization, this side is real. Independent of the parametrization, this side defines the CP-conserving direction [21] reference to order λ^3. Aligning it, with length equal to 1, along the real axis, the coordinates of the 3 vertices are then (0,0), (1,0) and (ρ, η). Note that, although the orientation of the triangle in the complex plane is phase-convention dependent, the triangle itself is a physical object: the length of the sides and the angles can be directly measured.

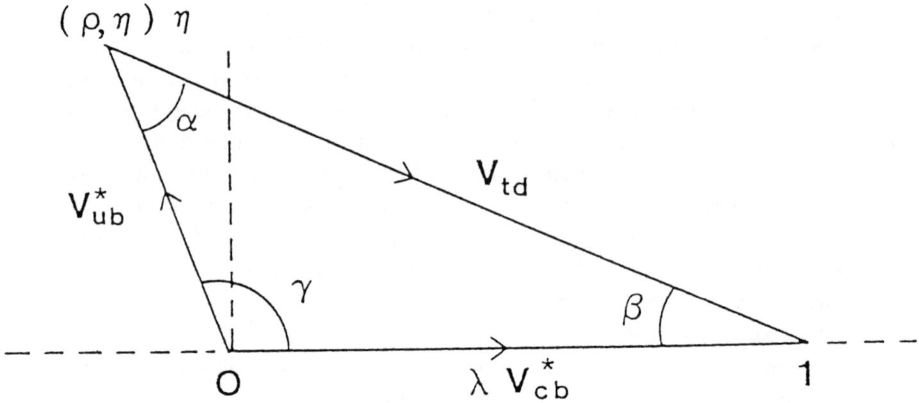

Figure 6: The (bd) Unitarity Triangle

The sides of the Unitarity Triangle are determined from the measured ratio $\Gamma(b \longrightarrow u)/\Gamma(b \longrightarrow c)$ and from $B_d^0 - \bar{B}_d^0$ mixing:

$$R_b = \left|\frac{V_{ub}^* V_{ud}}{V_{cb}^* V_{cd}}\right| \approx \left|\frac{V_{ub}}{\lambda V_{cb}}\right| \approx \sqrt{\rho^2 + \eta^2}$$
$$R_t = \left|\frac{V_{tb}^* V_{td}}{V_{cb}^* V_{cd}}\right| \approx \left|\frac{V_{td}}{\lambda V_{cb}}\right| \approx \sqrt{(1-\rho)^2 + \eta^2} \qquad (31)$$

The significance of the V_{td} coupling is seen in the box diagram of Figure 7, relevant to the $B_d^0 - \bar{B}_d^0$ mixing. A priori, the measurement of the two sides, performed through CP-conserving observables, could make possible to establish that the area is not vanishing and that Fig. 6 is indeed a triangle. With the present theoretical and experimental uncertainties, this is however not possible.

In section I we have discussed the value of ϵ_K,which specifies a hyperbola in the (ρ, η) plane. This constraint, together with the constraints from R_b and R_t, which result in the circles centered at (0,0) and (1,0), respectively, give an allowed range of values [22] for (ρ,η) as shown in Fig. 8. The final value [22] is given by the intersection of all constraints.

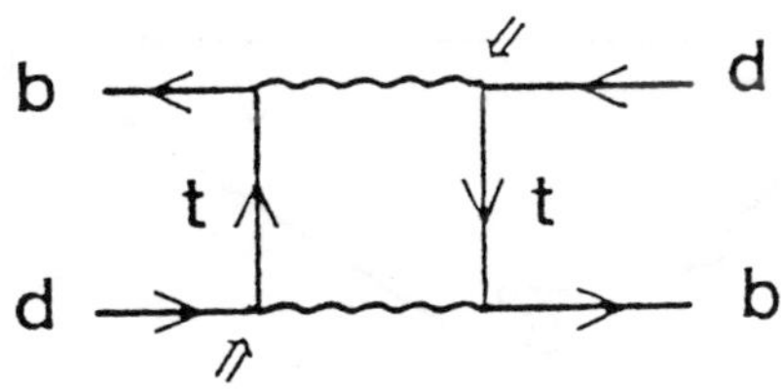

Figure 7: Quark diagram responsible of the $B_d^o - \bar{B}_d^o$ mixing. The arrows indicate the V_{td} coupling.

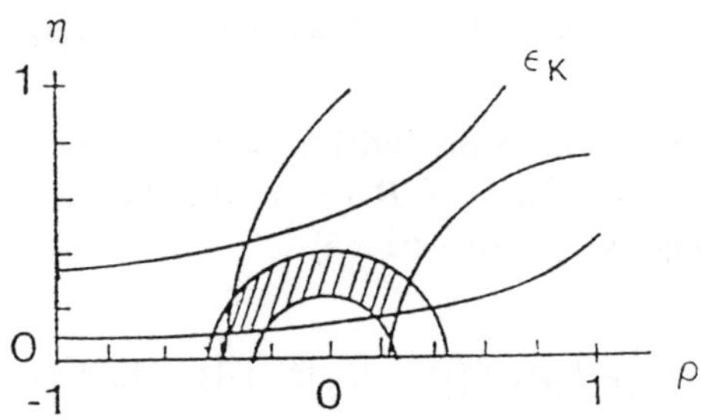

Figure 8: Constraints for (ρ, η) coming from the measured values of R_b, R_t and ϵ_K.

5 Principle of CP violating observables

i) With a large $B^o - \bar{B}^o$ mixing we can generate a CP-violating interference in the following way. There are quite a few non-leptonic final states which are reachable both from a B^o and a $\bar{B}^o$. For these flavour non-specific decays the B^o (or $\bar{B}^o$) can decay directly to a given final state f, or do it after the meson has been changed to its antiparticle via the mixing process; i.e., there are two different amplitudes, $A(B^o \longrightarrow f)$ and $A(B^o \longrightarrow \bar{B}^o \longrightarrow f)$, corresponding to two possible decay paths, as shown in Fig. 9

CP-violating effects can then result [23] from the interference of these two amplitudes. To build the associated asymmetry one needs to TAG the initial B-flavour.

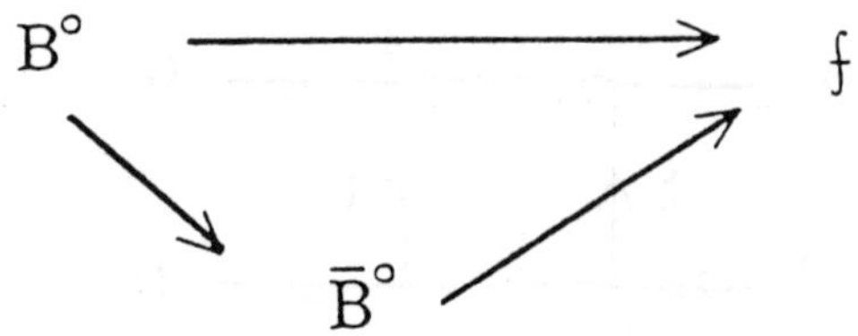

Figure 9: Two decay paths from B^o to f, with interfering amplitudes

The time-dependent decay probabilities for the decay of a neutral B meson created at the time $t_o = 0$ as a pure $B^o(\bar{B}^o)$ into the final state $f(\bar{f} \equiv CPf)$ are (we neglect the tiny $\Delta\Gamma_{B^o}$ corrections):

$$\Gamma[B^o(t) \longrightarrow f] \sim \frac{1}{2} e^{\Gamma t} |A_f|^2 \{[1 + |\bar{\rho}_f|^2] + [1 - |\bar{\rho}_f|^2] \cos(\Delta M t) - 2\,\mathrm{Im}(\frac{q}{p}\bar{\rho}_f) \sin(\Delta M t)\}$$

$$\Gamma[\bar{B}^o(t) \longrightarrow \bar{f}] \sim \frac{1}{2} e^{-\Gamma t} |\bar{A}_{\bar{f}}|^2 \{[1 + |\rho_{\bar{f}}|^2] + [1 - |\rho_{\bar{f}}|^2] \cos(\Delta M t) - 2\,\mathrm{Im}(\frac{p}{q}\rho_{\bar{f}}) \sin(\Delta M t)\} \quad (32)$$

where we have introduced the notation

$$A_f \equiv A[B^o \longrightarrow f], \quad \bar{A}_f \equiv -A[\bar{B}^o \longrightarrow f], \quad \bar{\rho}_f \equiv \bar{A}_f / A_f$$
$$A_{\bar{f}} \equiv A[B^o \longrightarrow \bar{f}], \quad \bar{A}_{\bar{f}} \equiv -A[\bar{B}^o \longrightarrow \bar{f}], \quad \rho_{\bar{f}} \equiv A_{\bar{f}} / \bar{A}_{\bar{f}} \quad (33)$$

CP-invariance demands the probabilities of CP conjugate processes to be identical. Thus, CP conservation requires $A_f = \bar{A}_{\bar{f}}, A_{\bar{f}} = \bar{A}_f, \bar{\rho}_f = \rho_{\bar{f}}$ and

$\mathrm{Im}(\frac{q}{p}\bar{\rho}_f) = \mathrm{Im}(\frac{p}{q}\rho_{\bar{f}})$. The violation of any of the first three equalities would be a signal of direct CP-violation, i.e., in the decay amplitudes. The fourth equality test CP violation generated by the interference of the direct decay $B^o \longrightarrow f$ and the mixing-induced decay $B^o \longrightarrow \bar{B}^o \longrightarrow f$, as shown in Fig. 9.

If the final state is a CP-eigenstate, $\bar{f} = \tau_f f$ with $\tau_f = \pm 1$ and there is only one transition decay amplitude[a], we have

$$\rho_{\bar{f}} = \bar{\rho}_f = \tau_f e^{2i\emptyset WD}$$

$$\frac{\Gamma[B^o(t) \longrightarrow f] - \Gamma[\bar{B}^o(t) \longrightarrow f]}{\Gamma[B^o(t) \longrightarrow f] - \Gamma[\bar{B}^o(t) \longrightarrow f]} = \tau_f \sin(2\emptyset)\sin(\Delta M t) \tag{34}$$

where $\emptyset$ is the convention-independent CP-phase wich results from the combination of the phase $\emptyset_{WD}$ of the decay amplitude and the phase of the mixing parameter $\frac{p}{q}$. This mechanism is thus called "the interplay between Mixing and Direct CP-Violation".

ii) Consider now two interfering decay amplitudes which contribute to a given process $B \longrightarrow f$:

$$\begin{aligned} A(B \longrightarrow f) &= M_1 e^{i\emptyset_1} e^{i\alpha_1} + M_2 e^{i\emptyset_2} e^{i\alpha_2} \\ A(\bar{B} \longrightarrow \bar{f}) &= M_1 e^{-i\emptyset_1} e^{i\alpha_1} + M_2 e^{-i\emptyset_2} e^{i\alpha_2} \end{aligned} \tag{35}$$

where $\emptyset_1$, $\emptyset_2$ are the corresponding weak decay phases and α_1, α_2 are the final state "strong" interaction phases. They are such that, in going to the CP-conjugated process, the weak phases change their sign, whereas the strong phases keep their sign. The strong phases originate from the beyond the Born approximation complexity of the decay amplitudes. The rate Asymmetry is thus given by [24]

$$\frac{\Gamma(B \longrightarrow f) - \Gamma(\bar{B} \longrightarrow \bar{f})}{\Gamma(B \longrightarrow f) + \Gamma(\bar{B} \longrightarrow \bar{f})} = \frac{-2M_1M_2\sin(\emptyset_1 - \emptyset_2)\sin(\alpha_1 - \alpha_2)}{|M_1|^2 + |M_2|^2 + 2M_1M_2\cos(\emptyset_1 - \emptyset_2)\cos(\alpha_1 - \alpha_2)} \tag{36}$$

so that one needs [24]:

1) Two, at least, interfering amplitudes.
2) Two different weak phases $\emptyset_1 \neq \emptyset_2$.
3) Two different strong phases $\alpha_1 \neq \alpha_2$.
4) To enhance the asymmetry, M_1 and M_2 should be of comparable size.

[a] Each transition amplitude is identified as the weak interaction matrix element of a given operator with a definite CP-phase

This "Direct CP-Violation" operates either for charged B's or neutral B's with flavour specific decays.

iii) There is a third method to generate CP-Violation, that of CP-Violation in $B^o - \bar{B}^o$ Mixing or "Indirect CP-Violation". Take the semileptonic decays of a $B^o \longrightarrow \bar{B}^o$ pair with equal sign dileptons: $N(l^+l^+)$ or $N(l^-l^-)$ is a signal of mixing, their difference is a signal of CP-Violation in this mixing.

One gets the semileptonic asymmetry of $B^o \longrightarrow Xl\nu_l$

$$a_{SL} = \frac{N(l^+l^+) - N(l^-l^-)}{N(l^+l^+) + N(l^-l^-)} = \frac{|p/q|^2 - |q/p|^2}{|p/q|^2 + |q/p|^2} = 4\,\mathrm{Re}(\epsilon_B) \tag{37}$$

for a small value of $\mathrm{Re}(\epsilon_B)$.

The off-diagonal matrix elements of the effective mixing matrix of $B^o - \bar{B}^o$ give rise to a difference of the masses and widths of the two physical eigenstates. In the Standard Model the two complex eigenvalues satisfy

$$|\frac{\Delta\Gamma_B}{\Delta M_B}| \sim |\frac{\Gamma_{12}}{M_{12}}| \sim \frac{m_b^2}{m_t^2} \tag{38}$$

To generate $\mathrm{Re}(\epsilon_B) \neq 0$ one needs both $\Delta\Gamma_B \neq 0$ and a missalignment of the (complex) values of Γ_{12} and M_{12}. The relative argument between M_{12} and Γ_{12} is expected to be of $0(m_c^2/m_b^2)$ in the Standard Model. All togeether, a_{SL} is expected to be less than 10^{-3} for B_d^o and less than 10^{-4} for B_S^o and thus outside of the capabilities of the experimental facilities. Present experimental values [25] give $|\mathrm{Re}(\epsilon_B)| = (-1 \pm 5) \times 10^{-3}$.

Contrary to the small value (37), Im (ϵ_B) has no servitude from $\Delta\Gamma_B$ for the rephasing invariant CP-Mixing parameter defined as [26]

$$|B_{1,2}> = \frac{1}{\sqrt{1+|\epsilon|^2}}(|B\pm> + \epsilon|B_{\mp}>) \tag{39}$$

where $|B_{1,2}>$ are the physical states of definite mass and lifetime and $|B_{\pm}>$ are the CP eigenstates

$$|B_{\pm}> = \frac{1}{\sqrt{2}}(I \pm CP)|B^o> \tag{40}$$

For a given CP operator, ϵ connects Mixing and CP. $CP_{12} \equiv < B^o|CP|\bar{B}^o >$ plays the role of a reference phase, given by the flavour mixing amplitude in the CP-conserving limit. With $\Delta\Gamma \sim 0$, $\mathrm{Re}(\epsilon) \sim 0$, one has

$$\frac{\mathrm{Im}(\epsilon)}{1+|\epsilon|^2} \simeq \frac{\mathrm{Im}(M_{12}^* CP_{12})}{\Delta m} \tag{41}$$

Ref. [26] discusses an interference experiment between the CP eigenstates $|B\pm>$ obtained from the time evolution of a CP-Tag. A CP eigenstate of the decay products induced by b$\longrightarrow c\ \bar{c}\ s$ does this preparation [21], because the (bs) unitarity triangle collapses to a CP conserving line to $0(\lambda^3)$. Fig. (6) then shows that, in the SM with top dominance for M_{12}, the CP violating quantity (41) is $-\sin(2\beta)$, with no suppression coming from $\lambda's$.

6 Rate asymmetries

By using the first method discussed in the last Section, one considers the process [27] $B \longrightarrow J/\psi K_s$ able to measure sin (2β), where β is the angle of the unitarity triangle shown in Fig. 6. The quark diagrams responsible of this process are a tree diagram and a penguin diagram, as shown in Fig. 10

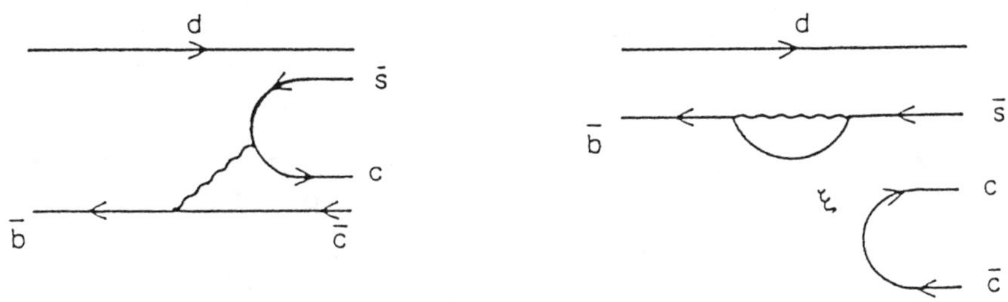

Figure 10: Tree and penguin diagrams relevant to the process $B \longrightarrow J/\psi K_s$

The penguin contribution is a small, estimated to be less than 1% of the tree amplitude, and furthermore one discovers by inspection that both diagrams contain the CP-phase, i.e., there is a unique (complex) hadronic amplitude to describe the exclusive process.

For $B_d's$ in the initial state, one has $\Delta\Gamma/\bar{\Gamma} \sim 5 \times 10^{-3}$ so $\Delta\Gamma$ can be neglected in the time distribution. The corresponding rate asymmetry for this definite final state $f \equiv J/\psi K_s$, eigenstate of CP, from definite flavour B^o, $\bar{B}^o$ states, leads to

$$\frac{\Gamma_f - \bar{\Gamma}_f}{\Gamma_f + \bar{\Gamma}_f} = -\sin(2\beta)\sin(\Delta m t) \tag{42}$$

independent of the (unique) hadronic amplitude!

This asymmetry is expected to be detected and measured at the Tevatron, B-factories, HERA-B and at the LHC experiments. Precision determinations with $\sigma(\sin 2\beta) \sim 0.02$ can be expected. At present, OPALi [28] and CDF [29] Collaborations give $\sin(2\beta) = 3.2^{+1.8}_{-2.0} \pm 0.5$ and $1.8 \pm 1.1 \pm 0.3$, respectively.

At B-factories, the preparation of entangled $B - \bar{B}$ states leads, through the earlier detection of $J/\psi K_s$ in one side, to tag the CP eigenstate $|B+>$ in the other side. The leptonic asymmetries [26] have then access to measure $\mathrm{Im}(\epsilon_B)$, i.e., $\sin(2\beta)$ too.

Taking into account that the $(c\,\bar{c})$ system in the quark diagrams of Fig. 10 is, by itself, composed of charge-conjugated quarks, one can envisage the consideration of the semiinclusive process [30] $B \longrightarrow K_s X(c\,\bar{c})$. No cancellation of the rate asymmetries (42) is expected when summing over the different hadronic states of the $(c\,\bar{c})$ spectrum.

With the aim to extract the angle α of the unitary triangle, one considers the exclusive process $B_d \longrightarrow \pi^+\pi^-$, where $f \equiv \pi^+\pi^-$ is a CP-eigenstate not aligned with the CP conserving direction. The quark diagrams, shown in Fig. 11, contributing to the process are again a tree diagram and a penguin diagram

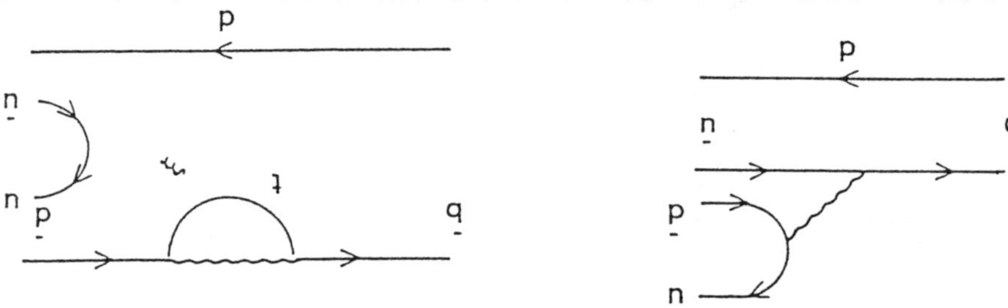

Figure 11: Tree and penguin diagrams relevant to the process $B_d \longrightarrow \pi^+\pi^-$

The tree amplitude is governed by the quark mixing matrix element $V_{ub}^* V_{ud} \sim A\lambda^3(\rho+i\eta)$ whereas the penguin amplitude contains $V_{tb}^* V_{td} \sim A\lambda^3(1-\rho-i\eta)$ so that both are of similar CKM strength. The effect of the loop and the gluonic exchange in the penguin amplitude has been estimated to give a relative contribution of the order of 10%-20%. An independent measure of these two contributions and their relative magnitudes is highly desirable.

Taking into account the contribution of the two different hadronic amplitudes, one induces a Time-Dependent CP-violating Rate Asymmetry

$$\frac{N - \bar{N}}{N + \bar{N}} = a\cos(\Delta m t) + b\sin(\Delta m t) \tag{43}$$

where "a" is obtained from the interference of the Direct CP violating Decay Amplitudes, whereas "b" comes from the interplay between Mixing and Decay.

Although the process $B_d \longrightarrow \pi\pi$ is highly CKM-forbidden, the main problem is not statistics, particularly in the hadronic machines. Its major source of background comes from the accompanying decays $B_d \longrightarrow K^+\pi^-$; $B_s \longrightarrow K^+\pi^-, K^+K^-$, which by themselves could have contributions to the asymetry and not only to the rate. The virtues of having particle identification in the detector are evident.

¿From the theoretical side, the absence of the penguin amplitude would lead to $a = 0, b = -\sin(2\alpha)$, where α as shown in the unitary triangle of Fig. 6. The process $B_d \longrightarrow \pi^+\pi^-$ is, however, theoretically difficult, because the presence of the penguin amplitude P affects the extraction of α as obtained from the tree amplitude T. To leading order in |P/T|small, one has

$$\begin{cases} a = 2|\frac{P}{T}|\sin(\delta_P - \delta_T)\sin\alpha \\ b = -\sin(2\alpha) - 2|\frac{P}{T}|\cos(\delta_P - \delta_T)\sin\alpha\cos 2\alpha \end{cases} \tag{44}$$

where $\delta_{T(P)}$ is the elastic-final-state-interaction phase shift for the T(P) hadronic amplitude.

If |P/T|were known independently, the value of α could be extracted from the measurement of both "a" and "b". The quantity "b" is very sensitive to α and much less to the phase shift difference $\delta_P - \delta_T$, whereas the quantity "a" can determine $\delta_P - \delta_T$. The correlation between α and $\delta_P - \delta_T$ is quite small.

To determine |P/T|experimentally, Gronau and London have used the isospin triangle relations [31] among the amplitudes M_{+0} for $B_u^\pm \longrightarrow \pi^\pm\pi^o$, M_{+-} for $B_d \longrightarrow \pi^+\pi^-$ and M_{oo} for $B_d \longrightarrow \pi^o\pi^o$

$$M_{+-} + \sqrt{2}M_{oo} = \sqrt{2M_{+o}} \tag{45}$$

If *all* the three decay modes are measured, the value of |P/T|can be extracted. However, $B_d \longrightarrow \pi^+\pi^-$ is colour allowed in the quark diagram, whereas $B_d \longrightarrow \pi^o\pi^o$ is colour suppressed and it needs a colour rearrangement. What is the price to be paid by this dynamics could be as to lower the branching ratio for $\pi^o\pi^o$ by a factor 10.

The channel $B_d \longrightarrow \pi^+\pi^-$ could have other theoretical problems, like the existence of an electroweak (beyond the gluonic) penguin amplitude and/or inelastic-final-state-interaction effects. These problems need further scrutiny[32].

7 Summary

The flavour structure is one of the main open questions in fundamental physics. The Standard Model incorporates the Cabibbo-Kobayashi-Maskawa mechanism to generate CP-violation. For 3 families, one single CP-phase appears, leading to constrained physics. A fundamental explanation of the origin of CP-violation is however lacking.

In K-physics, "indirect" CP Violation described by ϵ_K plays the most prominent role. Results for "direct" CP Violation in the 2π channels, governed by ϵ'/ϵ, to the 10^{-4} level are expected from both double ratio and Coherent $(K^o\bar{K}^o)$ methods. The rare CP-Violating Decay $K_L \longrightarrow \pi^o\nu\,\bar{\nu}$ is particularly attractive due to its small hadronic uncertainties.

The CP-odd asymmetries in the $B^o - \bar{B}_o$ system can lead, by the choice of appropriate final states, to a determination of the unitarity triangle when combined with side measurements. The three angles α, β and $\gamma = \pi - \alpha - \beta$ can be separately determined and some experimental methods to do so have been discussed here. But still one perceives that all possible methods for the determination of the unitarity triangle have not yet been studied. The hadronic physics involved in some matrix elements needs further scrutiny. The overall consistency will need of different experimental methods to extract the sides and angles of the unitarity constraint.

The next decade presents a brilliant future in front of us for this B-physics, with electron facilities at the B-factories and hadronic machines, with HERA-B, B-TeV and LHC experiments, devoted to CP Violation studies

Acknowledgments

This work has been supported by CICYT, under Grant AEN-96/1718.

References

1. J.H. Christenson, J.W. Cronin, V.L. Fitch, R. Turlay, *Phys. Rev. Lett.***13**(1964) 138.
2. J.W. Gaillard et al., Phys. Rev. Lett. 18(1967) 20; J.W. Cronin et al.,*Phys. Rev. Lett.***18** (1967) 25.
3. S. Bennet et al., *Phys. Rev. Lett.***19** (1967) 993; D. Dorfan et al., *Phys. Rev. Lett.***19** (1967) 987
4. Review of Particle Physics, Eur. Phys. J.C3(1998)1.
5. E. Chell, M.G. Olsson, *Phys. Rev.* D**48**(1993) 4076.
6. R. Adler et al., CPLEAR, *Phys. Lett.* B**363**(1995)243
7. G.D. Barr et al., NA31, *Phys. Lett.* B**317**(1993)233.
8. L.K. Gibbons et al., E731, *Phys. Rev. Lett.***70**(1993) 1203.
9. M. Kobayashi, M. Maskawa, Progr. Theor. Phys. **49**(1973) 652.
10. L. Wolfenstein, *Phys. Rev. Lett.***51**(1984) 1945.
11. S. Bertolini, M. Fabbrichesi, J.O. Eeg, SISSA-19-98-EP, hep-ph 9802405.
12. G. Pancheri, Acta Phys. Pol. **B29**(1998) 2763
13. J. Bernabéu, F.J. Botella, J. Roldán, in CP-Violation in Particle Physics and Astrophysics, p. 389., ed. J. Tran Thanh Van, Editions Frontières (1990).

14. H.J. Lipkin, Phys. Rev. **176**(1968) 1715.
15. I. Dunietz, J. Hauser, J.L. Rosner, *Phys. Rev.* D**35**(1987) 2166.
16. J. Bernabéu, F.J. Botella and J. Roldán, *Phys. Lett.* B**211**(1988) 226.
17. G. Buchalla, A.J. Buras, *Phys. Lett.* B**333**(1994) 221; *Phys. Rev.* D**54**(1996)6782.
18. A.J. Buras, A. Romanino, L. Silvestrini, *Nucl. Phys.* B**520**(1998) 3
19. S. Adler et al., BNL787- Collaboration, hep-ex/9708031.
20. C. Jarlskog, Phys. Rev. Lett. 55(1985) 1039; Z. Phys. C29(1985) 491; J. Bernabéu, G. Branco, M. Gronau, *Phys. Lett.* B**169**(1986) 243.
21. M.C. Bañuls, J. Bernabéu, FTUV 98/37(1998).
22. A. Ali, D. London, *Nuovo Cimento***A109**(1996) 957; A. Pich, J. Prades, *Phys. Lett.* B**346**(1995) 342.
23. A. Carter, A.I. Sanda, *Phys. Rev. Lett.***45**(1980) 952; *Phys. Rev.* D**23**(1981) 1567.
24. J. Bernabéu, C. Jarlskog, *Z. Phys.* C**8**(1981) 233
25. M.P. Jimack, H. G. Moser, *Nucl. Instrum. Methods* A**408**(1998) 36.
26. M.C. Bañuls, J. Bernabéu, *Phys. Lett.* B**423**(1998) 151.
27. I.I. Bigi, A.I. Sanda, *Nucl. Phys.* B**193**(1981) 85.
28. K. Ackerstaff et al., OPAL-Collaboration, Eur. Phys. J. C5(1998) 379.
29. F. Abe et al., CDF-Collaboration, *Phys. Rev. Lett.***81**(1998) 5513.
30. J. Bernabéu, C. Jarlskog, *Phys. Lett.* B**301**(1993) 275.
31. M. Gronau, D. London, *Phys. Rev. Lett.***65**(1990) 3381.
32. G. Kramer, W.F. Palmer, *Phys. Rev.* D**52**(1995) 6411.

PRECISION TESTS OF THE ELECTROWEAK INTERACTION AND THE HIGGS MASS.

MANEL MARTINEZ
Institut de Fisica d'Altes Energies, Edifici Cn, U. A. B.
E-08193 Bellaterra (Barcelona), Spain

The most relevant LEP electroweak measurements obtained during the LEP-I operation phase are briefly reviewed and their internal consistency as well as their agreement with the Minimal Standard Model predictions are discussed. Assuming the MSM, the precision LEP electroweak data together with other measurements performed at SLAC and FERMILAB allows already an interesting inference of the Higgs mass.

1 Introduction

Since its start-up in 1989, LEP has proven to be the most powerful machine to probe the theory of the electroweak interactions, the Minimal Standard Model (MSM). A full bunch of high precision measurements performed by the four LEP experiments allow now the extraction of many important parameters to an unprecedented level of accuracy.

With the start of the LEP-II program in fall 1995, the LEP-I program arrived to its end although, as of now, part of the LEP-I data has not yet reached final publication. The reason for that is that several measurements have been reanalyzed in the light of a better theoretical and experimental understanding of the systematic effects. Nevertheless, altogether the LEP-I measurements have been already the subject of many good and extensive reviews (see, for instance [1]) and therefore, in spite of their great importance, in this note we are going to omit the details about their experimental determination and discuss just the results, their meaning and their interpretation. The discussion about the electroweak results from LEP-II and from other experiments can be found in other contributions to these proceedings.

The results presented here for the LEP-I measurements come, for most observables, from all data taken between 1989 and 1995 by the four LEP experiments: about 15 million hadronic Z decays and about 1.7 million charged-lepton Z decays coming from the Z resonance scan. Nevertheless, in the Standard Model fits, the results coming from the already analyzed 4000 WW events at and above threshold at LEP-II, as well as the latest top and W mass measured by TEVATRON, the most recent polarizations asymmetries measured at SLD and the recent neutrino nucleon scattering results measured in different places are also used. Most of the results discussed are still preliminary although they have been already presented in public [2].

The Z observables are classified into three groups: Z lineshape and lepton F-B asymmetry measurements, tau polarization measurements and finally electroweak results with quarks, where the present results, analyzed together with the ones obtained at SLAC, have originated some discussion about possible systematic effects in the quark couplings to the Z as we shall see later.

They are all discussed in turn in the following sections.

2 The LEP-I measurements.

2.1 Z lineshape and lepton F-B asymmetry measurements.

The cross section for the production of a fermion pair $f\bar{f}$ can be written as

$$\sigma_{f\bar{f}}(s) = \int_{4m_f^2}^{s} ds' H(s,s')\hat{\sigma}_{f\bar{f}}(s') \tag{1}$$

where $H(s,s')$ is the so-called radiator function which takes care of initial state radiation corrections and the reduced cross section $\hat{\sigma}$ is written as

$$\begin{aligned} \hat{\sigma}_{f\bar{f}}(s) = \sigma^0_{f\bar{f}} \cdot & \frac{s\Gamma_Z^2}{(s-M_Z^2)^2 + (s\Gamma_Z/M_Z)^2} \\ & + (\gamma - Z) + |\gamma|^2 \end{aligned} \tag{2}$$

where $\sigma^0_{f\bar{f}}$ is called the peak cross section, $\Gamma_Z = \Gamma_h + \Gamma_e + \Gamma_\mu + \Gamma_\tau + \Gamma_{inv}$ stands for the total Z width and M_Z is the Z mass. This parameterization assumes the validity of QED for the photon exchange part and also takes from the Minimal Standard Model the interference between the photon- and Z-mediated amplitudes [a]. This interference is very small around the Z pole. In the case of Bhabha scattering, $f = e$, one can either subtract for the data the t-channel contributions, also taken from the theory or add them to the previous expression. The cross section at the peak can be written in turn in terms of the Z mass and width and the Z partial widths to the initial state Γ_e and the final state Γ_f:

$$\sigma^0_{f\bar{f}} = \frac{12\pi}{M_Z^2} \cdot \frac{\Gamma_e\Gamma_f}{\Gamma_Z^2} \tag{3}$$

[a] For some experiments this is only the case for hadronic final states.

where the partial widths can be written in terms of effective vector and axial couplings of the fermions to the Z:

$$\Gamma_l = \frac{G_F M_Z^3}{6\pi\sqrt{2}} ({g_{V_l}}^2 + {g_{A_l}}^2) \left(1 + \frac{3\alpha}{4\pi}\right) \tag{4}$$

Assuming lepton universality, only four parameters are needed to describe the s-dependence of the hadronic and leptonic cross sections: M_Z, Γ_Z, σ_h^0 and the ratio of hadronic to leptonic partial widths ($R_l = \Gamma_h/\Gamma_l = \sigma_h^0/\sigma_l^0$), where the lepton is taken to be massless. If lepton universality is not assumed, R_l is substituted by three analogous quantities, R_e, R_μ, R_τ. This choice of parameters, given the fact that they are directly related to geometrical characteristics of the lineshape, has two advantages: on the one hand, their correlation are small, and on the other hand they simplify the task of disentangling common uncertainties:

- M_Z and G_Z are the only measurements in the energy scale and they are strongly affected by common LEP energy uncertainties which, thanks to the use of the resonant depolarization technique for the determination of the LEP energy, have been reduced to 1.7 MeV and 1.3 MeV respectively [3]. The present LEP averages are $M_Z = 91.1867 \pm 0.0021$ GeV and $G_Z = 2.4939 \pm 0.0024$ GeV.

- σ_h^0 is the only lineshape observable in which the overall normalization, and hence the luminosity measurement uncertainties, enter. Presently this measurement is already dominated by the common systematical error due to the theoretical uncertainty in the low-angle Bhabha cross section, which presently is estimated to be of 0.11% [4]. Since R_l is a ratio of cross sections, the luminosity measurement cancels and there is, so far, no relevant source of common systematics affecting sizably it. The present LEP averages are $\sigma_h^0 = 41.491 \pm 0.058$ nb and $R_l = 20.765 \pm 0.026$.

The forward-backward asymmetry is defined as

$$A_{FB} = \frac{\sigma_F - \sigma_B}{\sigma_F + \sigma_B}, \tag{5}$$

where F and B indicates the forward or backward hemisphere. Normally it is obtained by fitting the measured angular distribution to the formula

$$\frac{d\sigma^l}{d\cos\theta}(s) = \frac{3}{8}\sigma^l(s) \cdot (1 + \cos^2\theta + \frac{8}{3}A_{FB}^l(s)\cos\theta) \tag{6}$$

In the case of e^+e^- final state, the t-channel contribution is either subtracted from the observed asymmetry or added to the previous expression.

Once the different $A^l_{FB}(s_i)$ are obtained, they are fitted together with the lineshape data to get the the lineshape parameters mentioned above and the peak asymmetry, $A^{0,l}_{FB}$:

$$A^{0,l}_{FB} = \frac{3}{4} A_e A_l$$
$$A_f = \frac{2 g_{V_f}/g_{A_f}}{1 + \left(g_{V_f}/g_{A_f}\right)^2}. \tag{7}$$

from which the effective weak mixing angle is measured. The main error in this measurement is still statistical. Experimental systematics can only come from simultaneous charge and forward-backward asymmetries in the detector, which are bound to be very small. The present knowledge of the beam energy contributes a non-negligible 0.0005 to $\Delta A^{0,l}_{FB}$. The LEP average for the peak asymmetry is $A^{0,l}_{FB} = 0.01683 \pm 0.00096$.

The use of the sketched procedure does introduce a model dependence into the parameterization of the lineshape and asymmetry by the treatment of the γZ interference contribution to the hadronic cross section, because quark couplings to the Z are not individually measured for each flavour. This term is therefore taken from the Standard Model. A more general approach is to parameterize the γZ interference term independently of the Z exchange amplitude and has been implemented by the LEP experiments using the so-called s-matrix ansatz [5]. Fitting the LEP-I data as well as the LEP data collected at centre-of-mass energies between 130 and 172 GeV in that approach, the Z mass as defined above turns out to be $M_Z = 91.1882 \pm 0.0031$ and the parameter which measures the hadronic γZ interference turns out to be $j^{tot}_{had} = 0.14 \pm 0.14$ in good agreement with the MSM expectation $j^{tot}_{had} = 0.22$ [2].

2.2 Tau Polarization

The measurement of the final state longitudinal polarization asymmetry for tau leptons:

$$< P_\tau >= \frac{\sigma_R - \sigma_L}{\sigma_R + \sigma_L} = -A_\tau, \tag{8}$$

where $\sigma_{R(L)}$ is the integrated cross section for right (left) handed tau's, provides a means to measure the tau couplings to the Z directly. Since the tau decay inside the detectors, their helicity information can be obtained by using their parity-violating weak decays as a spin analyzer. The main decays are used: to electron, muon, pion, rho, and a_1. Maximal sensitivity is obtained for the semileptonic modes (pion and rho) since then, just one neutrino escapes detection. In this case, the decay angle of the hadron in the tau rest frame (or equivalently its scaled energy) is measured and the integrated tau polarization is extracted from

$$\frac{1}{N}\frac{dN}{dx} = 1 + \alpha < P_\tau > \cos\theta^* \tag{9}$$

where α is a sensitivity coefficient linked to the spin of the hadron. For pions (s=0) $\alpha = 1$ and for rho and a_1 (s=1) $\alpha < 1$ but can be improved by studying the hadron helicity through the analysis of its decay products. The systematics are in this case comparable to the statistical errors and their eventual reduction would require a lot of effort in the further understanding of the calorimeters.

By measuring the tau polarization as a function of the the tau production angle, θ one can write

$$P_\tau(\cos\theta) = -\frac{A_\tau + A_e \cdot \frac{2\cos\theta}{1+\cos^2\theta}}{1 + A_e A_\tau \cdot \frac{2\cos\theta}{1+\cos^2\theta}} \ . \tag{10}$$

From this expression its is apparent that, while the integrated polarization measures A_τ, its forward-backward asymmetry measures the electron coupling, A_e.

The present LEP average results for these two measurements are $A_\tau = 0.1431 \pm 0.0045$ and $A_e = 0.1479 \pm 0.0051$.

This last measurement can be directly compared with the left-right initial state polarization asymmetry from SLD [6] since, in a rather model independent way, both determine the same quantity: $A^0_{LR} = A_e$. The SLD measurement $A_e = 0.1504 \pm 0.0023$ turns out to be in perfect agreement with the LEP average.

2.3 Electroweak results with quarks.

Given the lack of techniques to tag efficiently the light quark flavors, most of the LEP experiments compute inclusively the forward-backward asymmetry in all hadronic events by estimating the average quark charge via a momentum-weighted mean of charges of the hadrons belonging to each quark's jet. This way, the inclusive forward-backward asymmetry for the actual Z decay mixture of quark flavors, or Quark Charge Asymmetry $< Q_{FB} >$ can be extracted leading to a LEP average of $\sin^2\theta^{lept}_{eff} = 0.2321 \pm 0.0010$. This uncertainty is dominated by the Monte Carlo modelling for light quarks.

The measurement of the Z decay width into b hadrons is especially important because, within the Minimal Standard Model, it receives a vertex correction involving the top quark which is absent from any other final state. By taking the ratio $R_b = \Gamma_b/\Gamma_h$, most of the vacuum polarization corrections depending on the top quark and the Higgs mass cancel out, and one is left with the following approximate expression:

$$R_b \simeq R_d \cdot \left[1 - \frac{20}{13}\frac{\alpha}{\pi}\left(\frac{m_t^2}{M_Z^2} + \frac{13}{6}\log\frac{m_t^2}{M_Z^2}\right)\right] \tag{11}$$

Therefore, R_b has a singular rôle since its accurate measurement should provide an indirect determination of m_t independent on M_H, which is something that none of the rest of precision measurements can do. Anyway, the effect of the top quark vertex corrections is only of order 2% for a top mass of 150 GeV. Therefore only a precise measurement, to better than 1%, is useful to get information on the top mass.

The relatively large b mass (~ 4.7 GeV) and lifetime ($\sim 1.5 \times 10^{-12}$ s), makes possible the use of its decay kinematics to have the largest identification efficiency and purities among all the quarks. Several methods have been used to tag b events at LEP:

- Lepton tag: It uses high p, high p_t leptons from b decays. High purity can be achieved but one has to pay for the small semileptonic b branching ratio.

- Event shape tag: High mass, high momentum b mesons or baryons give raise to particular event shapes which have been used to tag b events with high efficiency although rather modest purity. Recognition has been optimized using Neural Network techniques.

- Lifetime tag: The long lifetime of the b quark can be used, using silicon microvertex detectors, to tag b events by looking for tracks not coming from the Z production vertex. This is currently the best performing method with both high purity and efficiency.

The main systematic errors come from the evaluation of the efficiency and the background of the selection. The best option is to try to use data to estimate both. In practice the techniques mentioned above can be used to tag only one hemisphere and look at the other one to measure the tag efficiency and the background. This is the basic idea of the "double tag" methods, in which these quantities are obtained from the measurement of the number of the tagged hemispheres N_t and double-tagged events N_{tt} through the following equations:

$$\begin{aligned} N_t &= 2N_{had}(R_b\varepsilon_b + R_c\varepsilon_c + (1 - R_b - R_c)\varepsilon_{uds}) \\ N_{tt} &= N_{had}(R_b\varepsilon_b^2 C_b + R_c\varepsilon_c^2 C_c \\ &\quad +(1 - R_b - R_c)\varepsilon_{uds}^2 C_{uds}) \end{aligned} \tag{12}$$

where ε_i stand for the efficiencies (ε_b is of $\mathcal{O}$ (25%) with lifetime tag, ε_c typically $\mathcal{O}$ (0.5%) so that the typical b purity is of $\mathcal{O}$ (95%) and ε_{uds} negligible) and C_i are the tagging efficiency hemisphere correlations (typically $\mathcal{O}$ (1)). To solve these equations, R_b and ε_b are taken to be the unknowns while the rest of quantities are computed using Monte Carlo simulation.

The c quarks can be tagged in two different ways:

- using b tag techniques (lepton tag, event shape) extending them to the lower p and p_t regions and fitting then simultaneously the b and c information,
- through the reconstruction on charmed meson decays. The cleanest one is $D^{*+} \to D^0\pi^+$ and then $D^0 \to K^-\pi^+$ and its charge conjugate, because the low $D^* - D^0$ mass difference produces the signature of a soft π acompaning a D^0 with opposite charge to the K.

In either case, the efficiencies and purities are much lower than for b quarks and the dependence on external input for the product and decay branching ratios very important. Usually the selection cuts are chosen to optimize the ratio between statistical and systematical uncertainties and, in this sense, these measurements are not systematically limited. At any rate, the weight of correlated systematical uncertainties is very important in the final errors and therefore, the proper averaging of the measurements from the different experiments and using different techniques is non-trivial. A detailed discussion, which also applies to the heavy flavour asymmetries to be presented later, can be found in [7]. The improvement in the tag purities in the final analyses of the data have reduced the contamination of b events into the c sample and vice versa leading to a correlation between the final averages of R_b and R_c of about -0.17.

The heavy quark forward-backward asymmetries provide presently the most precise determination of the effective weak mixing angle at LEP. What is measured is

$$A_{FB}^{0,q} = \frac{3}{4}A_e A_q. \tag{13}$$

where A_q is $\sim 0.66, 0.93$ for c- and b-type quarks, respectively and depends only mildly on $\sin^2\theta_{eff}^{lept}$. Therefore, the asymmetry is quite large and mainly sensitive to the $\sin^2\theta_{eff}^{lept}$ dependence of A_e. The main difficulties measuring the forward-backward asymmetry for quark final states are the flavour identification and the charge assignment.

For b quarks, the charge assignment is done in two different ways: either the charge of a high p, high p_t lepton from semileptonic decays which identify the b events is also used to extract the charge of the parent quark, or the lifetime information in one hemisphere is used to tag the event while the weighted mean charge (jet charge) in the other hemisphere measures the quark charge. The two methods lead to samples almost completely statistically independent. The systematical errors differ as well: in the first method, the knowledge of the lepton purities and of the semi-leptonic branching ratios is crucial; in the second, the charm background in the b sample is the main worry though, as in the case of the measurement of R_b, a lot of effort has been put by the LEP collaborations in reduing it as much as possible for the final analyses. The results obtained by the four LEP collaborations with the different methods explained can be seen in figure 1.

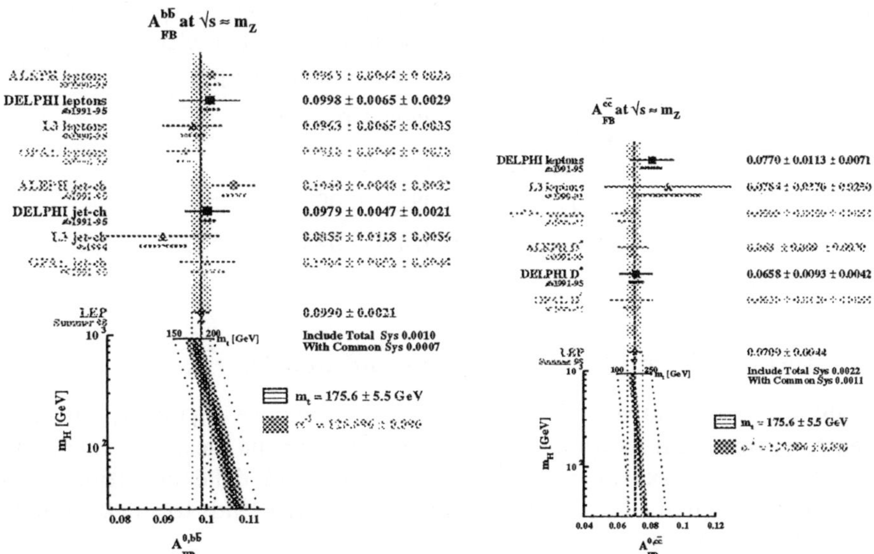

Figure 1. The b-quark and c-quark forward-backward charge asymmetry measured by the four experiments together with the mean and the MSM prediction as a function of the top quark mass. The error bars reflect the full uncertainty.

For c quarks, the charge is extracted either from the lepton charge, like for b's, in global fits using the low p_t leptons from semileptonic decays or, from the charge of the $D^{*\pm}$ meson. The results obtained by the four LEP collaborations with the different methods explained can be seen in figure 1.

The LEP results are combined at the "row" asymmetry level, as measured at the average LEP energy and therefore, they have to be corrected for QED, QCD and energy effects to obtain $A_{FB}^{0,q}$ as appearing in eq. 13. After the corrections, and including the relevant data from SLD as well, the results are

$$\begin{aligned} A_{FB}^{0,b} &= 0.0990 \pm 0.0021 \\ A_{FB}^{0,c} &= 0.0709 \pm 0.0044 \end{aligned} \tag{14}$$

3 Interpretation of the Z measurements.

3.1 Tests of lepton universality.

Lepton universality in the Z couplings can be tested by analyzing simultaneously the information coming from the lepton partial widths and the leptonic forward-backward and tau polarization asymmetries. Then, following eqs.4 and 7 the vector and axial couplings of the Z to each lepton specie can be disentangled. The results (see fig. 2) show perfect agreement with the hypothesis of lepton universality for both, the vector and axial couplings.

$$\begin{aligned} g_{a_\mu}/g_{a_e} &= 0.9996 \pm 0.0014 \\ g_{a_\tau}/g_{a_e} &= 1.0015 \pm 0.0015 \\ g_{v_\mu}/g_{v_e} &= 0.984 \pm 0.098 \\ g_{v_\tau}/g_{v_e} &= 0.974 \pm 0.043 \end{aligned} \tag{15}$$

3.2 Tests of Z couplings to leptons and quarks.

Assuming lepton universality and using all the above asymmetry measurements, leptonic and hadronic, one can test whether the measured Z couplings to leptons and quarks do agree with their Standard Model expectations or not. The combination of all the LEP leptonic asymmetries leads to a determination of the Z coupling to the leptons of

$$A_l = 0.1469 \pm 0.0027$$

As already explained, the quark asymmetries determined by LEP do not measure directly the quark couplings to the Z but instead their product with the

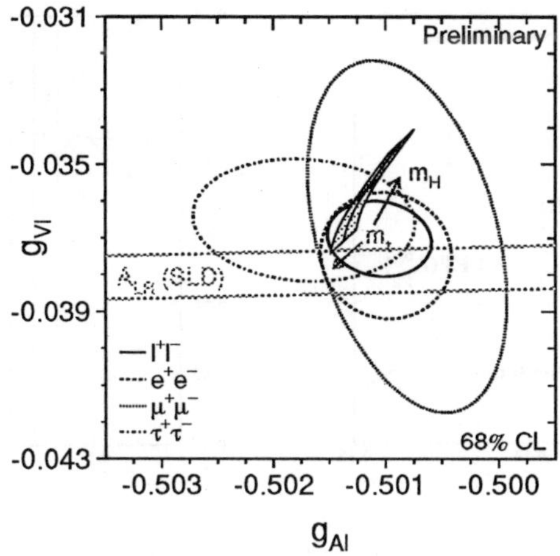

Figure 2. The 68% probability contours in the $g_{V_l} - g_{A_l}$ plane compared to the MSM predictions.

leptonic ones. This is illustrated in figures 3 where also the direct determination of A_b, A_c and A_l performed at SLD by using initial state polarization is shown. From these figures our it is clear that no significative discrepancy with respect to the SM prediction can be claimed from the LEP measurements. Also it is clear that the leptonic measurements lead to a value of A_l which in the SM requires a rather light Higgs, as we'll see in more detail later.

3.3 The effective sinus.

Figure 4 shows the compilation of all the values of $\sin^2\theta_{eff}^{lept}$ obtained from the asymmetry measurements at LEP presented in the previous sections. The SLD determination coming from the measurement of the left-right polarization asymmetry [6] is also shown. Its discrepancy with the average of the LEP measurements alone is of about 2.1 σ. The overall agreement of all the measurements is reasonably good and the mean value provides a very precise determination of the effective weak mixing angle, which is very sensitive to the top quark and the Higgs masses.

Figure 5 summarizes the information coming form the measurement of the LEP asymmetries and the LEP lineshapes and shows clearly the need for

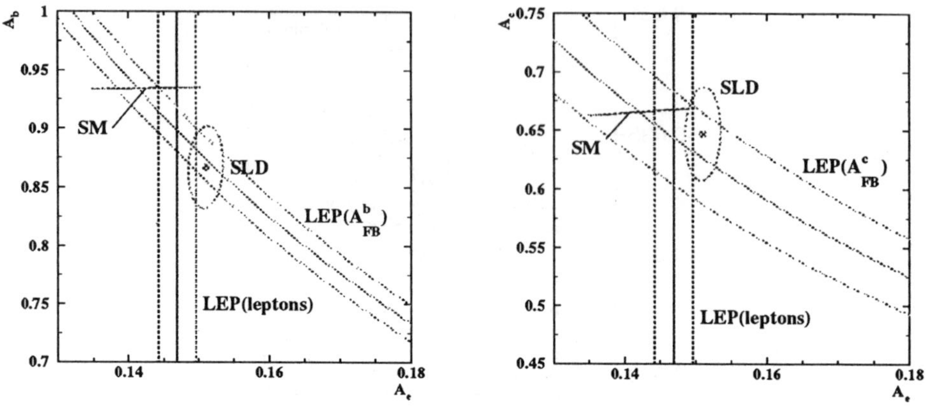

Figure 3. 1-sigma constraints in the A_b and A_c versus A_l plains coming from the leptonic and hadronic asymmetry measurements.

the existence of genuine electroweak radiative corrections beyond the running of the electromagnetic coupling constant to explain the measured values. It also shows that the uncertainty in $\Delta(1/\alpha(M_Z^2)) = 0.090$ that we'll assume through this note has an impact in the interpretation of $\sin^2\theta_{eff}^{lept}$ comparable to the the experimental error.

3.4 Quantities derived from ratios.

The ratios of partial widths, due to the cancellation of universal $\Delta\rho$ corrections between numerator and denominator, allow the direct determination of important parameters of the theory without relying, in principle, too much on the validity of the MSM. Extensions of the MSM which would manifest themselves mainly via effects in vacuum polarization would produce very similar predictions for ratios.

R_l depends little on m_t and M_H whereas it has a large dependence on the strong coupling constant $R_l \sim R_l^0(1+\alpha_s(M_Z)/\pi)$. Therefore it allows a direct determination of $\alpha_s(M_Z)$ with minimum theoretical uncertainties. From the LEP average of R_l and using the formulae suggested in [8], which relate R_l with the QCD prediction, known to $\mathcal{O}$ (α_s^3), one gets

$$\alpha_s(M_Z^2) = 0.122 \pm 0.004 \pm 0.001$$

where the second error accounts varying M_H in the range $M_H = 76^{+85}_{-47}$ GeV. This result is in good agreement with the world average $\alpha_s(M_Z^2) = 0.119 \pm 0.002$.

The invisible width, Γ_{inv} can be derived from the direct lineshape measurements by subtracting from the total width the one due to visible channels. If one assumes that all the invisible width is due to neutrino final states, then one can derive the number of light neutrino species N_ν by writing

$$\frac{\Gamma_{inv}}{\Gamma_l} = N_\nu \cdot \frac{\Gamma_\nu}{\Gamma_l} \tag{16}$$

and taking the ratio Γ_ν over Γ_l from the MSM: $\Gamma_\nu/\Gamma_l = 1.991 \pm 0.001$. It should be noted the small error in the Minimal Standard Model prediction for this ratio, which does not depend on α_s and in which the top and Higgs mass dependences largely cancel. Using the LEP average for $\Gamma_{inv}/\Gamma_l = 5.961 \pm 0.023$ one obtains

$$N_\nu = 2.994 \pm 0.011$$

Since the result favours three species without any doubt, this measurement can be interpreted then as the determination of the width for additional invisible decays of the Z, which yields:

$$\Delta\Gamma_{inv} = -1.6 \pm 1.9 \text{ MeV}$$

or, equivalently

$$\Delta\Gamma_{inv} < 2.8 \text{ MeV}$$

at the 95 % confidence level.

4 Standard Model fits.

The results presented in the previous sections, together with other relevant electroweak measurements performed at LEP-II and in other experiments can be interpreted in the context of the Minimal Standard Model allowing a check of its validity in describing all the observations, and enabling the inference of some of the fundamental MSM parameters.

The comparison is made through a fit of the measurements shown in table 1 to their MSM predictions in terms of the top quark mass m_t, the Higgs boson mass M_H and the value of $\alpha_s(M_Z)$.

The most up-to-date MSM calculations have been used and their estimated theoretical uncertainties [9] have been also propagated in the fitting procedure, being the uncertainty in $1/\alpha(M_Z^2) = 128.878 \pm 0.090$ [10], due to the contribution of light quarks in the photon vacuum-polarization, the one that dominates by far.

The results of the fit with different datasets are presented in table 2. It is clear from these results that the data prefers a rather light Higgs boson mass as can be seen also in figure 6. In that figure, the impact of taking $1/\alpha(M_Z^2) = 128.905 \pm 0.036$ as advocated in references [11] is also shown.

In addition, from table 2 one sees that there is a nice agreement between the direct measurement of m_t and M_W and their indirect determination through their effect in the precision electroweak measurements. This fact constitutes a successful deep test of the Standard Model predictions beyond the tree level structure. This agreement can be also seen in figure 7 which also illustrates the fact mentioned above: in a completely independent manner, the precision electroweak measurements and the direct determinations of m_t and M_W point to a rather light Higgs boson mass.

5 SUMMARY

The precision electroweak measurements at LEP-I have been a big success, even beyond the most optimistic expectations. Thanks to these measurements, the Z boson properties are known today with very high precision.

All together, these measurements allow an unprecedented test of the good health of the Minimal Standard Model in describing the electroweak data and, together with the LEP-II measurements and other relevant electroweak measurements performed around the world, open a significative window for the inference of the last missing piece of the puzzle:M_H.

References

1. "Precision Tests of the Electroweak Interaction and the Z Polc", M.Martinez, R.Miquel, L.Rolandi and R.Tenchini,CERN-EP/98-27, February 1998, to appear in Review of Modern Physics.
2. "A Combination of Preliminary Electroweak Measurements and Constraints on the Standard Model", The LEP Electroweak Working Group, CERN-EP/98, November 1998. Prepared from Contributions of the LEP and SLD Experiments to the 1998 summer conferences.
3. "Calibration of the centre-of-mass energies at LEP1 for precise measurements of Z properties", The LEP Energy Working Group, CERN-EP/98-40, CERN-SL/98-12, March 1988, to appear in Eur.Phys.Jour.C.
4. "New Results on the Theoretical Precision of the LEP/SLC Luminosity", B.F.L.Ward, S.Jadach, M.Meles and S.A.Yost, UTHEP-98-0501,

July 1998.
5. A.Leike,T.Riemann and J.Rose, Phys. Lett. B273 (1991) 513.
T.Riemann, Phys. Lett. B293 (1992) 451.
S.Kirsh, T.Riemann, Comp.Phys.Comm.88 (1995) 89.
6. "Measurements of A_{LR} and A_l from SLD", K.Baird for the SLD collaboration, talk presented at ICHEP 98, Vancouver, B.C., Canada, 23-29 July 1998.
7. "Input Parameters for the LEP/SLD Electroweak Heavy Flavour Results for Summer 1998 Conferences", The LEP Heavy Flavour Group, Internal note LEPHF/98-01, http://www.cern.ch/LEPEWWG/heavy/lephf9801.ps.gz.
8. T. Hebbeker, M. Martinez, G. Passarino, G. Quast, Phys. Lett. B331 (1994) 165.
P.A.Raczka and A.Szymacha, Phys.Rev.D54 (1996) 3073.
D.E.Soper and L.R.Surguladze, Phys.Rev.D54 (1996) 4566.
9. D. Bardin et al.,"Precision Calculation Working Group"(PCWG), CERN 95-03 Yellow report, 31 March 1995 and references therein.
"Upgrading of Precision Calculations for Electroweak Observables", D.Bardin and G.Passarino, CERN-TH/98-92.
10. S. Eidelmann and F. Jegerlehner, PSI-PR-95-1, January 1995 // H. Burkhardt and B. Pietrzyk, Phys.Lett.B356 (1995) 398.
11. M.Davier and A.Höcker, Phys.Lett.B419 (1998) 419.

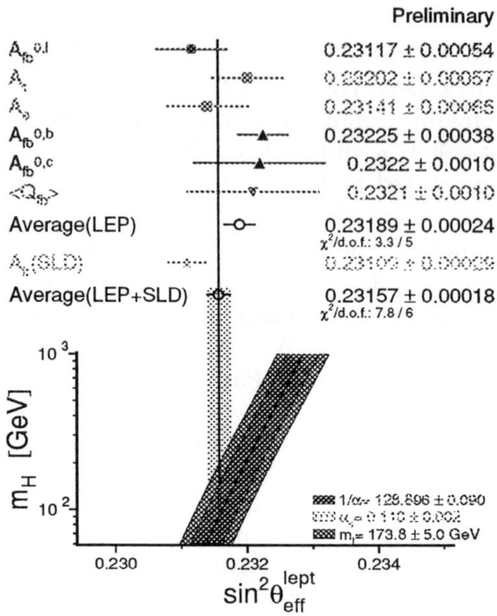

Figure 4. The LEP averages for $\sin^2\theta_{eff}^{lept}$ obtained from the different asymmetry observables discussed in the text, together with SLD measurement and the total mean.

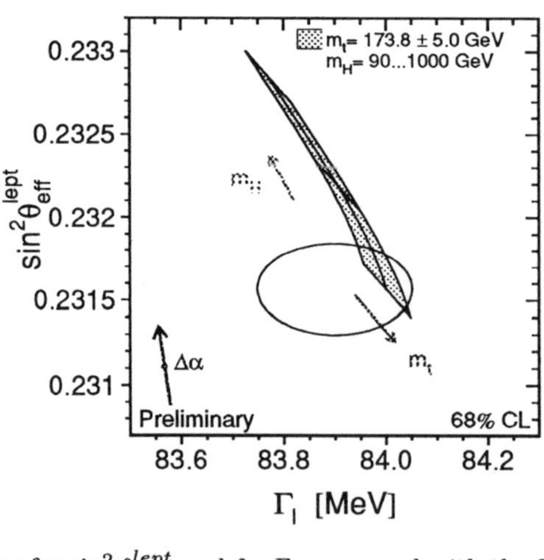

Figure 5. The average for $\sin^2\theta_{eff}^{lept}$ and for Γ_l compared with the Standard Model predictions. The arrow shows these predictions when the only electroweak radiative correction included is the running of the elctromagnetic coupling constant.

Table 1. Summary of measurements included in the combined analysis of Standard Model parameters. Section a) summarizes LEP averages (which in the case of R_b and R_c include also SLD) section b) summarizes the relevant additionnal SLD results and section c) the electroweak results from hadron colliders and νN-scattering. In parenthesis is indicated the size of the estimated systematic component of the quoted uncertainty. The MSM fit result in column 3 and the pulls in column 4 are derived from the fit including all data with the Higgs mass as a free parameter.

	measurement	Standard Model fit	pull
$\alpha(M_Z^2)^{-1}$	$128.878 \pm 0.090(83)$	128.878	0.0
a) LEP			
line-shape and lepton asymm.:			
M_Z [GeV]	$91.1867 \pm 0.0021(17)$	91.1865	0.1
Γ_Z [GeV]	$2.4939 \pm 0.0024(13)$	2.4958	−0.8
σ_h^0 [nb]	$41.491 \pm 0.058(57)$	41.473	0.3
R_l	$20.765 \pm 0.026(20)$	20.748	0.7
$A_{FB}^{0,l}$	$0.01683 \pm 0.00096(60)$	0.01613	0.7
τ polarization:			
A_τ	$0.1431 \pm 0.0045(27)$	0.1467	−0.8
A_e	$0.1479 \pm 0.0051(10)$	0.1467	0.2
b and c quark:			
R_b^0	$0.21656 \pm 0.00074(57)$	0.21590	0.9
R_c^0	$0.1735 \pm 0.0044(34)$	0.1722	0.3
$A_{FB}^{0,b}$	$0.0990 \pm 0.0021(10)$	0.1028	−1.8
$A_{FB}^{0,c}$	$0.0709 \pm 0.0044(22)$	0.0734	−0.6
$q\bar{q}$ charge asymm.:			
$\sin^2\theta_{eff}^{lept}(Q_{FB})$	$0.2321 \pm 0.0010(8)$	0.23157	0.5
M_W [GeV]	$80.37 \pm 0.09(6)$	80.37	0.0
b) SLD			
$\sin^2\theta_{eff}^{lept}(A_{LR})$	$0.23109 \pm 0.00029(18)$	0.23157	−1.7
A_b	$0.867 \pm 0.035(25)$	0.935	−1.9
A_c	$0.647 \pm 0.040(23)$	0.668	−0.5
c) $p\bar{p}$ and νN			
M_W [GeV] ($p\bar{p}$)	$80.41 \pm 0.09(07)$	80.37	0.4
$1 - M_W^2/M_Z^2(\nu N)$	$0.2254 \pm 0.0021(10)$	0.2232	1.1
m_t [GeV] ($p\bar{p}$)	$173.8 \pm 5.0(3.9)$	171.1	0.5

Table 2. Results of fits to LEP and other data for m_t, M_H and $\alpha_s(M_Z^2)$. No external constraint on $\alpha_s(M_Z^2)$ has been imposed.

	No M_W,no m_t	No M_W	All data
m_t (GeV)	$158\,^{+9}_{-8}$	171.0 ± 4.9	151.1 ± 4.9
M_H (GeV)	$32\,^{+41}_{-15}$	$82\,^{+95}_{-51}$	$76\,^{+85}_{-47}$
$\alpha_s(M_Z^2)$	0.120 ± 0.003	0.120 ± 0.003	0.119 ± 0.003
$\chi^2/(d.o.f.)$	13/12	15/13	15/15
$\sin^2\theta^l_{eff}$	0.23157 ± 0.00018	0.23159 ± 0.00020	0.23157 ± 0.00019
M_W (GeV)	80.332 ± 0.037	80.367 ± 0.029	80.371 ± 0.026

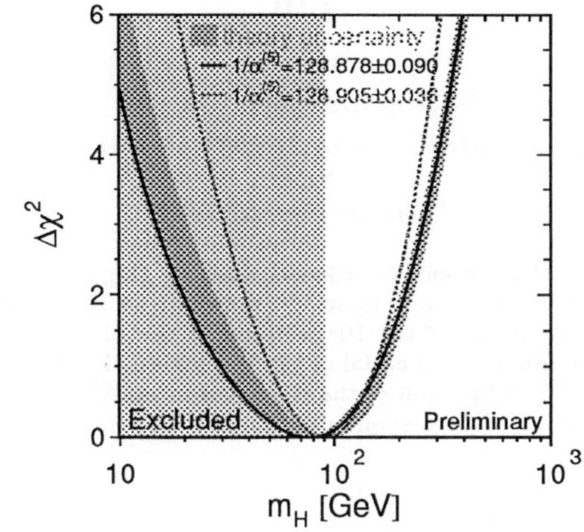

Figure 6. $\Delta\chi^2$ vs M_H using all data.

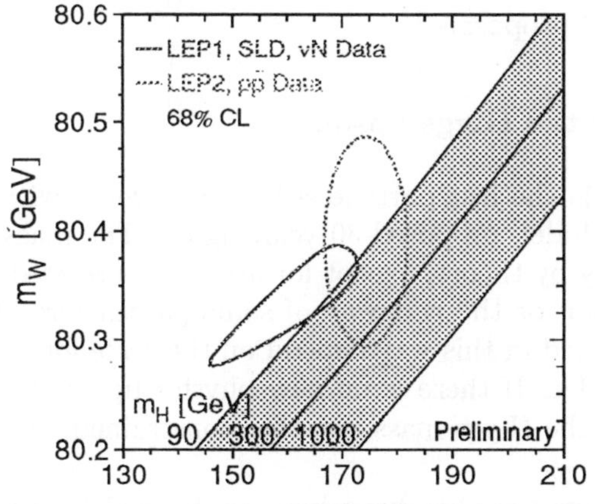

Figure 7. The direct and indirect determinations of m_t and M_W compared with the Standard Model predictions.

SEARCH FOR THE STANDARD MODEL HIGGS BOSON AT LHC

EDUARDO ROS
Universidad de Valencia-IFIC, Av.Dr.Moliner 50, 46111-Burjassot(Valencia), SPAIN
E-mail: edros@cern.ch

A short review of the present situation concerning Higgs boson searches at the LHC is presented. The various existing limits on the Higgs mass are reviewed and the various signatures of the Higgs boson at the LHC are briefly described. Finally the Higgs discovery potential of the two main LHC detectors, ATLAS and CMS, is presented as a function of the Higgs boson mass, assuming an integrated luminosity of 3.10^4 pb^{-1} corresponding to 3 years of running at low luminosity.

1 Introduction

The search for the Higgs boson is one of the main challenges of the LHC project. A large amount of work has been devoted by the two main collaborations, ATLAS and CMS, in order to find an optimal detector configuration to perform this search. A compilation of this work can be found in the corresponding technical proposals [1].

2 The mass of the Higgs boson

The Higgs boson is the last particle still undiscovered within the framework of the Standard Model, proposed 30 years ago [2]. The mass of this particle is not constrained by by theory, except for arguments related to the stability of the Higgs potential, or the runaway of some parameters of this potential [3]. The bounds obtained in this way depend on the scale for new physics beyond the Standard Model. If there is no new physics up to the Planck scale, the allowed range for the Higgs mass is rather constrained: between 140 and 180 GeV.

Extensive Higgs searches have been performed by LEP experiments [4], setting lower limits to the mass of the standard Higgs boson of the order 80 GeV. The limit value of 100 GeV might be reached at the end of the LEP200 program by the year 2000. Experimental upper bounds can also be set by inspecting the influence of the Higgs mass, via radiative corrections, on the measured values of the top-quark and W-boson masses. These upper bounds are currently of the order of 400 GeV, but will also improve in the near future.

3 Higgs boson production at the LHC

The main production mechanism for the Higgs boson at LHC is the gluon-gluon fusion, via a top-quark loop. The cross-section for this process is of the order of 30 pb for a Higgs mass of 100 GeV, and drops to 0.1 pb for a Higgs mass of 1 TeV. Other production mechanisms are the following:

- WW fusion, producing a Higgs in association with a quark pair. The cross-section for this process is typically one order of magnitude smaller than for the gg fusion process, except at very high Higgs masses.

- Higgs production in association with a top-quark pair, a W or a Z (all these processes are referred to as 'associate production'). The cross-sections for these processes are typically 2 or 3 orders of magnitude smaller than for the dominant process.

WW fusion and associate production mechanisms are useful to search for the Higgs boson only in those cases where the background for the dominant process, the gg fusion, is prohibitive.

The cross-sections for all these processes have been calculated using perturbative QCD at next-to-leading order [5]. The cross-section for the dominant process increases rapidly with beam energy. At LHC ($\sqrt{s}$= 14 TeV) the cross-section is about 100 times larger than at the Tevatron ($\sqrt{s}$= 1.8 TeV), assuming for example a Higgs mass of 150 GeV. This difference increases further for larger Higgs masses. Recent Tevatron upgrade studies set the potential search of the Higgs boson up to a mass limit of 120 GeV [6].

4 LHC luminosity

The LHC luminosity is expected to increase from an initial value of 10^{33} $cm^{-2}s^{-1}$ to a peak value of 10^{34} $cm^{-2}s^{-1}$. According to these performance values, it is expected to collect 10^4 pb^{-1} during the first 3 years of operation at low luminosity, and 10^{-5} pb^{-1} per year in the following. Taking into account the bunch crossing of 25 ns, it is expected that at high luminosity multiple interactions occur in each beam collision. Present calculations reveal a background of about 20 minimum bias events, equivalent to 700 low p_t tracks, on top of each interesting event. This soft background implies in some cases a degradation of the detector response. In the following the search for the Higgs boson is described under the assumption of 3 years of operation at low luminosity, avoiding in this way the problem of the degraded performance.

5 Higgs branching ratios

The Higgs boson decays into both massive fermion pairs and massive gauge boson pairs. The decay width into fermion pairs is:

$$\Gamma_{f\bar{f}} \sim G_F M_H m_f^2 \tag{1}$$

where G_F is the Fermi constant, M_H the Higgs mass and m_f the fermion mass. The decay width into gauge boson pairs is:

$$\Gamma_{B\bar{B}} \sim G_F M_H^3 \tag{2}$$

independent of the gauge boson mass. The decays into massive gauge bosons (W and Z) are dominant above their corresponding production thresholds. Decays into massless gauge bosons are also possible, although considerably suppressed. The decay into gluon pairs occurs via a t-quark loop and has the following width:

$$\Gamma_{gg} \sim \alpha_s^2 G_F M_H^3 \tag{3}$$

where α_s is the strong coupling constant. The decay into photon pairs is dominated by a W-boson loop and has a width of:

$$\Gamma_{\gamma\gamma} \sim \alpha^2 G_F M_H^3 \tag{4}$$

where α is the electromagnetic coupling constant. This decay is therefore marginal, but plays however an important role in Higgs searches as explained later.

All Higgs decay rates have been precisely calculated using perturbative QCD wherever relevant, and can be found for example in reference [5]. For Higgs masses below $\sim$ 120 GeV, fermionic decays are dominant ($\Gamma_{b\bar{b}} \sim 80\%$, $\Gamma_{\tau\bar{\tau}} \sim 80\%$). Above a Higgs mass of $\sim$ 200 GeV, Gauge boson decays become completely dominating, even above top threshold production ($\Gamma_{WW} \sim 2/3$, $\Gamma_{ZZ} \sim 1/3$). In the region between 120 and 200 GeV both fermionic and gauge boson decays are present. Just above W threshold production ($\sim$ 160 GeV), WW decays become completely dominant. Finally $\gamma\gamma$ decays are present at a maximum level of 0.2% around a Higgs mass of 120 GeV. In the mass region where fermionic decays are dominant, The Higgs is a narrow particle, with a few Mev width. Above threshold for gauge boson production, on the contrary, the width increases like M_H^3.

For a Higgs mass of 500 GeV the width value is 10% of the mass value. For a mass of 1 TeV the width value becomes comparable to the mass value.

6 The four-lepton signature

The most promising signature for the Higgs boson at LHC is the decay into Z-boson pairs, with the subsequent decay of the Z-bosons into electrons or muons. The expected number of events, for an integrated luminosity of 3.10^4 pb^{-1} reaches a maximum of 390 events for a Higgs mass of about 200 GeV. This number decreases both at lower masses (24 events for a Higgs mass of 150 GeV) and higher masses (18 events for a Higgs mass of 800 GeV). The drop at low masses is due to the drop in the branching ratio ($H \rightarrow ZZ$), whereas the drop at high masses is due to the smaller cross-section. The background in this channel is mainly irreducible, from Z-boson pairs produced in the continuum. This background can be easily separated after imposing cuts on the Z-boson p_t, and invariant mass cuts on the lepton pairs, in order to reconstruct the Z mass.

The 4-lepton invariant mass plays an important role in the analysis and has been studied in detail by both ATLAS and CMS. Monte Carlo simulations indicate that the invariant mass resolution is better then 2%. This implies that the width of the Higgs signal peak is dominated by the detector resolution for Higgs masses below 350 GeV, and by the natural width of the Higgs for masses above this value. In the case of ATLAS, the expected number of signal events after applying the relevant cuts is 94 for a Higgs mass of 300 GeV, and 21 if the Higgs mass is 600 GeV. The corresponding background events are 20 and 7, implying a significance (defined as the signal divided by the square-root of the number of background events) of 21 and 8, respectively, well above the level of 5, usually considered the threshold for a discovery. The analysis performed by both ATLAS and CMS reveals that the discovery of the HIGGS boson using the four-lepton signature is possible after 3 years of operation provided the Higgs mass is in the range 120-800 GeV.

7 The two-photon signature

The mass range 100-120 GeV is uncovered both by LEP searches and by the 4-lepton signature described in the previous section. The dominant Higgs decay in this mass range is the decay into b-quark pairs. If the Higgs mass is 100 GeV, the number of expected decays in this channel is rather large, about 600000 for an integrated luminosity of 3.10^4 pb^{-1}. The expected background from b-quark pairs produced in the continuum is however overwhelming, more than 2 million times larger than the signal. Since the invariant mass resolution is rather poor in this case, this signature cannot be used to achieve the Higgs discovery.

The most promising channel in the mass range around 100 GeV is in fact the decay into photon pairs H$\rightarrow \gamma\gamma$. A large effort has been devoted by both ATLAS and CMS to optimise the detection of the Higgs boson in this decay channel, which sets important constraints in the choice for electromagnetic calorimetry. Taking into account the branching ratio of about 0.2%, the expected number of signal events after 3 years of operation at low luminosity is about 1500. The irreducible background alone yields a background over signal ratio in the range 50 to 70. The reducible background from jet pairs is expected to be suppressed by imposing adequate calorimeter cuts. If $\pi^0\gamma$ separation works adequately, this irreducible background is in fact negligible. Due to the large value of the irreducible background from the continuum, the two-photon invariant mass plays a critical role in this analysis. For a Higgs mass of 100 GeV, ATLAS can achieve a two-photon invariant mass resolution of 1.4 GeV. CMS can achieve a resolution better than 1 GeV due to the optimisation in the electromagnetic energy resolution resulting from the choice of calorimetry (a high resolution crystal). After applying the relevant analysis cuts, ATLAS, for example, finds a signal of 252 events on top of a background of 14220 events, after 3 years of operation. The significance of this signal is therefore about 2, insufficient for a Higgs discovery. The result claimed by CMS is somewhat better. It is however doubtful that the two-photon channel alone is able to provide sufficient evidence for the presence the Higgs boson and therefore other decay channels have to be explored.

8 The b-quark pair signature

As explained before, the H$\rightarrow b\bar{b}$ decay cannot be used to detect the Higgs at LHC if the Higgs is produced via the dominant gg fusion mechanism. Other production mechanisms provide however a sufficiently clear signature. In the following, two of these mechanisms are discussed:

- The associate production with a W according to $p\bar{p} \rightarrow WH \rightarrow lb\bar{b}$, which provides sufficient signal (of the order of 1000 events for an integrated luminosity of 3.10^4 pb^{-1}) and a large, but not overwhelming, irreducible background consisting of a W accompanied by a pair of b-quarks from the continuum.

- The associate production with a t-quark pair according to $p\bar{p} \rightarrow t\bar{t}H \rightarrow lb\bar{b}b\bar{b}$, which also provides sufficient signal (again of the order of 1000 events) this time with a negligible irreducible background, but potentially large technical complications in the reconstruction of the final state.

In both cases a high p_t lepton in the final state is required to ensure the triggering on the events. In the first analysis ($lb\bar{b}$ final state, the b-quark pair invariant mass can be obtained with a resolution of about 11 GeV for a Higgs mass of 100 GeV. The main challenge of this analysis is to reduce sufficiently the background composed by a W accompanied by jets, top quark pairs and the resonant events with W and Z in the final state. The tagging of b-quarks plays an important role and has been studied in detail by both ATLAS and CMS. The ATLAS analysis, for example, concludes that, after all cuts are applied, the expected signal for an integrated luminosity of 3.10^4 pb^{-1} is 330 events, on top of a background of 17000 events, implying a significance of 2.5. In the second analysis ($lb\bar{b}b\bar{b}$ final state), the difficulty is to deal with a large multiplicity of b-quarks in the final state. In this case the complete reconstruction of the final state seems difficult, but a simple counting of events with at least 3 tagged b-jets and a high p_t lepton leads to a signal of 261 events over a background (reducible in this case) of some 10000 events. This implies also a significance of 2.5.

9 The high mass case

If the mass of the Higgs boson is very high (800 GeV or more) the event rate decreases rapidly due to the drop of the cross-section. The four-lepton yield is in this case insufficient to provide a significant signature for the Higgs boson. Four-lepton decays are however a small fraction of all possible $H \rightarrow ZZ$ decays (only 0.5%). If $Z \rightarrow \nu\bar{\nu}$ decays for one of the two Z-bosons are considered, the signal increases by a factor of 6. For a Higgs mass of 800 GeV, the rate in the four-lepton mode is only 18 events, increasing to approximately 100 in the $l\bar{l}\nu\bar{\nu}$ mode, after 3 years of operation at low luminosity. The $l\bar{l}\nu\bar{\nu}$ mode can be detected by searching for events with high p_t leptons and large missing energy (missing E_t larger than 200 GeV). The background is composed of ZZ events produced in the continuum and Z events with missing energy. In this way background over signal ratios of the order of 1 can be achieved.

If the Higgs mass increases further to 1 TeV, the rate of $l\bar{l}\nu\bar{\nu}$ events drops to about 30 and is again insufficient to achieve a Higgs discovery. In this case it is necessary to use all possible Higgs decays, with jets in the final state and at least one high p_t lepton in order to ensure the triggering. These decays include the channels $H \rightarrow WW \rightarrow l\nu jj$ and $H \rightarrow ZZ \rightarrow l\bar{l}jj$. The gain in event rate, compared to the 4-lepton case, is 150. The event yield for an integrated luminosity of 3.10^4 pb^{-1} consists of 500 WW and 90 ZZ events. The background, formed by W and Z bosons accompanied by jets is however too large to proceed without further cuts (the background over signal is of the order of 150000). In fact it is more advantageous in this case to refer to the $WW \rightarrow H$ production mechanism (WW fusion), where a pair of quarks are produced in addition to the Higgs. The tagging of these quarks (typically in the end-cap parts of the detector) in addition to the reconstruction of the Z and W bosons in the jet-jet modes, provides a sufficiently large signal over background ratio (of the order of 3) and a significance for the Higgs signal larger than 5, after 3 years of operation. The Higgs width is of no help in this case, since for these Higgs mass values the width is extremely large.

10 conclusion

Extensive studies have been performed by the Atlas and CMS collaborations in order to demonstrate that a signal for the standard model Higgs boson with significance larger than 5 can be obtained at the LHC after 3 years of operation at low luminosity, corresponding to an integrated luminosity of 3.10^4 pb^{-1}. If the Higgs mass is in the range 120-800 GeV, evidence for the Higgs boson will show up in the 4-lepton decay mode. The analysis in the low mass case (below 120 GeV) in more involved and implies the study of several signatures including photon pairs and b-quarks accompanied by high p_t leptons. The analysis of the high mass case (above 800 GeV) implies the study of signatures including high p_t leptons and missing energy or a high multiplicity of jets in the final state. The study of Higgs-boson properties (width, branching ratios, production cross-sections) will require many years of running at high luminosity. It is finally expected that the discovery of the Higgs boson will complete the Standard Model scenario of elementary particles and provide new insight into the the problem of the scale for new physics.

Acknowledgments

It is a pleasure to thank the organisers of the meeting, and in particular Bernardo Adeva, for providing a pleasant stay at La Toja.

References

1. ATLAS technical proposal, CERN/LHC 94-43 (1994),
   CMS technical proposal, CERN/LHC 94-38 (1994).
2. S.Weinberg, *Phys. Rev. Lett.* **19**, 1264 (1967).
3. See for example: M.Quiros in 'Proceedings of the XXIV Int.Physics on Fundamental Physics', Edited by A.Ferrer (1996).

4. See for example: P.Janot in 'Proceedings of the HEP Conference', Jerusalem (1997).
5. M.Spira, QCD effects in Higgs Physics, CERN-TH/97-68.
6. See for example: J.Huston in 'LHC workshop', CERN (1998).

W PHYSICS AT LEP II

LL.M. MIR

Institut de Física d'Altes Energies
Universitat Autònoma de Barcelona
Edifici Cn, E-08193 Bellaterra, Spain
E-mail: mir@ifae.es

Recent experimental results in W physics from the LEP Collaborations are presented.

1 Introduction

In summer 1996, the LEP centre-of-mass energy was raised to 161 GeV [1,4,7] just above the W-pair production threshold. W pairs in e^+e^- collisions were observed for the first time by the LEP experiments, which collected $\sim 10\text{pb}^{-1}$ of integrated luminosity each. Later that year, the energy was raised to 172 GeV [2,5,8,11], and another $\sim 10\text{pb}^{-1}$ were collected. A further increase in energy took place in 1997, allowing the experiments to collect $\sim 55\text{pb}^{-1}$ each, at a centre-of-mass energy of 183 GeV[3,6,9,12].

The 1998 run has already started and at least $\sim 150\text{pb}^{-1}$ of integrated luminosity per experiment are expected at a centre-of-mass energy of 189 GeV. Finally, in the two coming years, LEP will run at its highest reachable energy, 200 GeV, providing as much integrated luminosity as possible.

W bosons are produced in pairs through the so-called CC03 diagrams, shown in figure 1.

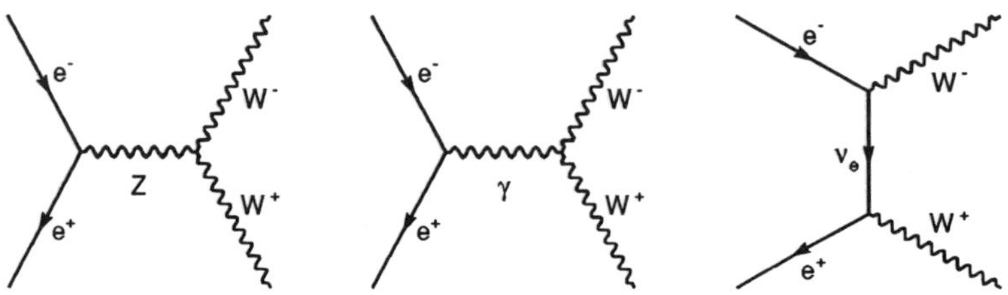

Figure 1. The CC03 diagrams

The W boson decays to a pair quark-antiquark (in 67% of the cases) and to charged lepton plus neutrino (in 33% of the cases). According to that, W-pair events can be classified into three different final states: fully hadronic, when the two Ws decay to quarks (45%); semileptonic, when one W decays to quarks and the other one to leptons (44%) and fully leptonic, when the two Ws decay to leptons (11%). The signatures of these three types of events are very different and, consequently, separate selection criteria have been developed for each of them. They are described in the following section.

2 Event Selection

The fully hadronic events are characterised by having four (or more) hadronic jets, high multiplicity and no missing mass. These events must be separated statistically from the main background, coming from $e^+e^- \to q\bar{q} \to$ multijets, especially from events with hard gluon radiation. Multivariate analyses, in particular neural network techniques, are applied after soft preselections to separate signal from background, reaching efficiencies and purities of 80%.

The systematic uncertainties which most affect this channel are the accuracy of the signal and background modelling, the miscalibrations of the detectors and the final state interactions, in particular, Bose-Einstein and colour reconnection effects.

The semileptonic events are characterised by having one energetic lepton plus two hadronic jets. So they feature high multiplicity, missing transverse momentum due to the escaping neutrino and the presence of an isolated energetic lepton with momentum anti-parallel to the missing momentum. The backgrounds are, in general, very low, and efficiencies higher than 80% (around 85% , 90% and 60%, for the electron, muon and tau channels respectively) are possible together with purities of 90%.

In addition to the accuracy of the modelling of signal and backgrounds and selection efficiencies, the uncertainties on the identification of the leptons can contribute to the systematic uncertainty.

Finally, the fully leptonic events are characterised by having two acoplanar energetic leptons. Low multiplicity, high missing transverse momentum and acoplanarity are their main signatures. Efficiencies of 70% with purities of 90% are reachable. The main backgrounds come from two-photon events giving tau pairs, leptonic decays of tau pairs, ZZ, Zee and Weν events and dileptons. The largest detector-related systematic effect comes from the photon vetoes against radiative dileptons.

3 Cross-Section Determination

As already mentioned in the introduction, the W bosons are produced in pairs through the CC03 diagrams, which lead to a four-fermion final state. Many other diagrams lead to the same four-fermion final states and interfere with the CC03 diagrams; however their effect can be reduced to at most few percent by requiring that all four fermions be within the detector acceptance and well separated from each other.

The cross-sections are presented as CC03 cross-sections, and the effect of the other diagrams is corrected for by comparing Monte Carlo CC03 with Monte Carlo containing the full set of four fermion diagrams (4f). Since the largest sensitivity to the mass comes precisely from these processes with two resonating Ws, this approach allows to extract the value of the W mass from the measured cross-sections. Furthermore, this CC03 result is insensitive to the value of the W mass used in the Monte Carlo generation.

The cross-sections are computed separately for each channel and the different channels are combined by means of a likelihood fit which uses as input the number of events in each channel and assumes the Standard Model branching ratios, the total cross-section being the only unknown. This result is consistent with the one obtained from the sum of the individual cross-sections. Finally, if lepton universality is assumed, a fit with the total cross-section and the hadronic branching ratio as unknowns can be performed. This hadronic branching ratio allows to constrain the Cabibbo-Kobayashi-Maskawa matrix element V_{CS}, leading to an improvement on the direct result from the measurement of the D meson decay.

Table 1 shows the WW measured cross-sections by the experiments at different energies as well as their combination. The first error corresponds to the statistical error and the second one to the systematic one.

Table 1. WW cross-sections at three different energies as measured by the LEP experiments. The first error is the statistical one and the second the systematic.

	σ_{WW}(pb)		
	161 GeV	172 GeV	183 GeV
ALEPH	$4.23 \pm 0.73 \pm 0.19$	$11.71 \pm 1.23 \pm 0.28$	$15.51 \pm 0.61 \pm 0.36$
DELPHI	$3.67^{+0.97}_{-0.85} \pm 0.19$	$11.58^{+1.44}_{-1.35} \pm 0.32$	$16.01 \pm 0.71 \pm 0.43$
L3	$2.89^{+0.81}_{-0.70} \pm 0.14$	$12.27^{+1.41}_{-1.32} \pm 0.23$	$16.66 \pm 0.66 \pm 0.30$
OPAL	$3.62^{+0.93}_{-0.82} \pm 0.16$	$12.3 \pm 1.3 \pm 0.3$	$15.52 \pm 0.62 \pm 0.35$
LEP	3.69 ± 0.45	12.0 ± 0.7	15.9 ± 0.4

Figure 2 shows the measured WW cross-sections at LEP at the different centre-of-mass energies, together with the theoretical predictions for different models.

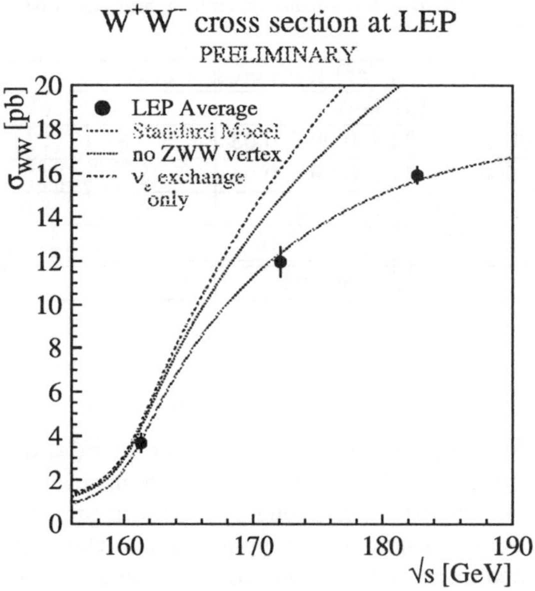

Figure 2. WW cross-sections at LEP

4 Mass Measurement

4.1 Mass Measurement at Threshold

At threshold, with $\sim 10\text{pb}^{-1}$ of integrated luminosity recorded per experiment, there are not enough events to obtain reconstructed mass distributions. However, the sensitivity of the cross-section to the mass can be used to extract the value of the W mass. This value is obtained by the LEP experiments using the theoretical prediction from `GENTLE` [13], so this estimate is strictly valid only within the Standard Model.

Table 2 shows the W mass extracted from the cross-section at different energies, obtained by the four LEP experiments. Their combination is also shown. The first error corresponds to the statistical error, the second one to the systematic error and the third comes from the LEP energy uncertainty.

Table 2. W mass from cross section at different energies as measured by the LEP experiments. The first error is statistical, the second systematic and the third comes from the LEP energy uncertainty.

$M_W (GeV/c^2)$ from cross-section		
	161 GeV	172 GeV
ALEPH	$80.14 \pm 0.34 \pm 0.09 \pm 0.03$	$81.17^{+1.15}_{-1.62}$
DELPHI	$80.40 \pm 0.44 \pm 0.09 \pm 0.03$	$81.18^{+1.26+0.38}_{-2.02-0.61} \pm 0.03$
L3	$80.80^{+0.48}_{-0.42} \pm 0.03$	$80.5^{+1.4}_{-2.4} \pm 0.03$
OPAL	$80.40^{+0.44+0.09}_{-0.41-0.10} \pm 0.10$	$80.5^{+1.4+0.5}_{-2.2-0.6}$
LEP	$80.40^{+0.22}_{-0.21} \pm 0.03$	

4.2 Mass Measurement by Direct Reconstruction

A larger sample of W pairs was collected at higher energies allowing the W mass to be measured from direct reconstruction of the invariant mass of its decay products. Since the centre-of-mass energy is known precisely, energy and momentum conservation constraints can be applied to the event reconstruction, leading to a much improved invariant mass resolution. This is true both in semileptonic W pairs, which are forced into a two-jet configuration excluding the lepton candidate, and in W pairs decaying to four (five) jets, where the relatively poor jet energy resolution is compensated by these kinematic constraints. To further improve the mass resolution the two reconstructed masses per event can be forced to be equal.

The hadronic channel presents an additional difficulty due to the presence of combinatorial background: the four jets can be coupled into di-jets in three different ways, of which only one is the correct. A jet pairing algorithm, which selects just one of the combinations, must be applied. The efficiency of the pairing algorithms is usually around 70%.

Figures 3 and 4 show the invariant mass distributions for hadronic and semileptonic events.

It can be seen that the ideal Breit-Wigner shape is distorted because of the presence of initial state radiation, phase space boundaries, detector resolution effects, misassignment of particles to Ws, backgrounds, event selection effects, etc. Modified Breit-Wigner functions can be used to fit the mass distributions, but linearity is not guaranteed, and the result should be corrected by means of a calibration curve.

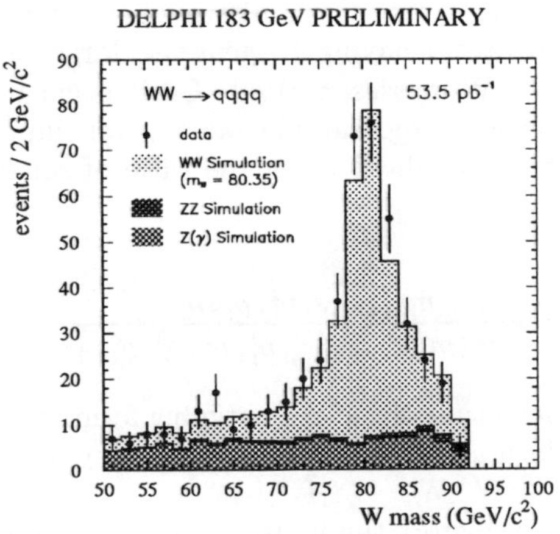

Figure 3. Mass distribution for hadronic data (points with error bars), background (dark area) and signal+background Monte Carlo (light area).

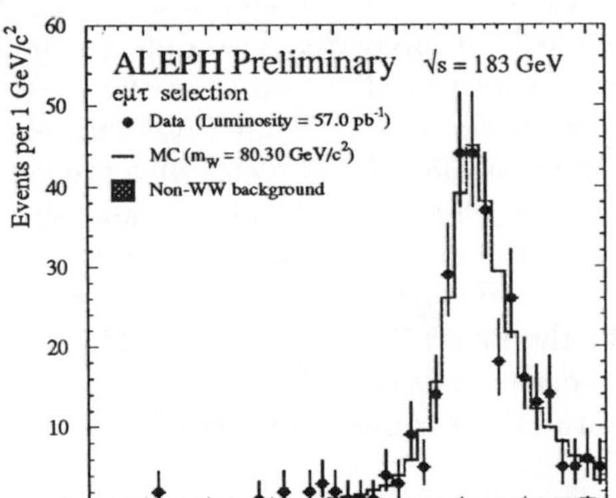

Figure 4. Mass distribution for semileptonic data (points with error bars), background (shaded area) and signal+background Monte Carlo (solid line histogram).

Another approach would be to compare the measured invariant mass distributions with the corresponding Monte Carlo distributions, generated with different input W masses. To avoid having to generate large Monte Carlo samples at many different input W masses, a Monte Carlo event re-weighting technique can be used: only one large sample, at a given input mass, is generated, and the events are re-weighted using the ratio of squared matrix elements:

$$w_i(m_W, \Gamma_W) = \frac{|\mathcal{M}(m_W, \Gamma_W, p_i^1, p_i^2, p_i^3, p_i^4)|^2}{|\mathcal{M}(m_W^{MC}, \Gamma_W^{MC}, p_i^1, p_i^2, p_i^3, p_i^4)|^2} \tag{1}$$

where p_i^j denotes the four-momentum of the jth outgoing fermion for a particular event i, Γ_W is the W width and $\mathcal{M}(m_W, \Gamma_W, p_i^1, p_i^2, p_i^3, p_i^4)$ is the matrix element of the process $e^+e^- \to W^+W^- \to f_1\bar{f}_2f_3\bar{f}_4$.

Final state interactions can distort the hadronic mass distributions, and hence the W mass measurement, if they are not properly simulated in the Monte Carlo references. These interactions come from the fact that the separation of the W decay vertices is of the order of $1/\Gamma_W \sim 0.1$ fm, distance much smaller than the typical hadronization scale of the decay products, of the order of ~ 1 fm. The final state particle production is not dictated by the separate decays of the W bosons because the identities of the individual W decay products are not well defined anymore. Two different interactions are studied: colour reconnection, which predicts shifts in the mass of the order of 10 to 50 MeV, depending on the models, and Bose-Einstein correlations between identical bosons (in particular pions) which also predict shifts of the order of 20-50 MeV. It should be stressed that in both cases the final effect is very likely to be dependent on the experimental procedure.

Table 3 shows the W mass extracted from direct reconstruction at different energies, obtained by the four LEP experiments. Their combination is also shown. The first error corresponds to the statistical error, the second one to the systematic error and the third comes from the LEP energy uncertainty.

4.3 Combined result

Figure 5 shows the W mass measured at LEP, both from the cross-section and direct reconstruction, as well as their combination. The 90 MeV final error breaks down as follows: 70 MeV experimental error, 40 MeV from final state interactions and 30 MeV coming from the uncertainty in the LEP energy.

Figure 6 shows the W mass measured at LEP together with the result from pp-colliders and their combination. This world average value for the W

Table 3. W mass from direct reconstruction at different energies as measured by the LEP experiments. The first error is statistical, the second systematic and the third comes from the LEP energy uncertainty.

$M_W(\mathrm{GeV}/c^2)$ from direct reconstruction		
	172 GeV	183 GeV
ALEPH	$80.80 \pm 0.32 \pm 0.11 \pm 0.03$	$81.30 \pm 0.13 \pm 0.07 \pm 0.03$
DELPHI	$80.22 \pm 0.41 \pm 0.06 \pm 0.03$	$80.23 \pm 0.16 \pm 0.15 \pm 0.03$
L3	$80.71^{+0.34}_{-0.35} \pm 0.09$	$80.32 \pm 0.16 \pm 0.09$
OPAL	$80.32 \pm 0.30 \pm 0.09$	$80.34 \pm 0.14 \pm 0.08$
LEP	$80.53 \pm 0.17 \pm 0.05$	$80.30 \pm 0.08 \pm 0.09$

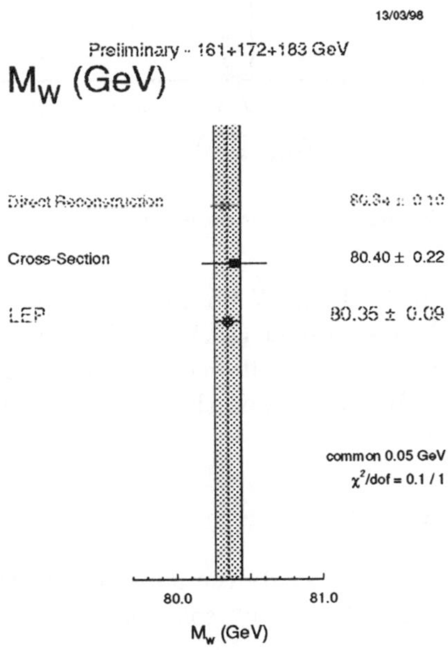

Figure 5. W mass measurement at LEP.

mass is compared with the Standard Model prediction for m_W obtained using all electroweak data from LEPI as SLD as input.

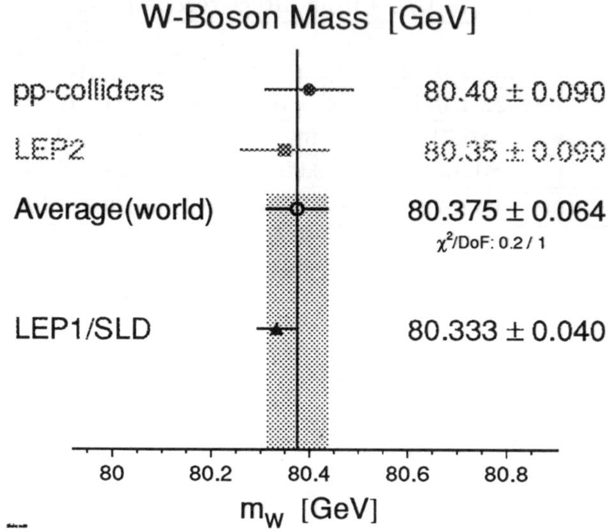

Figure 6. W mass measurements.

Figure 7 shows the mass of the W boson and the top quark from direct (LEPII and pp-colliders) and indirect (LEPI, SLD and neutrino data) measurements, together with the region allowed for three different values of the mass of the Higgs boson. It can be seen that low values for the mass of the Higss boson are favoured.

5 Triple Gauge Couplings

LEPII allows for the first time to study the non-abelian structure of the electroweak Standard Model directly in e^+e^- collisions (via the two first reactions of figure 1).

The most general structure of the Standard Model which is consistent with Lorentz invariance and $U(1)_{EM}$ gauge invariance contains 14 couplings. These couplings naturally split into three classes: six which conserve Charge Conjugation and Parity separately (g_1^γ, g_1^Z, κ_γ, κ_Z, λ_γ and λ_Z), two which conserve CP but violate Charge Conjugation and Parity (g_5^γ and g_5^Z) and six which violate CP (g_4^γ, g_4^Z, $\tilde{\kappa}_\gamma$, $\tilde{\kappa}_Z$, $\tilde{\lambda}_\gamma$ and $\tilde{\lambda}_Z$)[14].

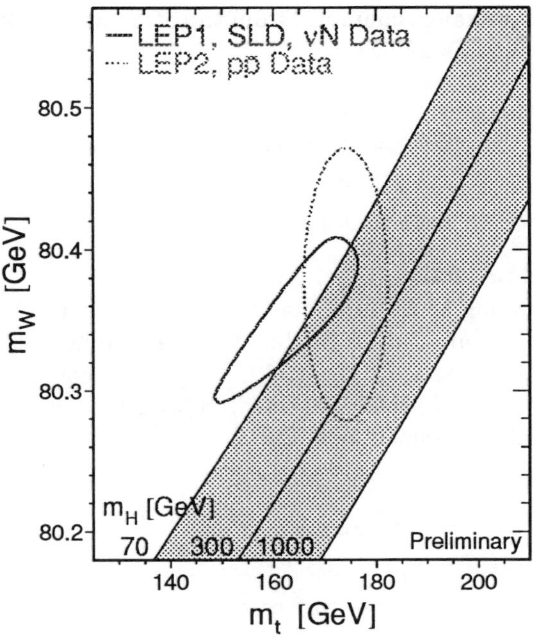

Figure 7. W mass versus top mass from direct and indirect data.

The statistics that will be accumulated by the LEP experiments will not allow a simultaneous determination of the 14 couplings, so, since it is very unlikely to observe C, P or CP violation at LEPII, due to the indirect measurements obtained from LEPI data, only the C and P conserving couplings will be considered. These couplings are related with the charge, magnetic dipole moment and electric quadrupole moment of the W boson through the following relations:

$$Q_W = e g_1^\gamma \tag{2}$$

$$\mu_W = \frac{e}{2m_W}(g_1^\gamma + \kappa_\gamma + \lambda_\gamma) \tag{3}$$

$$q_W = -\frac{e}{m_W^2}(\kappa_\gamma - \lambda_\gamma) \tag{4}$$

Any theory incorporating new Physics beyond the Standard Model and including the electroweak theory as an effective low-energy limit, may introduce deviations of the triple gauge couplings from the Standard Model values. These deviations are defined as follows:

$$\Delta g_1^Z \equiv (g_1^Z - 1) \tag{5}$$
$$\Delta \kappa_\gamma \equiv (\kappa_\gamma - 1) \tag{6}$$
$$\Delta \kappa_Z \equiv (\kappa_Z - 1) \tag{7}$$

and the Standard Model prediction for λ_γ and λ_Z is zero.

The triple gauge coupling contribute, via loop corrections, to observables which can be precisely measured at LEPI; but there are some "blind directions" in the multidimensional space of the T.G.C.s where the constraints from the LEPI analyses are very weak; these are the directions that are studied by the LEP experiments and are defined as follows:

$$\alpha_{W\phi} = \Delta g_1^Z \cos\theta_W \tag{8}$$
$$\alpha_W = \lambda_\gamma \tag{9}$$
$$\alpha_{B\phi} = \Delta\kappa_\gamma - \Delta g_1^Z \cos^2\theta_W \tag{10}$$

together with the constraints $\Delta\kappa_Z = \Delta g_1^Z - \Delta\kappa_\gamma \tan^2\theta_W$ and $\lambda_\gamma = \lambda_Z$. Notice that the Standard Model expectation for these couplings are $\alpha_{W\phi} = \alpha_W = \alpha_{B\phi} = 0$.

To evaluate the T.G.C.s one could, in principle, apply an unbinned maximum likelihood fit with probabilities evaluated from $\frac{d\sigma(\Omega,\bar{\alpha})}{d\Omega}$ using the measured kinematic variables; this way, the full angular information available is exploited but, since it does not include neither background nor experimental effects, the behaviour of the fit is not linear and some degradation in the resolution is expected.

Trying to include the experimental effects using Monte Carlo distributions (as it is done for the measurement of the mass) via a five-dimensional binned likelihood fit, would require to bin according to the experimental distributions and this would lead to an unaceptable large number of bins. One can always fit only the $\cos\theta_{\mathrm{W}}$ distribution, but this always mean loss of information, especially for the semileptonic channel where the leptonic quantities give information on the W helicity.

The approach followed by the LEP collaborations has been the use of "optimal observables", variables which project the relevant kinematic information to a certain T.G.C. onto a one-dimensional distribution. Since the T.G.C.s contribute only linear terms to the amplitude for a given process, the differencial cross-sections expanded in couplings contain no terms beyond second order:

$$\frac{d\sigma(\Omega,\bar{\alpha})}{d\Omega} = S_0(\Omega) + \sum_i S_{1i}(\Omega)\alpha_i + \sum_{i,j} S_{2ij}(\Omega)\alpha_i\alpha_j \tag{11}$$

where α_i are the couplings and Ω the measured phase space variables.

From the expression above, one can see that it is possible to derive limits on couplings by measuring the total cross-section; this would imply the use of only a small amount of all the accessible information and would be useful only if the values of the couplings were indeed quite large. To increase the sensitivity the W production and decay angles (see figure 8) can be used.

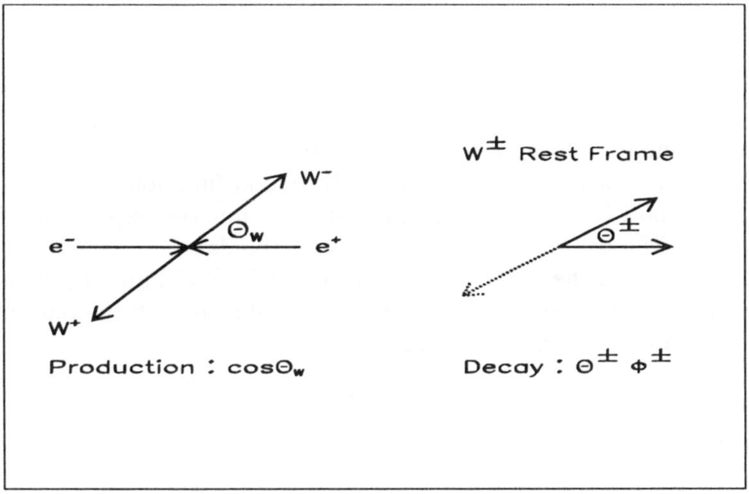

Figure 8. W production and decay angles.

The production angle is defined as the angle formed by the incoming electron and the produced W^-. In principle, all channels can be used, but the hadronic channel presents more difficulties due to the combinatoric background and to the fact that the charge of the W is, in principle, unknown. The decay angle for a given W is defined as the angle formed by that W and charged lepton (leptonic decay) or the down-type quark (hadronic decay). Since in hadronic decays the flavour of the quarks are not determined, there is a two-fold ambiguity per hadronic decay in the event.

For a given T.G.C. α_i, the optimal observable is defined as:

$$OO_i = \frac{\frac{d}{d\alpha_i}\left(\frac{d\sigma(\Omega,\bar{\alpha})}{d\Omega}\right)\Big|_{\alpha_i=0}}{\frac{d\sigma(\Omega,\bar{\alpha})}{d\Omega}\Big|_{\alpha_i=0}} = \frac{S_{1i}(\Omega)}{S_0(\Omega)} \tag{12}$$

For each detected event the quantity OO_i is calculated and a likelihood fit is performed:

$$\log \mathcal{L} = \sum_{i=1}^{N} \log \mathcal{P}(OO_i, \bar{\alpha}) \tag{13}$$

The total number of events observed (cross-section information) can be included by adding a Poisson term to the likelihood function.

This method is bias free and take into account the experimental effects provided the Monte Carlo describes correctly the data.

Figure 9 shows the results obtained by the LEP Collaborations for $\alpha_{W\phi}$, α_W and $\alpha_{B\phi}$. Figure 10 shows these results combined with the ones obtained by the D0 Collaboration at Tevatron.

6 Summary

W precision Physics has started to be a reality with the data collected by the LEP experiments so far, and there are still three years of running ahead. (By the time of writing these proceedings, around 175 pb^{-1} of WW events at 189 GeV centre-of-mass energy have been collected per experiment).

In what cross-sections are concerned, the four LEP experiments measure consistently the WW production cross-section and in good agreement with the Standard Model prediction. The mass is measured with a total error of 90 MeV. From now on, the systematic uncertainties will have to be kept well under control. Finally, the triple gauge coupling measurements start to be competitive.

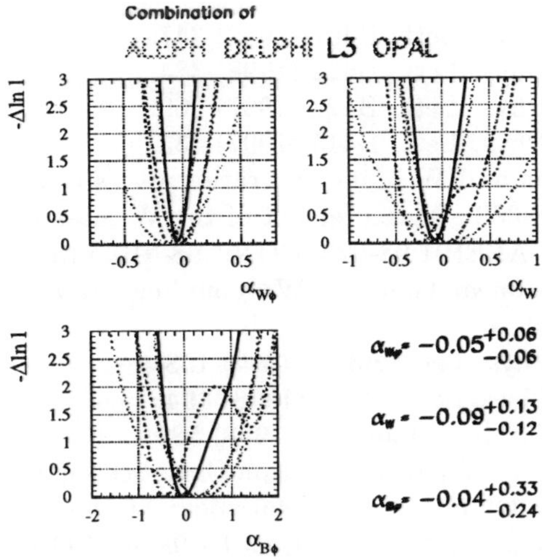

Figure 9. T.G.C's from the LEP Collaborations

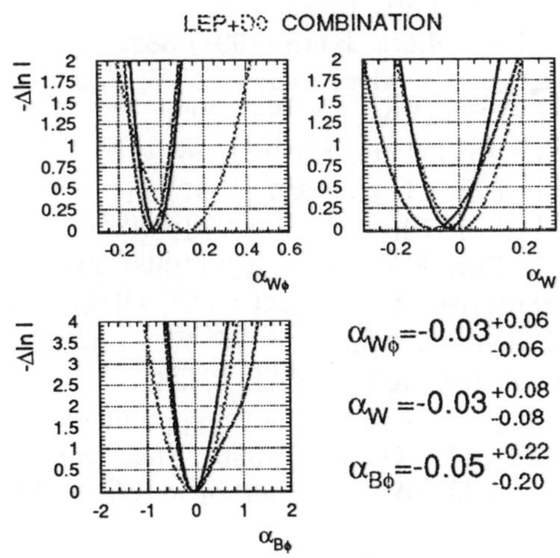

Figure 10. T.G.C's from the LEP and D0 Collaborations

References

1. R. Barate et al. Phys. Lett. B401 (1997) 347.
2. R. Barate et al. Phys. Lett. B422 (1998) 384.
   R. Barate et al. Phys. Lett. B422 (1997) 369.
   R. Barate et al. Phys. Lett. B415 (1997) 435.
3. 'WW cross section and W branching ratios at $\sqrt{s} = 183$ GeV', ALEPH 98-019, CONF 98-009. 'Measurement of the W mass in e^+e^- Collisions at 181-184 GeV', ALEPH 98-020, CONF 98-010. 'Isolated photons and constraints on Trilinear Gauge WWγ couplings at LEP2', ALEPH 98-012, CONF 98-002.
4. P. Abreu et al., Phys. Lett. B397 (1997) 158.
5. P. Abreu et al., Phys. Lett. B423 (1998) 194.
   P. Abreu et al., E. Phys. Jour, C2 (1998) 581.
6. 'Measurement of the W-pair Production Cross-section at $\sqrt{s} = 183$ GeV', DELPHI 98-20, CONF 120. 'Measurement of the W-pair Production Cross-section at $\sqrt{s} = 183$ GeV', DELPHI 98-20, CONF 120. 'Measurement of the mass of the W boson using direct reconstruction', DELPHI 98-27, CONF 123. 'Measurement of Trilinear Gauge Boson Couplings in e^+e^- Collisions at 183 GeV', DELPHI 98-21, CONF 121
7. M. Acciarri et al., Phys. Lett. B398 (1997) 223.
8. M. Acciarri et al., Phys. Lett. B441 (1997) 330.
   M. Acciarri et al.. Phys. Lett. B418 (1998) 389.
9. 'Preliminary Results on the Measurement of W-Pair Production in e^+e^- Interactions at $\sqrt{s} = 183$ GeV', L3 Note 2236.
10. K. Ackerstaff et al., Phys. Lett. B389 (1996) 416.
    K. Ackerstaff et al., Phys. Lett. B397 (1997) 147.
11. K. Ackerstaff et al., Eur. Phys. Jour. C1 (1998) 395.
    K. Ackerstaff et al., Eur. Phys. Jour. C2 (1998) 597.
12. 'W^+W^- event properties at$\sqrt{s} = 183$ GeV', OPAL PN342. 'Measurement of the W boson mass and the W^+W^- production cross-section in e^+e^- collisions at 183 GeV', OPAL PN331. 'Measurement of triple gauge boson couplings using $W^+W^- \rightarrow q\bar{q}q\bar{q}$ and $W^+W^- \rightarrow l\nu l\nu$ events at $\sqrt{s} = 181 - -184$ GeV', OPAL PN336. 'Measurement of triple gauge boson couplings from W^+W^- production at $\sqrt{s} = 172 - -183$ GeV', OPAL PN329.
13. D. Bardin Phys. Lett. B37 (1994) 148.
14. 'Physics at LEP2', CERN 96-01, Editors:G. Altarelli et al.

HERA-B: STATUS AND PERSPECTIVES

CRISTÓBAL PADILLA
DESY - Notkestrasse, 85, D-22603 Hamburg, Germany
E-mail: Cristobal.Padilla@desy.de

HERA-B, an experiment to measure CP violation in the $B^0 \to J/\psi K^0_S$ decay channel at the HERA proton storage ring at DESY, is a pioneering hadronic B factory using fixed target techniques. In addition to the CP violation measurement, the experiment addresses a wide range of topics in heavy quark physics, complementing the physics accessible at e^+e^- colliders. The detector design is discussed, making special emphasis in the needs for the CP violation measurement. Recent results from runs using a partially installed detector are presented.

1 Introduction

One of the most interesting observations in particle physics is the violation of the CP symmetry in decays of neutral kaons, discovered 30 years ago. Neutral B mesons provide one of the few other systems in nature where CP violation might be observable. This system can provide decisive tests of the implementation of CP violation in the standard model of electroweak interactions.

The CP violation effects are expected to be larger in neutral B mesons than in kaons, but the decay channels which can exhibit CP asymmetries are extremely rare, typically suppressed by four to five orders of magnitude. Experimental cuts to select clean signatures and to identify the B flavours reduce the useful rates further. An experiment will therefore require the production of large numbers of B mesons, *i.e.* a machine acting as a B factory.

The theoretical implementation of the CP violation is usually done through the CKM ansatz [1]. In this framework, CP violation is present when the unitarity triangle [2,3] has non-zero area. One of the angles of this triangle can be measured looking at CP violation in the $B^0 \to J/\psi K^0_S$ decay channel. HERA-B [4,5,6] is an experiment mainly designed to detect CP violation in B mesons using this decay. This lecture describe the main parameter designs of the experiment as well as its status and future schedule. Section 2 describes the principles of the CP violation measurement in the $B^0 \to J/\psi K^0_S$ decay channel. Section 3 gives an overview of the detector experimental physics reach. Section 4 explains why and how the detector is designed, giving special emphasis to the crucial components: the target and trigger systems. Section 5 gives a description of the tests and measurements performed so far. Finally, sec. 6 gives the summary an overview to the future of the experiment.

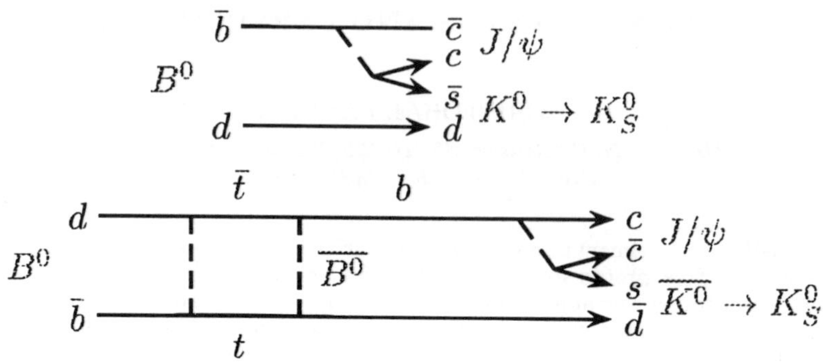

Figure 1. Diagrams contributing to the $B^0 \to J/\psi K^0_S$ decay.

2 CP violation in $B^0 \to J/\psi K^0_S$

The decay of B^0 mesons into $J/\psi\, K^0_S$ is generally considered the "golden decay" channel to detect CP violation in the B system. Even though its branching ratio is very small, it is very attractive due to its clear experimental signature, which results in good trigger conditions, and its small theoretical uncertainties coming from the absence of penguin contributions. The only two diagrams contributing to this decay are shown in fig. 1, the "direct" decay and the decay "through a $\overline{B}^0$", which results in small hadronic uncertainties. CP violation arise from the interference of both diagrams.

2.1 Principle of the Measurement

The measurement of the CP asymmetry will be done looking at the differences in decay rate to $J/\psi\, K^0_S$ (CP eigenstate) from B^0 and $\overline{B}^0$, given by

$$\text{Pure } B^0 \text{ beam} \longrightarrow n(t) \propto e^{-t}\,(1 + \sin xt \sin 2\beta) \tag{1}$$

$$\text{Pure } \overline{B}^0 \text{ beam} \longrightarrow \bar{n}(t) \propto e^{-t}\,(1 - \sin xt \sin 2\beta) \tag{2}$$

where t is measured in units of B lifetime, $x \approx 0.67$ [7,8] is the mixing parameter defined as $\Delta M/\Gamma$, where ΔM stands for the mass difference of the mass eigenstates and Γ for the inverse of the lifetime, and β is one of the angles of the unitarity triangle. Figure 2 shows this triangle in the (ρ,η) plane of the

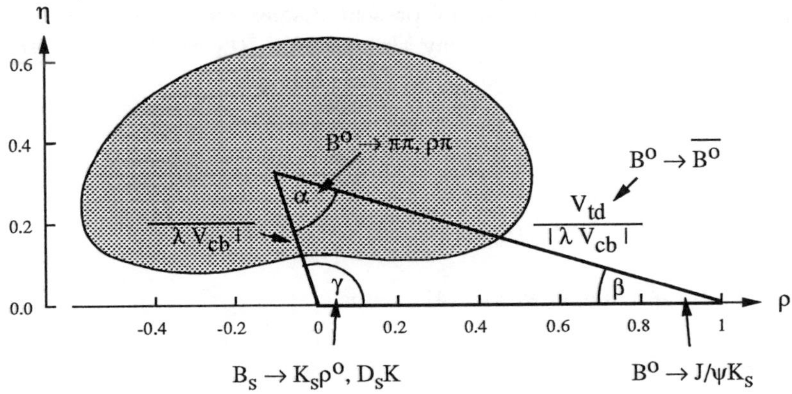

Figure 2. Unitarity triangle in the (ρ,η) plane of the Wolfenstein parametrization of the CKM matrix.

Wolfenstain parametrization. Also shown is a typical one sigma contour for the position of the tip as constrained by the data.

To detect CP violation in $B^0 \to J/\psi\, K^0_S$, one needs samples of pure B^0 and $\overline{B}^0$ mesons and measure their decay rates. Nevertheless a reconstructed $J/\psi\, K^0_S$ decay could come from either a B^0 or $\overline{B}^0$ and therefore the initial flavour of this "signal B" needs to be determined finding the b-quark contain of the second B in the event, the called "tagging B".

If B^0 and $\overline{B}^0$ are produced equally and detected with equal efficiency the CP violation can be obtained form the measurement of the asymmetry $a(t)$ as a function of t given by

$$a(t) = \frac{n(t) - \bar{n}(t)}{n(t) + \bar{n}(t)} = \sin\, xt\, \sin\, 2\beta \tag{3}$$

or its time integrated equivalent

$$A_{\text{int}} = \frac{\int n(t)\mathrm{d}t - \int \bar{n}(t)\mathrm{d}t}{\int n(t)\mathrm{d}t + \int \bar{n}(t)\mathrm{d}t} = \frac{x}{1+x^2}\, \sin\, 2\beta \approx 0.46\, \sin\, 2\beta. \tag{4}$$

In the experiment we must require a minimum separation between the primary vertex and the B decay vertex, in order to suppress background. This requirement, for the purpose of the present discussion, can be approximated by a fixed cut on the proper decay time, $t > t_0$ ($t_0 \approx 1$). The integrated asymmetry is then

$$A_{\text{int}}(t_0) = M_{\text{int}}(t_0)\ \sin 2\beta \tag{5}$$

where

$$M_{\text{int}}(t_0) = \frac{\sin xt_0 + x\ \cos xt_0}{1+x^2} \tag{6}$$

instead of $x/(1+x^2)$as it was in (4).

2.2 *Errror in* sin 2β

From eqs. (4) and (5) the statistical error can be estimated as

$$\Delta \sin 2\beta = \frac{1}{M_{\text{int}}(t_0)} \cdot \frac{1}{\sqrt{N_{\text{B}^0}(t_0)}} \tag{7}$$

where $M_{\text{int}}(t_0)$ is given in eq. (6) and $N_{\text{B}^0}(t_0)$ is defined as the number of B^0 and $\overline{\text{B}}^0$ that have not decayed after the time t_0.

As it has been already mentioned, this measurement requires the tagging of the second B. However, this tagging is never perfect and, any error in the tagging will dilute the CP asymmetry and increase the error in sin 2β. Two main factors have to be considered:

- Mixing of the tagging B: the effects of mixing in the tag are only relevant if B^0 or B^0_s are used. In this case, if no lifetime cuts are applied for the tagged B, the probability of changing flavour before tagging is

  $$f_M = \frac{x^2}{2\ (1+x^2)}, \tag{8}$$

  which can be directly translated into the number of events obtained as a function of time

  $$n_{\text{obs}}(t) = n(t)\ (1-f_M) + \bar{n}(t)\ f_M \bar{n}_{\text{obs}}(t) = \bar{n}(t)\ (1-f_M) + n(t)\ f_M. \tag{9}$$

  For B^0 one finds $f_M \approx 0.16$ and for B^0_s, $f_M \approx 0.5$ (with $x_\text{s} \gg 1$). The detectable integrated CP asymmetry is therefore reduced by a dilution factor D_M:

  $$A_{\text{obs}}(t_0) = D_M\ M(t_0)\ \sin 2\beta \tag{10}$$

  with $D_M = (1-2f_M) = 1/1+x^2 \approx 0.67$.

- Wrong tags due to particle misidentification: this effect is similar to the one already discussed. Calling f_T the probability of mistagging due to such reasons, the detectable integrated CP asymmetry is reduced by another dilution factor $D_T = (1-2f_T)$ (it goes from 0.5 up to 0.95 depending on the used tag), and finally:

$$A_{\rm obs}(t_0) = D_T\, D_M\, M(t_0)\, \sin 2\beta. \tag{11}$$

Taking into account all the above effects, the total error in sin 2β can be written as

$$\Delta \sin 2\beta = \frac{1}{M_{\rm int}(t_0)} \cdot \frac{1}{P} \cdot \frac{1}{\sqrt{N^{\rm rec}_{\rm B^0}(t_0)}} \tag{12}$$

where $N^{\rm rec}_{\rm B^0}(t_0)$ is the number of reconstructed $\rm B^0$ mesons and

$$P = D_M\, D_T\, \sqrt{\epsilon_{\rm tag}} \tag{13}$$

is the tagging power, with D_M and D_T the mixing and wrong tag dilution factors and $\epsilon_{\rm tag}$ the total tagging efficiency.

For values of $\sin 2\beta \sim 0.5$ we expect an observable CP asymmetry of $A_{\rm obs} \sim 0.15 - 0.2$ taking reasonable values for the tagging power (0.31 according to simulations). In this case, to achieve and error in $\sin 2\beta$ of the order of 0.13, $\sim$ 1500 reconstructed $\rm B^0$ mesons are needed.

2.3 $\rm B^0 \to J/\psi\, K^0_S$ *at HERA*

Figure 3 shows the scheme of the decay $\rm B^0 \to J/\psi\, K^0_S$ after a colision of a proton with a fixed target, as it is foreseen in the HERA proton storage ring. This scheme influences the design parameters of the HERA-B detector.

HERA produces and collides beams of protons and electrons of energies 820 GeV and 30 GeV, respectively. HERA-B will only make use of the protons, which will collide with a fixed target. Proton bunches are spaced 96 ns, which means a colliding frequency of ~10 MHz. In such a collision, sometimes, a $\rm b\overline{b}$ quark pair is produced. Sometimes one of the quarks hadronizes to a $\rm B^0$ or to a $\rm \overline{B}^0$ meson. The $\rm B^0$ meson has an average momentum of 120 GeV and travels of the order of 9 mm before decaying. If the decay is of the type $\rm B^0 \to J/\psi\, K^0_S$, the J/ψ decays inmediatelly, sometimes to a pair of leptons, which have an average momentum of 34 GeV. The $\rm K^0_S$ travels, on average, 1.1 m before decaying, sometimes to a pair of pions, with average momentum of 15 GeV. This decay chain corresponds to the "signal B", which will contribute to the measurement of CP violation as described in sec. 2.1.

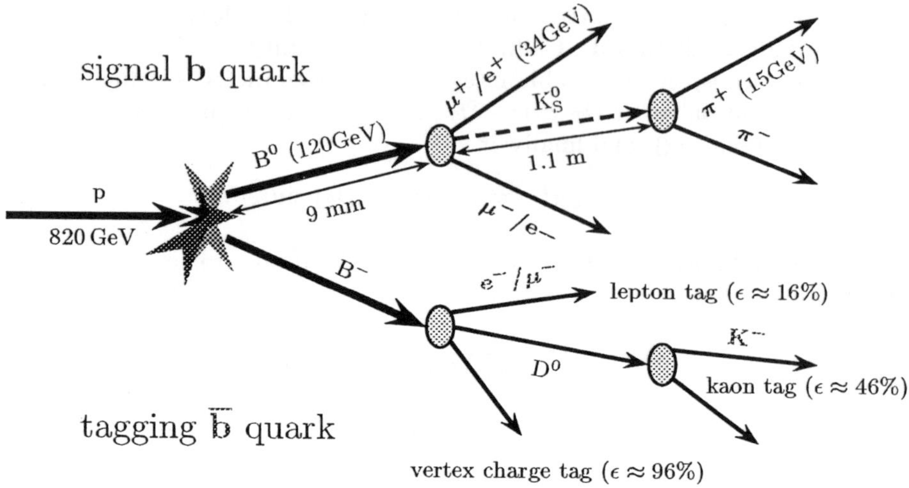

Figure 3. Scheme of the "golden decay" channel, $B^0 \rightarrow J/\psi\, K^0_S$, in a fix target experiment at HERA.

At the same time, the other charged meson can serve as "tagging B". This is the case only when this meson decays to a lepton, to a kaon or when the vertex can be reconstructed. In fig. 3, the efficiencies of each of these possible tags are also indicated.

Figure 4 gives some numbers for the production and reconstruction efficiency of the scheme presented in fig. 3. From the total number of proton-target collisions, only one in each million produces a $b\overline{b}$ quark pair. A $\overline{b}$ (b) quark hadronizes to a $B^0\,\overline{B}^0$ meson 40% of the cases. Taking into account the two possibilities, one can estimate that 80% of the 1 per million collisions are useful. The branching ratio of the deday $B^0 \rightarrow J/\psi\, K^0_S$ is $\sim 5 \cdot 10^{-4}$ [9]. The branching ratio of $J/\psi \rightarrow e^+e^-/\mu^+\mu^-$ is $\sim$ 12% and the branching ratio of $K^0_S \rightarrow \pi^+\pi^-$ is $\sim$ 69%. All these factors dramatically reduce the number of usable collisions.

The event is usable only if the "tagging B" can also be reconstructed. Taking into account the trigger and reconstruction efficiency of the three possible tags represented in fig. 3, a total efficiency of collisions of $4 \cdot 10^{-12}$ can be estimated.

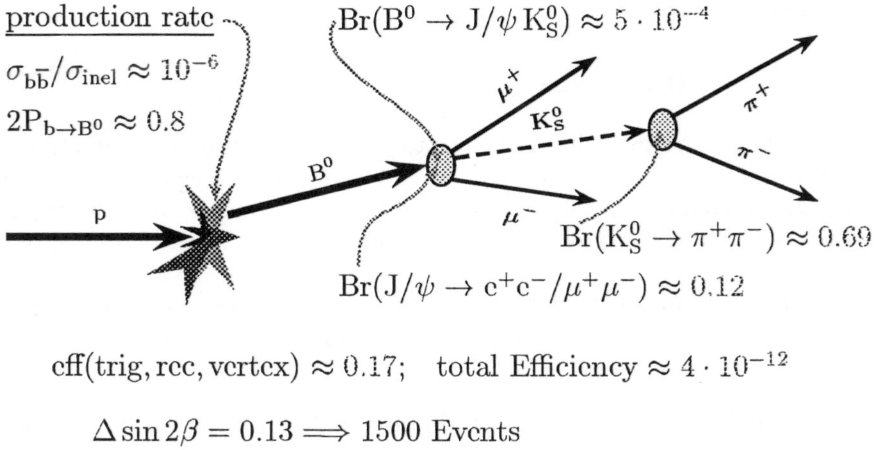

Figure 4. Production rate and reconstruction efficiency of the "golden decay" channel $B^0 \to J/\psi K^0_S$ in a fixed target experiment at HERA.

The $b\bar{b}$ quark pair production cross section in hadronic interactions has been measured by a variety of experiments [10,11,12,13,14,15,16] at different energies. The average value at 820 GeV is $\sigma_{b\bar{b}} \approx 12$ nb which means that $4{\cdot}10^{14}$ interactions are needed to get the 1500 reconstructed $B^0 \to J/\psi K^0_S$ and achieve and error of 0.13 in $\sin 2\beta$. The assumed number of seconds that an experiment can run in a year is 10^7. Thus, in order to get the needed number of interactions at HERA (with a bunch crossing frequency of 10 MHz) in this time, 4 interactions per bunch crossing are needed. This poses serious requirements to the target and detector design as described in section 4.

3 Other physics goals of HERA-B

Even if the HERA-B detector is optimized to trigger and reconstruct the J/ψ coming from neutral B mesons, it is clear that a lot of other channels will be recorded as well. Table 1 gives an overview of the possible channels that can be exploted with the expected number of events per year and the possible physical measurements that can be perfomed with them, which include measurements of β with other additional channels, B^0_s mixing, spectroscopy and decays of B mesons and baryons, rare and exotic B decays and QCD and heavy quark production studies.

4 The HERA-B detector

As stated in sec. 2, the HERA-B detector is optimized for the detection of the "golden decay" channel, $B^0 \to J/\psi K^0_S$. Figure 5 shows again the diagram of this decay channel specifying the detector requirements and the trigger scheme needed to identify each of its products. The design of the detector is based in this figure.

Fugure 6 shows the HERA-B detector. It is a huge magnetic forward spectrometer with outer dimensions of $20 \times 9 \times 7\,\mathrm{m}^3$. The main design choices are:

- Solid angle coverage from 10 mrad polar angle to about 200 mrad, corresponding to about 90% solid angle coverage in the centre of mass system.
- Use of a single normal-conducting dipole magnet for momentum analysis, with a field integral of 2.1 T m.

Table 1. Available channels that will be triggered at HERA-B, with the number of events expected per year and the possible physics interests.

Channel	# events per year		
$B \to J/\psi X$	$\approx 2.6 \cdot 10^5$	$\approx$ 17000 reconstructed B decays 1500 $J/\psi K^0_S$ } B^0 5500 $J/\psi K^{*0}$ } B^0 7000 $J/\psi K^+$ } B^+ 1500 $J/\psi K^{*+}$ } B^+ 1500 $J/\psi \phi$ } B_s 500 $J\psi \Lambda$ } Λ_B	– lifetimes $\Rightarrow \lvert V_{cb} \rvert$ – unbiased B sample – CP violation – B_c, Λ_b – rare decays – QCD ($gg \to b\overline{b}$) – other CP decays – $B \to ll$
$B_1 B_2 \to l_1 l_2 X$	$\approx 1.4 \cdot 10^6$	– $B^0\overline{B}^0$ mixing – $B^0_s\overline{B}^0_s$ mixing ($l_1\, l_2\, D_s^{(*)} \nu\nu$)	
$B \to l_1\, C X$	$\approx 3.8 \cdot 10^5$	– $D^0\overline{D}^0$ mixing ($B \to [D^0 \leftrightarrow \overline{D}^0]\, l\, \nu \to l^\pm\, l^\pm\, X$) – V_{cb} – V_{ub}, B_c, $B \to ll$, rare decays – source for unbiased B mesons	
$B \to l X$	$\approx 3.2 \cdot 10^6$	– unbiased tag of the second B meson	

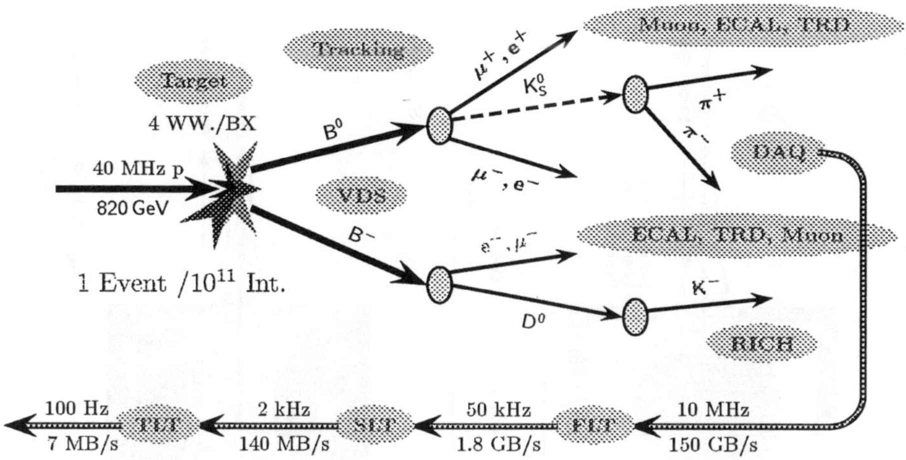

Figure 5. Detector components and trigger scheme required to detect the $B^0 \to J/\psi K^0_S$ decay.

- A silicon vertex detector system starting at 1 cm radius from the beam with seven layers of double-sided silicon strip detectors [17].

- A main tracker system, whose granularity and technology vary with distance from the beam in order to limit the occupancy of each detector cell and minimize the number of channels. Below a radius of 6 cm, silicon strip detectors will be used, followed by micro-strip gaseous detectors in the intermediate region out to about 20 cm, and by honeycomb drift cells at larger distances from the beam.

- A Ring Imaging CHerenkov (RICH) counter [18] as the only technology to identify a tagging kaon with its momentum between few GeV to about 50 GeV.

- A Transition Radiation Detector (TRD) [19] using a fibre and straw detector cells in the very forward region, in order to improve electron identification in the congested small-angle region.

- An electromagnetic Pb/scintillator and W/scintillator calorimeter [20] which fulfills the requirements concerning energy resolution in a cost ef-

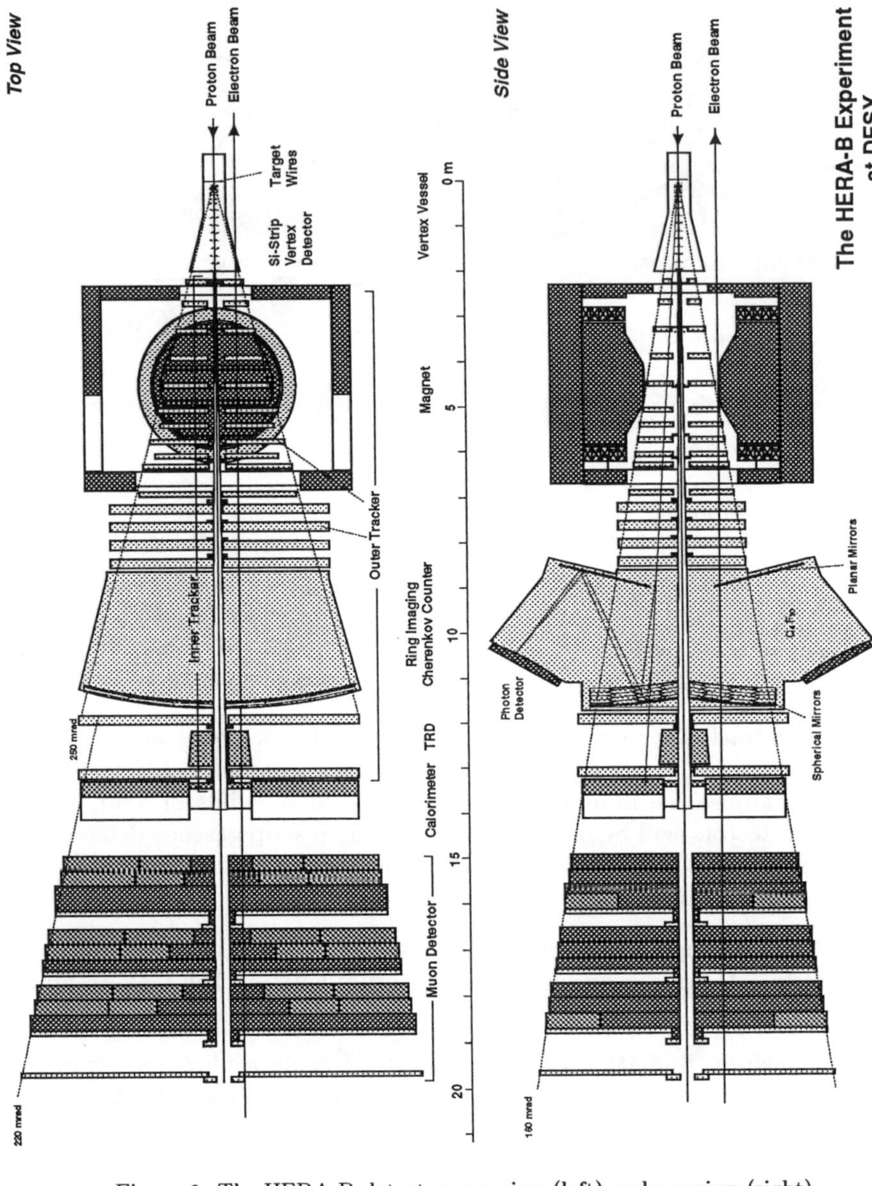

Figure 6. The HERA-B detector: z-x view (left) and z-y view (right).

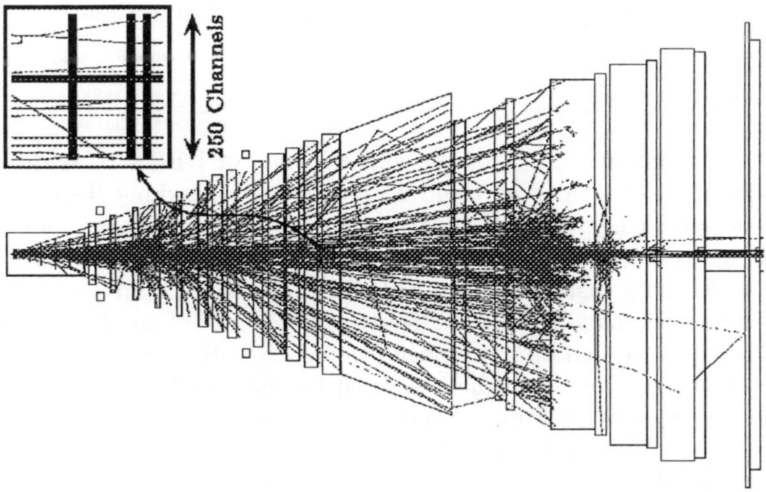

Figure 7. Typical picture of an event after a collision of a 820 GeV proton beam with the target in the HERA-B detector.

fective fashion, and allows matched granularities with a minimal Molière radius close to the beam.

- A conventional muon system with four chamber layers at different depths in the absorber.

As mentioned in sec. 2.3, an average of four interactions per bunch crossing are needed to get the desired statistics. Figure 7 shows the result of one of such bunch crossings. On average, 60 primary tracks are produced but the interaction of the particles with the detector produce up to 200 tracks before the calorimeter. These tracks are mainly distributed in the forward region of the detector, resulting in large occupancies there. The detail of fig. 7 shows that the granularity of the designed detector can stand such environment and still resolve all these tracks. Table 2 gives the number of readout channels per component and the occupancy in hits per bunch crossing per channel. The total number of electronic channels (which have to be readout every 96 ns) exceeds half a million.

There are two fundamental pieces of the detector that allow it to work: the internal target, able to produce the needed number of events and the trigger system, able to reduce the data input rate to a manegable level. These two parts, will be described in more detail in the following sections.

4.1 The internal target

The target system [21], one of the key elements of HERA-B, is designed to provide interaction rates up to 40 MHz withouth disturbing the ep operation of the storage ring. It consists of thin wires (cross sections $\sim 50 \times 500\,\mu m^2$) which are positioned around the proton beam at a distance of 4 to 10 r.m.s beam widths, *i.e.* inside the beam halo but outside the beam core. As a result, only those protons which have already left the core of the beam and would therefore be absorbed after a few additional revolutions anyway, will interact with the target. The beam halo scraped by the target wires is continuously replenished by protons diffusing outward, a process mainly driven by the non-linear effects of the ep beam-beam interactions.

Details of the operation and performance of the target as well as the interaction with the other HERA experiments will be discussed in sec. 5.

Table 2. Main detector components of the HERA-B detector.

detector	*technology*	*number of channels*	*hits per BX*
Vertex detector	Si-strip	136 k	≈ 0.05
Tracker			
inner (2-6 cm)	Si-strip	40 k	≈ 0.02
inner (6-19 cm)	microstrip gas-chamber (GEM)	135 k	≈ 0.04
outer (>19 cm)	honeycomb-DC	120 k	≈ 0.15
High-p_T trigger $B^0 \to \pi^+\pi^-$	gas pixel/straw	26 k	≈ 0.05
RICH Kaon iden.	C_4F_{10} radiator PMT photon detector	32 k	≈ 0.1
Small-Angle TRD Electron iden.	fibre radiator straw chambers	15.7 k	≈ 0.1
EM calorimeter Electron iden.	W/Pb scint. shashlik	5.8 k	≈ 0.2
Muon system Muon iden.	gas pad + pixel prop. tubes	31.3 k	< 0.01

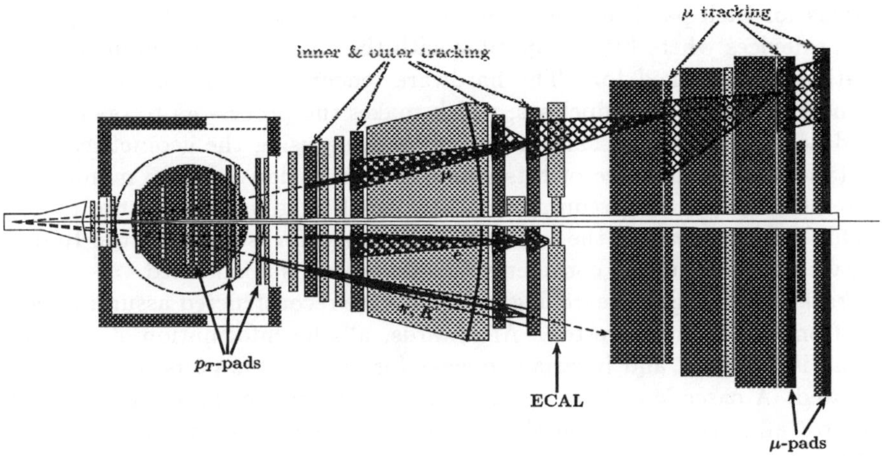

Figure 8. Track search scheme in the FLT. Detector components used in the track seeding and the track search are pointed out explicitely. The hatched areas indicate the search areas used in the particle finding procedure.

4.2 The trigger system

The trigger system of HERA-B consists of three levels with a rate reduction of 200, 25 and 20, respectively. The latency is of 12 μs for the First Level Trigger (FLT), 5-50 ms for the Second Level Trigger (SLT) and 100 ms for the Level Three (L3). The complete reconstruction of events that pass all levels is done at Level Four (L4) with a latency of 2 s, which does not aim at further output data reduction, but rather on classification of events for further analysis.

All processing in the FLT and SLT is based on Regions of Interest (ROI). The initial ROIs are established in the first step of the FLT and are then successively refined trhough the first two trigger levels. The definition of L3 is not completed. It will likely be based on a refit of all data related to a track found in the ROI searches plus the usage on non-ROI detector data like partial pattern recognition in the vertex detector.

The FLT algorithm searches for two lepton track candidates with an invariant mass compatible to that of a J/ψ or two high-p_T hadrons compatible of originating from a B meson. Figure 8 shows how this process works. Track seeds are originated by the μ-pad chambers, the calorimeter or the high-p_T chambers. Specific hardware processors (called pretriggers) exist to produce the seeds for each particle kind. Each track seed produces a ROI to specific tracking devices where hits compatible with the track parameters under consideration are searched for. The hardware tenchnique used in this process is asynchronous parallel pipelines, which makes the processors to operate on local data. The ROIs are shrunk during tracking using the geometrical limits obtained from the detector hits found and limits derived from momentum requirements under the assumption that a track originated from the target. If no hit compatible with the track being searched for is found in the ROI of the tracking device under consideration, the search is abandoned.

Tracks surviving the searching procedure are reconstructed assuming they come from the primary vertex. Afterwards, all the information of a bunch crossing is collected and invariant masses for all possible pairs of tracks are calculated. A cascade of look-up tables is used to derive the trigger decision on almost all information availabre for the two tracks; *e.g.* mass cuts can be p_T dependent. Also, counters of particle types are maintained. In this way a trigger can be generated if a certain number of high-p_T electrons, muons or hadrons is found.

The possibility of having reconstructed J/ψ mesons on FLT allows for an efficiency for CP-violation in the decay channel $B^0 \rightarrow J/\psi\, K^0_S$ of 50% at a minimum biass reduction level of 1000. The mass resolution in the muonic decay mode of the J/ψ is $100\,\mathrm{MeV/c^2}$.

The SLT makes a refinement of the tracks reconstructed at FLT using all the tracking devices and looking for secondary verteces in the Silicon vertex detector. The demands on processing power are satisfied by general purpose processors (partially Digital Signal Processors -SHARCS-) whith low overhead switching [22] which provides the high I/O bandwith into the processing nodes.

The L3 and L4 are no longer limited by data transfer. They are based in processor farms that execute the appropiate software codes.

5 Running since 1996.

Since 1996, prototypes of all the components of HERA-B were installed in the experimental area and data was taken at the same conditions that the detector will run in its final design *i.e.* with target running up to 40 MHz.

Figure 9 shows the installation, which consists on:

- Vertex Detector System(VDS)-vessel with 3 double sided Si-strip detector with 6000 readout channels and 8 target wires.

- Aluminum beam pipe of 0.5 mm.

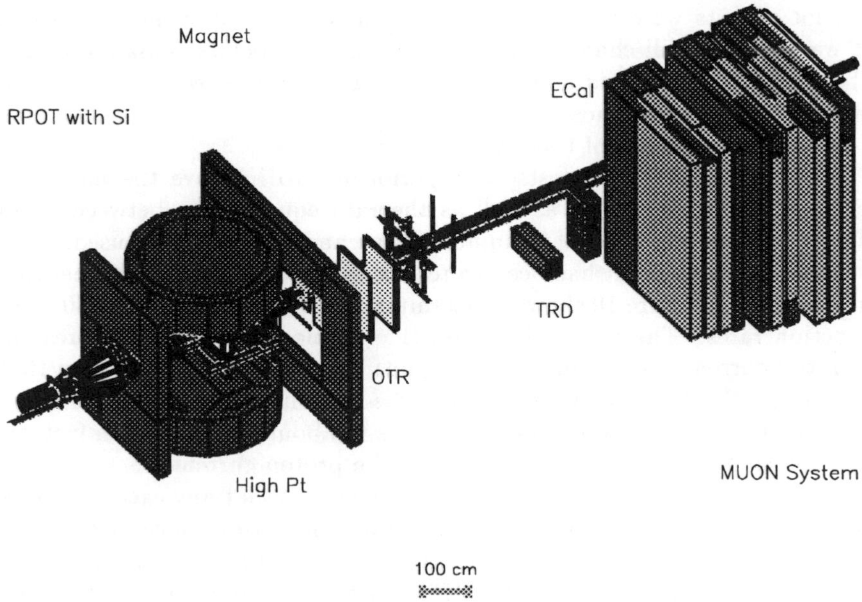

Figure 9. Scheme of the detector installation during the 1996 and 1997 running periods.

- Magnet and electron compensation coil.
- High-P_T gas pixel chambers.
- Micro Strip Gas Chambers (MSGC) with GEM foil (to divide the gas amplification in two zones) and scintillating fibre test modules.
- Three superlayers of the Outer TRacker (OTR), consisting of honeycomb drift chambers.
- A small piece of the calorimeter with 320 electronic channels.
- A module of the Transition Radiation Detector (TRD).
- Half of the muon absorber with 24 tube and 7 pad chambers.
- A small prototype of the RICH.

A lot of tests were performed with the installed components, something that was used to still change the design parameters before mass-production of the whole system starts and to get running experience with the detectors in the HERA-B environment.

Three major groups of tests were made:

Target tests: A lot of tests were performed to improve the automatic steering of the target wires as well as the rate equalization between them. Different materials (Carbon, Aluminum, Lead and Copper) were used.

One important point has been to test the impact of the target operation on the machine. Figure 10 shows an example of a target run at high (30 MHz) interaction rates. The interaction rate, the position of the target wires and the proton currents on the machine are plotted *vs.* time. It can be seen that, when the HERA-B target is moved into the proton beam, the proton beam lifetime is reduced from a few hundred hours to about fifty hours. In fact, the target consumes between 1 and 2 mA/h of the proton current.

A successful operation of the target requires also that any excessive backgrounds are produced which could deteriorate the performance of the other HERA experiments. In order to study this, the background rates of the experiments H1 [23], ZEUS [24] and HERMES [25] were monitored during the target test runs. In fig. 10 it is also shown the background rates measured by these two experiments. A clear correlation with the target rate is visible. However, the additional background created by the target wires was a factor of 3-5 below the limits above which the performance of the detectors will be seriously affected.

Standalone subdetector tests: At the same time the target was running, all the components installed during the last years were also performing crucial measurements in what will be the environment in which the full components will have to run. Even if a lot of the subdetectors were already tested in test-beams, the test in the HERA-B environment, using prototypes of the final electronics and the readout system, served to debug major parts of the components and allowed to proceed in the mass-production of the complete system.

The results of these tests can be summarized as follows:

- Radiation tests for all the subcomponents: Even if these tests were already performed in other environments, such as X-rays or any other irradiation facilities, it was for the first time that prototypes of the subdetectors were receiving the flux of particles that is expected in normal running conditions, something that will imply large occupancies in the detectors as can be seen in fig. 7 and is listed in tab. 2. Specially impor-

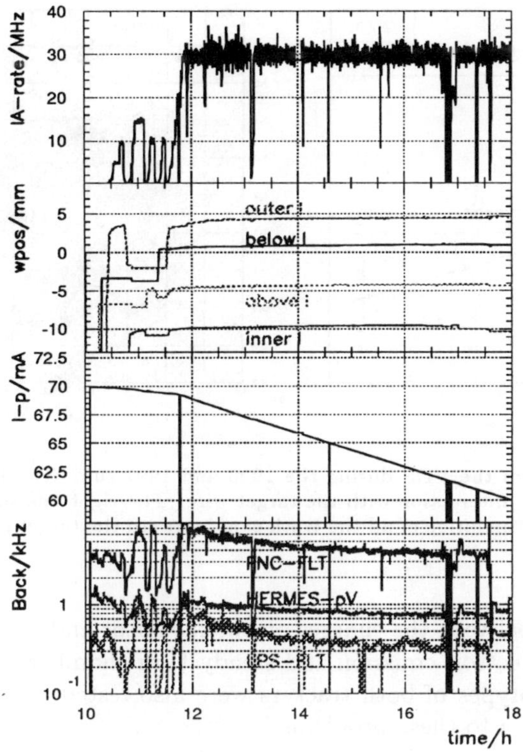

Figure 10. Example of a target run at high rates. From top to bottom: interaction rate *vs.* time, position of the wires *vs.* time, proton currents in HERA *vs.* time and background rates to the other HERA experiments *vs.* time.

tant were the studies performed with both trackers: MSGC and OTR, since the first was showing sparks caused by heavily inonizing particles and the second one was showing Malter currents after some irradiation. This later effect is partly connected to insulating layers developing on the chamber walls (which consist of folded polycarbonate foil). These insulating layers charge up during high rate operation, leading to strong

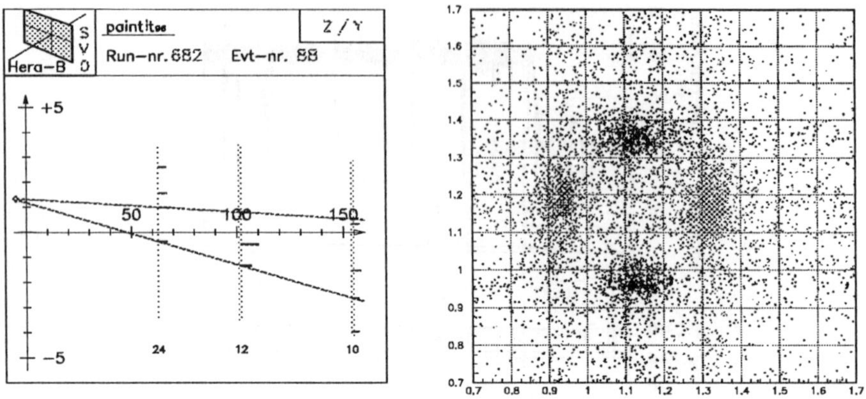

Figure 11. Results from the VDS during the 1996 and 1997 run. The left figure shows the event display of an interaction with the target wire. The right one shows the the x-y coordinates of the extrapolation of all the track candidates to the target plane.

electric fields between the charges on the insulator and the electrode and turn to field electron emission and steady high chamber currents. Some modified prototypes of both trackers were also tested during running to find the solution to these problems.

- Basic detector parameters tests: All subdetectors measured basic performance parameters, as position resolution, noise level, efficiencies,... These measurements served, in a lot of cases, to still change something in the production of the components or in the readout electronics.

  In particular, the three planes of the Silicon strips detectors installed in 1996 were used for tracking and pattern recognition studies, as well as estability of the detector efficiency. Figure 11 shows an event in which two tracks coming from the target wire cross the three detector planes. The lower track gives a hit in each detector plane. The x-y coordinates of the extrapolation of all the track candidates into the target plane are also shown. Four clusters can be identified corresponding to tracks originated in each of the four target wires in this plane. The track reconstruction resolution and the movement of the target wires contribute to the width of these clusters. The points outside the target wires are coming from conbinatoric background, since all tracks with, at least, two hits in the

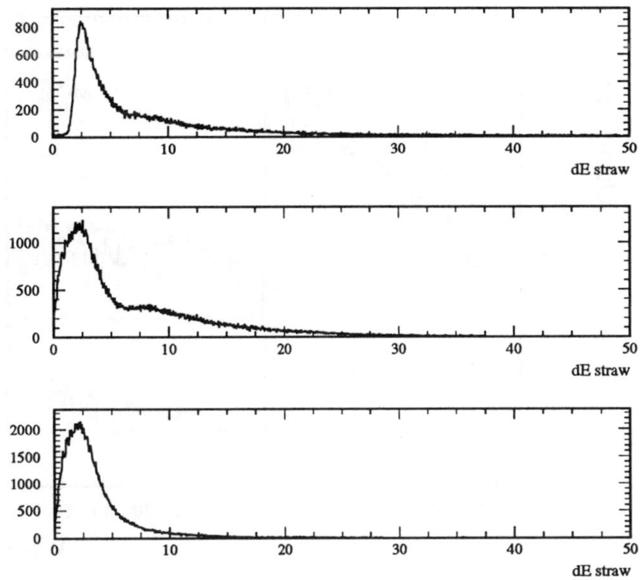

Figure 12. Observation of the transition detector signal from electrons: energy spectra in a straw tube measured at 10 MHz interaction rate (top) and Monte Carlo simulations from electrons (middle) and hadrons (bottom).

VDS are included in the plot.

Figure 12 shows, as another example, the observation of the transition detector signal measured by the TRD straw tubes. The data shows a larger tail in the energy spectra that is interpreted as the superposition of the two lower plots, coming from Monte Carlo simulations of electrons and hadrons. The one from electrons shows a peak a $\sim$8 KeV which correspond to the detection of photons coming from the transition radiator material.

- Measurement of occupancies: Occupancies up to 40 MHz inteaction rate were measured with all subcomponents and compared with Monte Carlo predictions. In all cases, results agree on the 10-20% level. Given the details of the Monte Carlo simulation used and the early stage of the experiment, this is a very promising result which allows to quote the

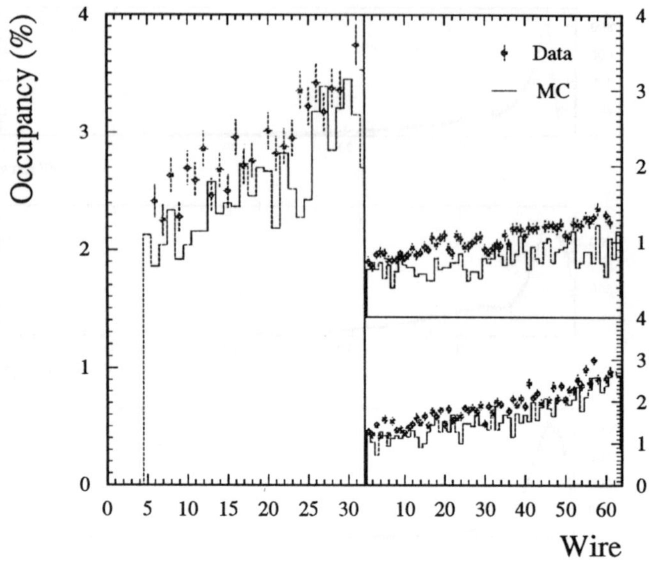

Figure 13. Occupancy measured in the three sectors of the OTR and comparison with Monte Carlo predictions. The target interaction rate was 10 MHz.

degree of accuracy of the general detector simulation.

Figure 13 gives an example for a section of the OTR. The modules were subdivided in two different granularities (from left to right) and in different sections (top and bottom) for the section closer to the beam pipe. The measured occupancies for each wire and the comparison with Monte Carlo simulations for a run in which the target interaction rate was 10 MHz are displayed. The agreement is in the order of 10%. At 40 MHz, the hotest wire was measuring an occupancy of $\sim$ 15%, which corresponds to the value in the hotest region given by the general simulation of the complete detector.

Common DAQ readout tests: From July 1997, all subdetectors installed were synchronized and readout in common. This was the first time the DAQ system was completely tested with real data. A lot of pieces of the readout, as the Fast Control System which distributes the synchronization and trigger signals to the front end electronics of all subdetectors, the SHARC boards,

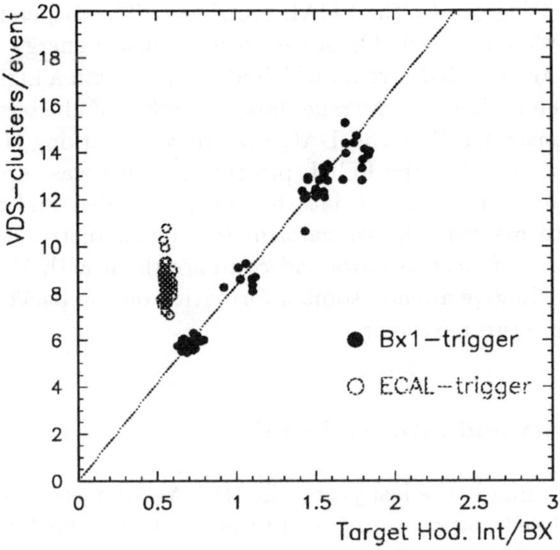

Figure 14. Number of cluster per event measured in the VDS *vs.* number of interactions per bunch crossing for two types of triggers.

which serve as Second Level Buffer, the SHARC-PCI bus interface, which serves to tranfer the data to the SLT farm, a prototype of the SLT switch as well as software pieces, as the run control and monitoring tasks, were tested for the first time. Since then, the system has been evolving aproaching to its final design.

This test, more than one year before the experiment will start to take data to measure CP violation, allowed the detector to develop a lot of monitoring tools which will serve in the future. The data taken was used to test the reconstruction chain and test the offline software, which, in the future, will have to run in the L4 farm in online mode.

A prototype board of the calorimeter pretrigger was also commisioned and operated. Figure 14 shows the number of clusters per event measured in the VDS as function of the number of interaction per bunch crossing estimated from the target monitoring hodoscopes. The points tagged as "BX1-trigger" are coming from minimum biass events. The ones tagged as "ECAL-trigger" (only available at one value of number of interactions per bunch crossing) come from events triggered by the ECAL pretrigger. The dispersion corresponds to runs with different thresholds of the electon cluster energy in the calorimeter pretrigger. As expected, events which already contain a high energy cluster in the calorimeter, show, on average, larger number of clusters in the VDS. The result demonstrates that the DAQ system was working, the detectors were syncrhonized and that the ECAL pretrigger board was running correctly.

Substantial ammount of data has been recorded and analysis are being performed to measure the p_T calorimeter cluster distribution as well as the invariant mass of clusters pairs and compare them with Monte Carlo simulations. This will give already some information on the background expected in the detector setup next year.

6 Summary and future schedule

After the running in the last years, the HERA-B detector has the goal of being fullly completed and commissioned to be ready for the 1999 physics running period.

It is clear that the measurement of CP violation is very challenging and this lecture explained the large particle occupancies that the detector subcomponents have to stand as well as the demands on the electronics, trigger and DAQ system to perform such a measurement.

In contrast to colliding experiments, HERA-B has the advantage that can be staged quite easily and pieces of the detector can be installed, commisioned and operated as they become ready. During the last running years, small parts of all subcomponents were already tested in the real environment, with target interaction rates up to more than 40 MHz.

The tests performed so far allewed to finsih the R&D that fixes all detector design details. They also allowed to develop several hardware, software and monitoring tools.

Production, assembly and commissioning of all detector components will happen in the next months with the goal to start the CP violation physics measurement not later than end of the millennium.

Acknowledgments

I would like to thank B. Adeva for the invitation to the school and for the good organization which allowed to create a pleasant atmosphere where students were participating quite actively.

References

1. J. Bernabeu, *these proceedings.*
2. L.L. Chau and W.Y. Keung, Phys. Rev. Lett. **53**, 1802 (1984).
3. C. Jarlskog and R. Stora, Phys. Lett. **B208**, 268 (1988).
4. H. Albrecht *et al.*, *HERA-B Letter of Intent* DESY-PRC 92/04 (1992).
5. T. Lohse *et al.*, *HERA-B Proposal* DESY-PRC 94/02 (1994).
6. E. Hartouni *et al.*, *HERA-B Design Report* DESY-PRC 95/01 (1995).
7. H. Albrecht *et al.*, ARGUS Collaboration, Z. Phys. **C55**, 357 (1992).
8. M. Artuso *et al.*, CLEO Collaboration, Phys. Rev. Lett. **71**, 1680 (1993).
9. R.M. Barnett *et al.*, Phys. Rev. **D54**, 1 (1996).
10. P. Bordalo *et al.*, Z. Phys. **C39**, 7 (1988).
11. M.G. Catanesi *et al.*, Phys. Lett. **B231**, 328 (1989).
12. R. Jesik *et al.*, *Int. Conf. on High Energy Physics, Dallas, 1992.*
13. K. Kodama *et al.*, Phys. Lett. **B305**, 359 (1993); Phys. Lett. **B305**, 961 (1992).
14. C. Albjar *et al.*, Phys. Lett. **B256**, 121 (1991).
15. M. Schub, *28th Rencontre de Moriond, Les Arcs, France (1994).*
16. F. Abe, Phys. Rev. Lett **71**, 500 (1993); Phys. Rev. Lett **71**, 2396 (1993).
17. K.T. Knöpfle, Nucl. Inst. and Meth. **A368**, 192 (1995).
18. T. Hamacher, Nucl. Inst. and Meth. **A368**, 234 (1995).
19. B. Dolgoshein, Nucl. Inst. and Meth. **A368**, 239 (1995).
20. P. Krizan *et al.*, Nucl. Inst. and Meth. **A351**, 111 (1994).
21. C. Hast *et al.*, Nucl. Inst. and Meth. **A324**, 224 (1994).
22. M. Medinnis, Nucl. Inst. and Meth. **A368**, 161 (1995).
23. I. Abt *et al.*, H1 Collaboration, Nucl. Inst. and Meth. **A386**, 36 (1996).
24. M. Derrick *et al.*, ZEUS Collaboration, Z. Phys. **C63**, 391 (1994).
25. D.P. Barber *et al.*, HERMES Collaboration, Nucl. Inst. and Meth. **A338**, 166 (1994).

HIGGS SEARCH AT LEP

J. CUEVAS

Dpto. de Física, Universidad de Oviedo, Calvo Sotelo S/N, 33007- Oviedo, SPAIN
E-mail: Javier.Cuevas@cern.ch

This talk describes a search for the Higgs boson using data from e^+e^- collisions collected at center-of-mass up to an energy of 183 GeV by the LEP detectors. Two processes $e^+e^- \to hZ$, and $e^+e^- \to hA$ are mainly used. All relevant final states, hl^+l^-, $h\nu\bar{\nu}$, $hq\bar{q}$, $\tau^+\tau^- q\bar{q}$, and $b\bar{b}b\bar{b}$ were analyzed. No discovery has been performed at LEP, but the limit on the Higgs mass is being significantly improved. These data allows also to set new limits in the parameter space of the MSSM.

1 Introduction

The Standard Model is very successful in describing the interaction of elementary particles. However the mechanism which originates particle masses remain uncertain. Such a mechanism will be characterized by the observation of at least a scalar particle. Whether this object is elementary (Standard Model or Minimal Supersymmetric extension of the Standard Model) composite (as predicted in the Technicolor scenario) or too heavy to be observed as a particle remains an open question. In the Higgs mechanism, corresponding to the first scenario, $W^\pm$ and Z^0 gauge bosons acquire mass via spontaneous symmetry breaking.

In the first part the theoretical and experimental arguments in favor of this mechanism and the predicted properties of the neutral Higgs boson(s) within the SM and MSSM will be discussed. In the second part the strategy for the neutral Higgs boson searches at LEP2 will be reviewed. Finally prospects for higher energies at LEP will be presented.

2 The SM Higgs

In this scenario one assumes that there is a single scalar complex iso-doublet Φ which gives masses to $W^\pm$ and Z^0 bosons (and leaves the photon massless) through the Higgs mechanism. The masses are given in terms of the v.e.v. of Φ and of the gauge couplings. Three components of Φ create the longitudinal components of $W^\pm$ and Z^0 while the last degree of freedom is identified to the Higgs boson. The fermion masses (with the exception of the neutrinos) are generated through the Yukawa couplings. The Higgs boson then couples to the fermions proportionally to their masses. This fact is crucial in the experimental searches. The mass of the Higgs boson is given by $m_H^2 = 2\lambda v^2$ where v is the v.e.v. of Φ simply related to the W boson mass and λ the quartic dimensionless coupling which has to be positive for the scalar potential of Φ to be bounded from below. λ is not explicitly known but can be severely constrained imposing the condition of vacuum stability and the requirement that the SM stays valid, with no new physics up to a high scale [1]. This sets a lower limit to $m_h \approx 130\ GeV$ which can be relaxed if new physics appears to an energy scale well below the Planck scale.

3 MSSM Higgs sector

In the minimal supersymmetric extension of the Standard Model (MSSM), the Higgs sector consists of five physical states: two scalars, h^0, H^0 whose mixing involves an angle α, a pseudoscalar A^0 and two charged bosons $H^{\pm}$. At tree level, two parameters, for instance m_A and $tan\beta = \frac{v_1}{v_2}$ (where v_1 =v.e.v. of the doublet giving mass to up-quarks, v_2 =v.e.v of the doublet giving mass to down-quarks and leptons) are enough to describe the Higgs sector.

At loop level and focusing on h^0, its tree level mass is increased by radiative corrections and reads:

$$m_{h^0}^2 = m_z^2 cos^2 2\beta + 3(m_t^2/4\pi^2 v^2) ln(m_{\tilde{t}_1} m_{\tilde{t}_2}/m_t^2) + \tilde{A}_t^2 F(m_{\tilde{t}_1}, m_{\tilde{t}_2}, \tilde{A}_t) \quad (1)$$

where $m_{\tilde{t}_1}, m_{\tilde{t}_2}$ are obtained by diagonalizing the stop mass matrix and $\tilde{A}_t$ is called the mixing parameter. The increment depends on the 4th power of m_t, hence the importance of its accurate knowledge, and logarithmically on the stop masses $m_{\tilde{t}_1}, m_{\tilde{t}_2}$. m_{h^0} and the Higgs sector depends on the other parameters of the MSSM through the mixing parameter $\tilde{A}_t$.

One important point is to consider the complementarity between the couplings $h^0 Z^0 Z^0 \propto \sin(\alpha - \beta)$ and $h^0 A^0 Z^0 \propto \cos(\alpha - \beta)$. An interesting case is when m_{A^0} is large, the h^0 mass is unaffected and stays light, while all other Higgs bosons become mass degenerate with A^0, h^0 is Standard Model like and $\sin(\alpha - \beta) = 1$. In this framework the upper bound on the Higgs mass, initially m_Z has to be lower than $\approx 125\ GeV$ for any $tan\beta$ and stop mixing [2].

4 Higgs mass from precision measurements

There is an indirect source of information on the Higgs mass coming from precise measurements performed at LEP, SLC, and FNAL.

The SM allows the computation of loop corrections which can be precisely measured from various independent parameters at LEP/SLC: m_Z, $\sigma_{f\bar{f}}$, $\Gamma_{f\bar{f}}$, A^f_{FB}, A^e_{LR}, P_τ, m_W, and σ_{WW}. The main correction comes from the running of α up to the Z mass (see figure 1). The second main correction comes from loops involving the top quark and depends on m_t^2. The top mass is now fairly well known, $m_t = 174.1 \pm 5\ GeV$ and the correction can be precisely computed so that one can separate the effect from the Higgs contribution. The loop effect depends on $log(m_H)$ then providing a rather imprecise determination on that parameter. Nevertheless with the more precise experimental results and internal consistency between different measurements and with theoretical calculations meaningful for Higgs masses up to 1 TeV [3] an upper limit on the Higgs mass can safely be given [4] in the SM:

$$m_{h^0} < 280\, GeV \quad at\ 95\%\ CL \tag{2}$$

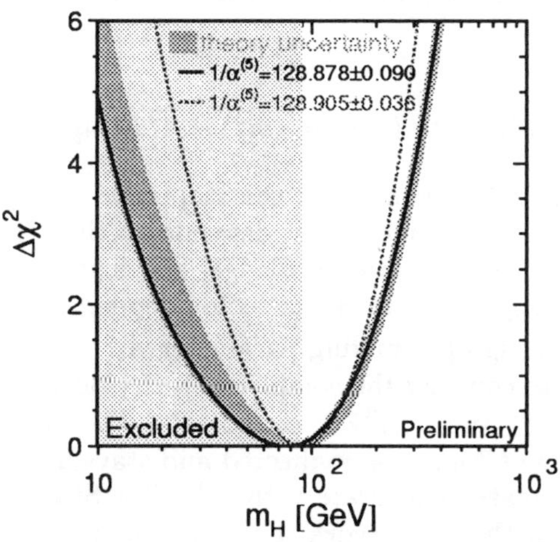

Figure 1. χ^2 distribution obtained from precision measurements at LEP, SLC and FNAL

In conclusion one can say that precision measurements put severe restrictions on the mass domain allowed for the Higgs boson within the Standard

Model. If new physics appears at finite mass scales, this domain can be anyway extended [5].

5 Higgs boson searches at LEP

5.1 Higgs production

In the Standard Model at the energies relevant at LEP 2, the Higgs boson is mainly produced through the Higgs-strahlung process, $e^+e^- \to HZ$. It should be noted that there is a sharp energy threshold for Higgs production in e^+e^- colliders (see figure 2) and there is no possible compensation for this lack of energy by increasing luminosity as in a hadron collider. Other contributions from W^+W^- and Z^0Z^0 fusion processes account for a small part of the total production.

In the MSSM, the h^0 and A^0 bosons are expected to be produced predominantly via two processes, the Higgs-strahlung process $e^+e^- \to h^0Z^0$ (as for SM), and the ***pair production*** process $e^+e^- \to h^0A^0$. For these two processes, the cross-sections σ_{hZ} and σ_{hA} are related at tree level to the SM cross-sections [6]

$$e^+e^- \to h^0Z^0: \quad \sigma_{hZ} = \sin^2(\beta-\alpha)\,\sigma_{hZ}^{SM} \tag{3}$$

$$e^+e^- \to h^0A^0: \quad \sigma_{hA} = \cos^2(\beta-\alpha)\,\bar{\lambda}\,\sigma_{\nu\bar{\nu}}^{SM} \tag{4}$$

where σ_{HZ}^{SM} and $\sigma_{\nu\bar{\nu}}^{SM}$ are the cross-sections for the corresponding SM processes: $e^+e^- \to H_{SM}Z$ and $e^+e^- \to \nu\bar{\nu}$, and $\bar{\lambda}$ is a kinematic factor depending on m_{h^0} and m_{A^0} typically having values between 0.5 and 0.7 for the center-of-mass energies under consideration.

5.2 Higgs decays

In the SM, the Higgs boson decays predominantly into b quarks when $m_H <$ 130 GeV. Above this mass, couplings to W and Z pairs become dominant. At LEP200 energies H decays in to $b\bar{b}$ in $\approx 85\%$ of the time. In MSSM this decay is also dominant, unless $h \to 2A$ becomes kinematically accessible.

A Higgs boson may also decay invisibly. In the SUSY scenario it can go into neutralino, gravitino or sneutrino.

6 Experimental Search

The lower limit on the SM Higgs boson up to $\sqrt{s} = 172\,GeV$ using the data taken up to this energy (current PDG [7] limit) is:

$$m_H > 77.5\,GeV \quad at\ 95\%\ CL \tag{5}$$

A center-of-mass energy of 183 GeV was reached in 1997, and about 55 pb^{-1} were collected in each experiment, ($57pb^{-1}$ in Aleph [8], $54pb^{-1}$ in Delphi [9], $54pb^{-1}$ in Opal [10], and $57pb^{-1}$ in L3 [11])

The process $e^+e^- \to A^0h^0$ which appears in MSSM (also in more general two Higgs Doublet Models) is searched for in the final states $A^0h^0 \to b\bar{b}b\bar{b}$ and $A^0h^0 \to b\bar{b}\tau^+\tau^-$. The process $h^0A^0 \to A^0A^0A^0 \to b\bar{b}b\bar{b}b\bar{b}$ has also been searched for.

The experimental situation at LEP200 is well known. Primordial backgrounds are not much larger than the signal (see fig 2). Furthermore W do not decay appreciably into b. The $q\bar{q}$ component is dominated by radiative return to the Z resonance. This type of effect has a definite pattern and can be removed. On the other hand the Z^0Z^0 final state when one Z^0 goes to $b\bar{b}$ and for $m_h \approx m_Z$, the situation presently explored, is an irreducible background.

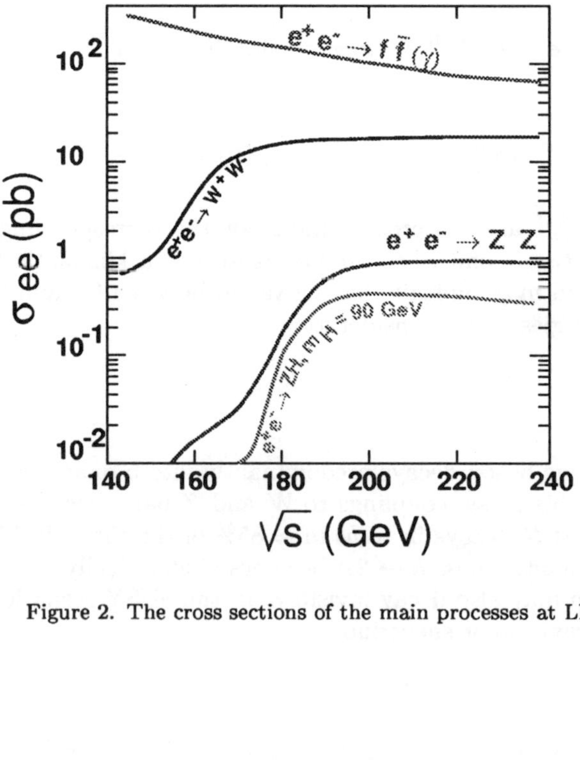

Figure 2. The cross sections of the main processes at LEP200

B-tagging is very useful and very powerful. Since Higgs bosons decays preferentially to $b\bar{b}$ pairs, the tagging of jets originating from b-quarks plays a crucial role in Higgs boson searches. Jet-wise b-tagging algorithms has been developed for all the four LEP collaborations using the main distinctive features of the b-quark decays, its relatively long lifetime, high decay multiplicity, and high mass of the b-flavored hadrons. Semileptonic b-decays are identified using electrons and muons of high transverse momentum with respect the jet directions. This information is usually combined using Artificial Neural Networks or likelihood methods into a single discriminant variable per jet. This jet-wise variable is the one used in the search for the Higgs boson in the various channels. This combined b-tagging provides a substantial gain in rejection against W^+W^- pairs as compared to the b-tagging using only impact parameter information [9].

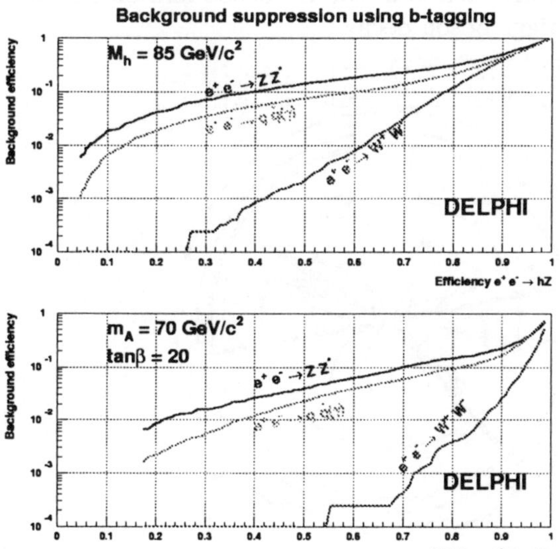

Figure 3. b-tag rejection of the various backgrounds versus signal efficiency.

An illustration of the b-tag rejection of the various backgrounds versus signal efficiency in DELPHI [9] is shown in figure 3. As an example an efficiency of 60% to the h^0Z^0 signal in 4-jets can be kept while the W-pair background is rejected by a factor ≈ 100. In all channels (except $H\nu\bar{\nu}$) a constrained fit is performed to extract the Higgs mass and often to reject background processes as well.

6.1 SM Higgs: The Four Jet Channel

The process $e^+e^- \to Z^0H^0 \to q\bar{q}b\bar{b}$ is not as distinctive as the leptonic and missing energy final states, but it accounts for approximately $60-65\%$ of the SM Higgs boson production cross-section. It is characterized by four energetic hadronic jets, large visible energy and the presence of b-hadron decays. The main backgrounds as mentioned above are $(Z/\gamma^*) \to q\bar{q}$ and four-fermion processes W^+W^- and Z^0Z^0. The suppression of these backgrounds relies on the kinematic reconstruction of the Z^0 boson and on the identification of b-quarks from the Higgs boson decay.

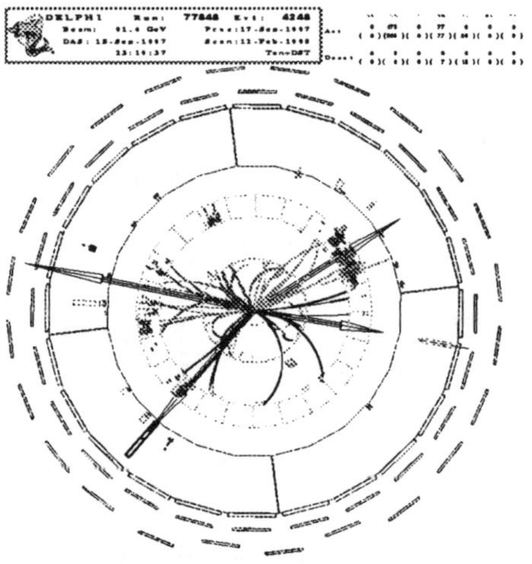

Figure 4. Transverse view of a DELPHI $Hq\bar{q}$ candidate.

Four-jet preselection is done, aimed to reduce the QCD background while keeping most of the Higgs boson signal. Typically $85-90\%$ is kept for the Higgs signal after this step while the QCD background is reduced by a factor approximately 10. After preselection, cut based, neural network, or likelihood analyses are performed [8 9 10 11].

At $\sqrt{s} = 183$ GeV, typical efficiencies of $\approx 40\%$ are obtained for a Higgs boson mass of 80 to 85 GeV. The expected remaining background being dominated by QCD and Z^0Z^0 processes (≈ 4 events for the luminosity collected at this center-of-mass energy). A transverse view of a candidate in DELPHI is shown in figure 4.

6.2 SM Higgs: The missing energy final state

The $e^+e^- \to \nu\bar{\nu}b\bar{b}$ process accounts for approximately 18% of the SM Higgs boson production cross-section with a small contribution (1.3% relative for $m_H = 85\ GeV$) from the W^+W^- fusion process. A large missing mass and two energetic, acoplanar b-flavored jets characterize this final state. The dominant backgrounds are miss-measured $(Z/\gamma^*) \to q\bar{q}$ events, four-fermion processes with neutrinos in the final state, such as $Z^0Z^{0*} \to \nu\bar{\nu}q\bar{q}$, $W^+W^- \to l^\pm\nu qq'$, $W^\pm e^\mp\nu \to qq'e^\mp\nu$ with the charged lepton escaping detection and, in general events in which particles go undetected down to the beam pipe. These backgrounds are largely reduced via b-tagging. The process $Z^0Z^{0*} \to \nu\bar{\nu}b\bar{b}$ remains an irreducible background. OPAL [10] finds for this channel an efficiency of $\approx$ 40% for m_H = 85 GeV. The total expected background is 1.53 ± 0.13 events, 0.29 events coming from QCD, and 1.27 from 4-fermion processes, and no event remains at this level.

6.3 SM Higgs: The leptonic final states

The Hl^+l^- ($l = e$ or μ) channel represents 6.7% of the Higgs-strahlung cross-section. The signal events are characterized by two leptons with an invariant mass close to m_Z and a recoil mass equal to the Higgs boson mass. Although the branching ratio of this channel is small, the experimental signature is very clear and the Higgs boson mass can be reconstructed with high resolution.

Identification of electrons and muons and b-tagging are the more important points in the separation of the signal which respect the dominant backgrounds. Typical efficiencies of $\approx 60\%$ are obtained for both $H\mu^+\mu^-$ and He^+e^- channels.

6.4 SM Higgs: The final states with τ leptons

The $q\bar{q}\tau^+\tau^-$ final state accounts for $\approx$ 9% of the total Z^0H^0 production rate. The process $HZ \to H\tau^+\tau^-$ corresponds to a 3.4% of the total Higgs-strahlung process, and $H \to \tau^+\tau^-$, $Z \to q\bar{q}$ corresponds to an additional 5.5%. The events in this channel are expected to have two oppositely charged low multiplicity jets in association with missing energy from the τ decays. The dominant backgrounds are the four-fermion processes $Z^0Z^{0*} \rightarrow q\bar{q}l^+l^-$ and $W^+W^- \rightarrow qq'l\nu$. The process $Z^0Z^{0*} \rightarrow b\bar{b}\tau^+\tau^-$ is an irreducible background. L3 [11] has obtained an efficiency varying between 15 and 20% for Higgs mass between 75 and 90 GeV. 0.7 background events are expected in the $H \rightarrow q\bar{q}, Z \rightarrow \tau^+\tau^-$ and 1.7 in the $H \rightarrow \tau^+\tau^-, Z \rightarrow q\bar{q}$, 1 event is left in data at this level.

6.5 *MSSM Higgs: The Four b Channel*

The signature for events from the process $A^0h^0 \rightarrow b\bar{b}b\bar{b}$ is a four jet topology with a high b-quark content. The dominant background process are $(Z/\gamma^*) \rightarrow q\bar{q}gg$ and Z^0Z^0 and W^+W^- production. Typical efficiencies of $\approx 55 - 60\%$ were obtained with Artificial Neural Networks, cut-based or likelihood analysis for $m_h = 75\ GeV$ and $tg\beta = 20$ at $\sqrt{s} = 183\ GeV$. For instance the ALEPH collaboration [8] has obtained an efficiency of 60.5%, with a background of 2.4 ± 0.1 events, 1.1 from $q\bar{q}$, 1.0 from ZZ and 0.3 from WW processes. Two events were selected in the data.

6.6 *MSSM Higgs: The $\tau^+\tau^-b\bar{b}$ final state*

The $h^0A^0 \rightarrow b\bar{b}\tau^+\tau^-$ final state where either h^0 or A^0 decays into the τ pair is topologically similar to the $H^0Z^0 \rightarrow q\bar{q}\tau^+\tau^-$ final state previously described, the main difference being the loss of the Z^0 mass constraint. Events in this final state are characterized by a pair of energetic and isolated narrow jets (originating from the τ decays) and a pair of energetic hadronic jets, accompanied by missing energy and transverse momentum. For $m_h = m_A = 75\ GeV$ ALEPH [8] obtains a signal efficiency of 28.6% corresponding to 0.27 signal events expected, with 0.07 background events expected. No events were selected in the data.

7 Global efficiencies and results

As there is no evidence for a signal, the results of the searches presented in the previous sections translate into exclusion limits in the masses of the neutral Higgs bosons in the SM and MSSM.

An important element to obtain the best limit is the optimal choice of the cut on the selective variable(s), the so-called working point, of the combination of the various channels, and, at the final stage of the four experiments.

For each analysis, the working point is selected using the expected confidence in the signal hypothesis if there is no signal, which is obtained using the simulation which takes into account all ZH or hA channels then avoiding any bias coming from the data.

Figure 5 shows the distribution of the reconstructed Higgs boson mass found in the HZ channel after all selection criteria for data and signal events. The mass limits expected and obtained at 95% level by each of the LEP experiments from their data up to 183 GeV are given in GeV, in table 1.

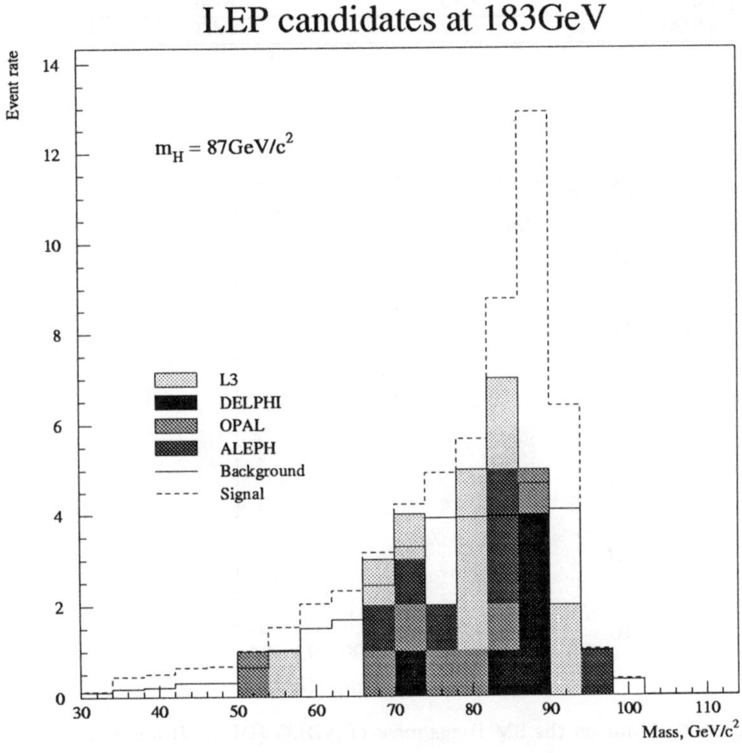

Figure 5. LEP200 Higgs candidates at center-of-mass energy of 183 GeV, and the expected signal of a 87 GeV Higgs boson.

The improvement on the limit from adding the lower energy data (161 and 172 GeV data) is approximately 0.1 GeV. Four methods are used to set up *frequentistic* confidence levels. Three of them are based on *gedanken* experiments and one on analytical calculation. They produce about equal performance and they are continuously compared.

Table 1. SM Higgs mass limits at $\sqrt{s} = 183$ GeV, obtained by each of the LEP experiments.

Experiment	expected mass (GeV)	obtained mass (GeV)
ALEPH	85.5	87.9
DELPHI	86.5	85.7
L3	85.0	87.6
OPAL	86.2	88.3

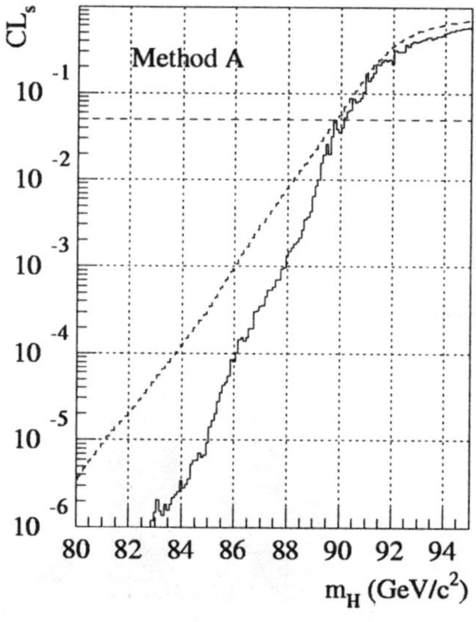

Figure 6. Limit on the SM Higgs mass of ADLO (DELPHI method).

The limits obtained by combining the results of the four LEP experiments with these four different statistical methods, are given in table 2

The lowest of the four limits, is the official exclusion of LEP [12]

$$m_H > 89.8\, GeV \quad at\ 95\%\ CL \tag{6}$$

while 90.4 GeV was expected. Figure 6 gives the limit on the SM Higgs mass obtained by method A (DELPHI's statistical method). The gain obtained by

Table 2. Mass limits at $\sqrt{s} = 183 GeV$, obtained by combining the results of the four LEP experiments with four different statistical methods.

	expected mass (GeV)	obtained mass (GeV)
Method A	90.0	90.1
Method B	89.9	90.1
Method C	90.4	89.8
Method D	90.5	90.1
Spread	$\pm$ 0.3	$\pm$ 0.15

combining the four experiments is +3 GeV with respect to the expected limit from individual experiments. This result can also be obtained on the basis of one ***lucky*** experiment, that is an experiment in which the background has fluctuated negatively yielding a better mass limit than expected.

In the case of SUSY Higgses one has to combine the results of the two production channels already mentioned hZ and hA. In this case one can try to exclude the MSSM scenario in terms of two parameters, e.g. m_h and $tg\ \beta$. The question is then to decide what to do with the other parameters like the squark top masses and the mixing parameter $\tilde{A}_t$ which intervene, at loop level. Following the recommendation of the LEP2 workshop [13], one usually assumes an average squark mass of 1 TeV and $\tilde{A}_t$ either small ***no mixing case***, or such that it maximizes the loop corrections ***maximal mixing***.

The corresponding results of the four LEP collaborations for their data up to 183 GeV are shown in figure 7. The figure shows the theoretically excluded regions, for the two extreme cases of mixing, as well as the expected and observed experimental exclusion contours. One sees that in the case of no mixing, a domain of low values of $tg\ \beta$ (between 0.8 and 2.1) is already excluded. For $tg\ \beta \geq 0.8$ LEP [12] excludes:

$$m_h > 78.8\, GeV, \quad m_A > 79.1\, GeV \tag{7}$$

The region below $tg\ \beta = 0.8$ is a difficult one. Close to the m_h lower bound m_A is small, possibly below the $b\bar{b}$ threshold, and the decay $h \rightarrow AA$ is open. The limit on m_h is weaker at large $tg\ \beta$. This effect is due to the hA channel.

To check whether a more general scan of the parameters under consideration can reveal ***weak*** points, which can lead to a reduction on the Higgs masses, a thorough scan of the parameter space (m_A, $tg\ \beta$, m_0, $m_{1/2}$, μ, A, ...) has been performed. The result obtained is that the limits reached are quite robust and that there are no significant changes except for very narrow windows of parameters.

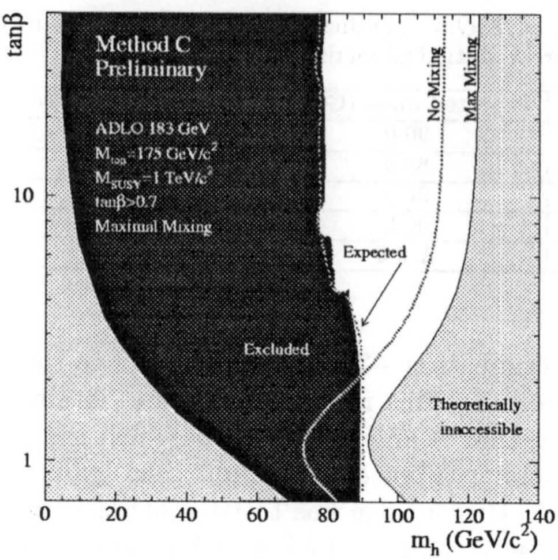

Figure 7. Limit on the MSSM Higgs masses reached by combining the four LEP experiments.

8 Other Higgs searches at LEP

Charged Higgs bosons in the MSSM should be heavier than m_W. Their pair production has been looked for by the four LEP collaborations in the modes $c\bar{s}$ and $\tau\nu$. The mass limits obtained [12] are given in table 3.

Other searches have also been performed, e.g. the invisibly decay of the h^0 into either a pair of LSP neutralinos or to a pair of majorons. The result on the mass limit of an invisible Higgs boson produced with a SM like cross section are of ≈ 80 GeV as obtained by each of the four LEP collaborations.

Table 3. $H^\pm$ mass limits at $\sqrt{s} = 183$ GeV, obtained by each of the LEP experiments.

	expected mass (GeV)	obtained mass (GeV)
ALEPH	57	59
DELPHI	56	56.6
L3	58.8	57.5
OPAL	62	59

9 Prospects for the future

Figure 8 shows the exclusion and discovery limits one can expect, as a function of the integrated luminosity per experiment, from the combination of their results [14], for two CM energies, 189 GeV (the one that has been reached during 1998) and 200 GeV which represents what one can ultimately hope for with the number of RF cavities available, and provided their mean accelerating field can be raised from their design value of 6 MeV/m to $\approx 6.8\ MeV/m$.

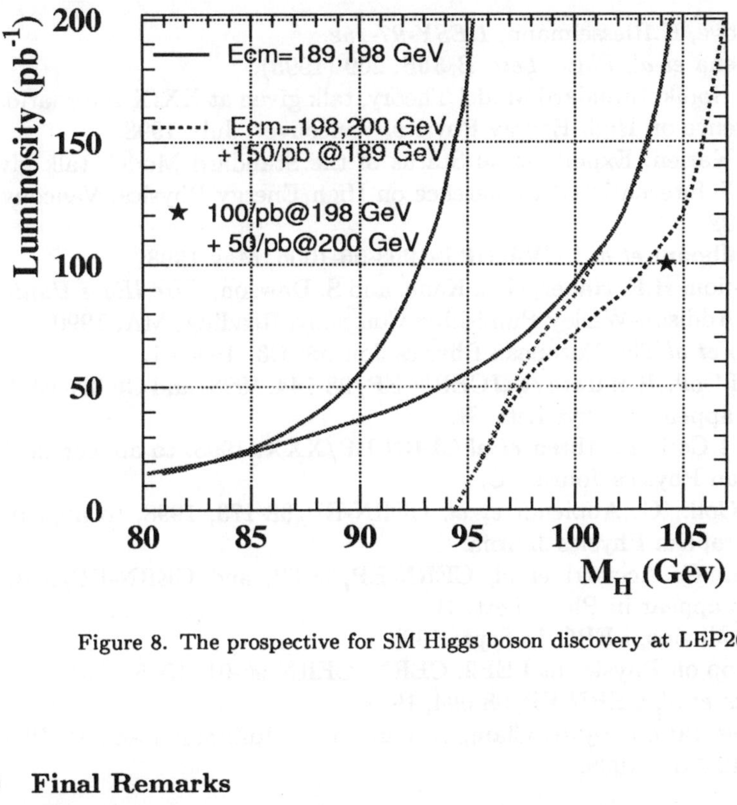

Figure 8. The prospective for SM Higgs boson discovery at LEP200.

10 Final Remarks

At the time of writng, more than 150 pb^{-1} have been collected by each of the four LEP experiments at $\sqrt{s} = 189\ GeV$. The preliminary results on the Higgs boson search have been presented at the LEP experiments committee [15]. No discovery has been reported. Individual mass limits on the SM Higgs of ≈ 94 GeV have been presented. In the MSSM, the preliminary limits, are $m_h > 82.4$ GeV and $m_A > 83.3$ GeV.

Acknowledgments

I thank the organizers of the meeting for their kind invitation. It has been a pleasure to attend this meeting. I thank all the colleagues from the LEP collaborations who provided me with information and explanations of their results.

References

1. T.Hambye, K.Riesselmann, *DESY-97-152.*
2. M. Carena *et al*, *Phys. Lett.* B **355**, 209 (1995).
3. See W. Hollik, Standard Model Theory, talk given at XXIX International COnference on High Energy Physics, Vancouver, July 1998.
4. See D. Karlen, Experimental Status of the Standard Model, talk given at XXIX International Conference on High Energy Physics, Vancouver, July 1998.
5. R. Casalbouni *et al* UGVA-DPT-1998-05-1006, May 1998.
6. J.F.Gunion, H.E. Haber, G.L. Kane, and S. Dawson, *THe Higgs Hunter's Guide*, Addison-Wesley Publishing Company, Reading, MA, 1990.
7. C. Caso *et al* The European Physics Journal C3 (1998) 1.
8. ALEPH Coll. R.Barate *et al* CERN-EP/98-144, 1998, and CERN-EP/98-145, to appear in Phys Lett. B.
9. DELPHI Coll. P. Abreu *et al* CERN-EP/XXX, 1998, to appear in The European Physics Journal C.
10. OPAL Coll., G. Abbiendi *et al*, CERN-EP/98-173, 1998, to appear in The European Physics Journal C.
11. L3 Coll., G. Acciarri *et al*, CERN-EP/98-52, and CERN-EP/98-072, 1998, to appear in Phys. Lett. B.
12. F. Di Lodovico, LEPC 15 Sept. 1998.
13. Workshop on Physics at LEP2, CERN. CERN 96-01, 1995.
14. E. Gross *et al*, CERN EP/98-094, 1998.
15. See presentations by R. Clare, E. Lancon, V. Ruhlmann and D. Plane, LEPC 12 Nov. 1998.

RECENT DEVELOPMENTS IN PHYSICS FAR BEYOND THE STANDARD MODEL *

L.E. IBÁÑEZ

Departamento de Física Teórica C-XI and Instituto de Física Teórica C-XVI, Universidad Autónoma de Madrid. Cantoblanco, 28049 Madrid, Spain

String theory is the leading candidate for a unified theory of the standard model and gravity. In the last few years theorists have realized that there is a unique structure underlying string theory. In this unification a prominent role is played by the duality symmetries which relate different theories. I present a brief overview of these developements and discuss their possible impact in low-energy physics. One of the lessons learned is that the string scale, usually asumed to be of order of the Planck scale, could be arbitrarily low, even close to accelerator energies.

1 Far Beyond the Standard Model (FBSM)

You have already heard many times that the Standard Model (SM) is fine and describes everything (except, maybe, neutrino masses) very well. And you have also heard almost the same number of times that in spite of that there are plenty of aspects of the SM that we do not yet understand. I am not going to repeat the well known list of questions without answers in the SM, it has not changed very much in the last decade. Instead of that I am going to talk today about one particular problem: the problem of combining the SM interactions with gravity. As is well known naive quantization of Einstein's gravitation leads to a non-renormalizable theory plagued with ultraviolate divergences. This problem has been with us for several decades and since fifteen years ago it is believed that string theory is the best candidate for a solution. In addition we need a theory in which chiral gauge interactions like those of the SM can coexist with quantum gravity. Also this aspect seems to be present in strings so let me first of all discuss some key ingredients of these theories.

2 String Theories

The origin of ultraviolate divergences in perturbative gravity (or any field theory for that matter) is related to the existence of interactions between particles which take place at a point. Insisting in pointlike interactions leads to

*CONTRIBUTION TO THE PROCEEDINGS OF THE XXVI INTERNATIONAL MEETING ON FUNDAMENTAL PHYSICS, LA TOJA, SPAIN (JUNE 1998).

ultraviolate divergences. Whereas in the field theories of the $SU(3)\times SU(2)\times U(1)$ standard model those divergences can be absorved in the renormalization process, that is not the case for gravity. String theories bypass this problem by avoiding point-like interactions and allowing the particles to interact in an extended region of space-time.

String theory [1] is a radical departure from the standard understanding of physics because asumes that elementary (pointlike) particles are not the fundamental blocks of nature. Rather the fundamental objects are strings, extended one-dimensional objects. Whereas the movement of a classical particle is described by giving the dependence of coordinates on e.g., proper time τ, $X^\mu(\tau)$, the movement of a string in space-time is described by giving $X^\mu(\tau,\sigma)$, where σ is the coordinate along the string. Strings maybe either open (with two boundary points X_a, X_b between which the string stretches) or closed ($X_a = X_b$). Interactions of strings take place by joining and splitting. Thus e.g., a closed string propagating in space can in a given moment split into two closed strings. In such a process of splitting the interaction takes place in an extended region in the variables (τ,σ) and this is at the basis of the absence of ultraviolate divergences in string theory.

An important property of string theory is that it predicts the existence of gravity. This comes about as follows. Let us consider the quantization of a closed string. One can consider this system as a set of an infinite number of harmonic oscilators with two type of vibration modes, right-handed (say) and left handed around the closed string. If a_n^μ and $\tilde{a}_n^\mu$ $(n = 0, \pm1, \pm2...)$ are the two type of oscilators one finds for the masses of the string modes:

$$m^2 \; = \; \frac{1}{\alpha'}(N \; + \tilde{N} - 2) \tag{1}$$

where $N = \sum_n a^\mu_{-n} a_n{}^\mu$ and $\tilde{N} = \sum_n \tilde{a}^\mu_{-n}\tilde{a}^\mu_n$ are the corresponding number operators. Here $\frac{1}{\alpha'}$ is the string tension and the -2 comes from the (regulated) zero point vacuum energy. There is an additional constraint (coming from the non-existence of a privileged point on the string coordinate σ) which imposses $N = \tilde{N}$ for the eigenvalues of physical states. Now, for $N = \tilde{N} = 1$ there is a *massless state with two vector indices* , given by $g^{\mu\nu} = a_1^\mu \tilde{a}_1^\nu|0>$. Thus we have a massless spin two state in the spectrum, which is no other but the graviton [a] . Unfortunatelly for $N = \tilde{N} = 0$ we have also a tachyonic state with mass^2 equal to $-2/\alpha'$. This is signaling an instability in this (purely

[a] More specifically, it is the symmetrized state which gives rise to the graviton. The anti-symmetrized state gives rise to an additional antisymmetric field $A^{\mu\nu}$ and the trace to a massless scalar, the dilaton ϕ .

bosonic) string. Furthermore this theory has no fermions in the spectrum and hence it cannot accomodate the observed quark and leptons.

Both problems are solved if we supersymmetrize the string variable $X^{\mu}(\tau,\sigma)$ and include fermionic coordinates $\psi^{\mu}(\tau,\sigma)$. There are two types of closed superstrings of this type which go under the names of Type IIA and Type IIB [2]. They differ in the structure of the supersymmetry in both theories. The pairs X^{μ}, ψ^{μ} may be considered themselves as fields in a two-dimensional field theory on the variables (τ,σ), the world-sheet. Both Type IIA and Type IIB have $N = 2$ supersymmetry on the worldsheet but the two SUSY generators have oposite chirality in the Type IIA case and the same chirality for Type IIB. In both cases one can show that, in a explicitely supersymmetric formalism the tachyon disappears from the massless spectrum. Furthermore they are finite theories in perturbation theory without any divergence, neither ultraviolate nor infrared. Thus they seem to provide us for the first time with consistent theories of quantum gravity. However they have a puzzling property: they are formulated in ten (one timelike and nine spacelike) dimensions. Extra dimensions are not necesarily a problem since, as already Kaluza and Klein showed in the twenties, they can be curled up into a compact space at very short distances in such a way that we only are able to see experimentally the standard four dimensions. In the case of Type II strings, six dimensions are assumed to be compactified at very short distances R in such a way that only if we have energies higher than $1/R$ we would be able to see the six extra dimensions. Consistency with known gravitational interactions suggest to identify :

$$1/R \propto \frac{1}{\sqrt{\alpha'}} \propto M_{Planck}. \tag{2}$$

The idea is that the SM particles (plus the graviton) correspond to string modes which are (aproximately) massless compared to such huge mass scales.

Even compactifying the six extra dimensions, Type II theories do not seem very promising (at least in their pre-1995 formulation) to include not only gravity but the standard model. We know that one of the most important properties of the SM is its chirality: left-handed and right-handed quarks and leptons transform differently under the SM group. Type IIA theory is non-chiral and does not seem possible to include a chiral theory like the SM inside it. Type IIB theory is on the other hand chiral but it turns out that it does not lead upon compactification to non-abelian chiral theories like $SU(3) \times SU(2) \times U(1)$, it leads at most to chiral $U(1)$ theories.

Before 1984 only a third type of supersymmetric string, also defined in ten dimensions, was known in addition to Type II theories. It was the so called Type I theories. This theory (unlike Type II) contain both closed and open strings. Closed (unoriented) strings give rise to the gravitational sector of the theory and open strings give rise in general to non-Abelian gauge theories. This is a very interesting property since one then has the hope of embedding the gauge group $SU(3) \times SU(2) \times U(1)$ into the open string sector. Unfortunatelly it was realized [3] that the theory has anomalies and hence it is inconsistent at the quantum level. Thus before 1984 the situation concerning string theory was very puzzling: Type II theories had a consistent quantum theory without anomalies but where unable to embedd the SM. On the other hand Type I strings had the potential to embed the SM but were anomalous! One thus can understand why before 1984 only very, very few theorists were working in the field of string theory.

In 1984 everything changed when it was realized that the conclusion that Type I string is inconsistent was in fact not true. Green and Schwarz showed [4] that for a very particular gauge group, $SO(32)$, all anomalies cancelled via a new and ellegant mechanism. Soon after Gross, Harvey, Martinec and Rohm [5] showed the existence of two more string theories in ten dimensions, the heterotic strings. They are theories that, like Type I, are chiral and contain gauge fields in their massless spectra. However they only contain closed strings. The gauge groups are either $SO(32)$ or $E_8 \times E_8$ and are also anomaly free. These developements caused the well known increase of popularity of this field. Now we had three string theories which were candidates to unify gravitation and the SM into a finite theory. Of course, being ten-dimensional a process of compactification down to the four physical dimensions is required. Most phenomenological attempts to embed the SM were based in the heterotic $E_8 \times E_8$ string since, as shown by Candelas, Horowitz, Strominger and Witten [6] , there are particular classes of compact six-dimensional varieties (known as Calabi-Yau manifolds) such that, upon compactification of the $E_8 \times E_8$ string give rise at low energies to an E_6 gauge theory with one $N = 1$ supersymmetry (usefull to solve the hierarchy problem) and a number of fermion generations given by one half the Euler characteristic of the Calabi-Yau manifold. The gauge group E_6 had been used as a grand unification group in the past so everything looked like if we were really close to the final unified theory of the standard model and gravity.

This optimism was cooled down when people tried to get a more specific contact with low energy phenomenology. It was soon realized that although E_6 is a nice group, is still far away from what we really want : the SM with three quark-lepton generations, apropriate Higgs doublets, a reasonable fermion mass spectrum, sufficiently stable proton, etc. A number of techniques (orbifolds, fermionic construction etc.) were developed to get miriads of new possibilities for heterotic string compactifications leading at low energies to plenty of possibilities for gauge group, fermion content and phenomenology. A few of the new string vacua obtained with the new techniques had massless spectrum quite close to that of the minimal supersymmetric SM (MSSM), which is already by itself a remarkable achievement [7] . In spite of this success, the presence of such miriads of aparently consistent string vacua leads to an obvious question: how is the physical vacuum (presumably corresponding to the SM) chosen by the dynamics of the theory? With unbroken supersymmetry all string vacua are degenerate, which leads us to the next question: how is supersymmetry broken in string theory ? Both questions seem to require a knowledge of non-perturbative effects in string theory.

3 T-duality

By 1985 the five existing supersymmetric $D = 10$ string theories had already been discovered: Type IIA, Type IIB, $SO(32)$ Type I, $SO(32)$ heterotic and $E_8 \times E_8$ heterotic. As we said, much effort was dedicated to the study of the heterotic strings and also to the study of two-dimensional conformal field theory (CFT) in general, the latter being relevant to study the properties of the string variables $X(\tau, \sigma)$ in the two-dimensional world-sheet spanned by τ and σ. Soon a peculiar symmetry of closed strings was noticed, which now goes under the name of T-duality. Consider now for simplicity the case of the purely bosonic string with coordinates $X^\mu(\tau, \sigma)$ which we discussed at the beginning of previous section. The simplest example of compactification is one in which only one dimension is curled up forming a microscopic circle of radius R. One can see that the mass^2 of the string modes has now the form[1]:

$$m^2 = (\frac{p^2}{R^2} + \frac{w^2 R^2}{\alpha'^2}) + \frac{1}{\alpha'}(N + \tilde{N} - 2) \tag{3}$$

where $p, w = 0, \pm 1, \pm 2, ...$ In this formula the first term in the rhs. is not special of string theories, it just corresponds to the fact that in quantum mechanics with one dimension of finite size, there are quantized momemtum (p) states, the Kaluza-Klein modes in our case. The second term in the rhs. (proportional to R^2) is purely stringy, as is obvious from the fact that α' appears. It represents the possibility that a closed string winds w times around the circle of radius R. The above formula has an interesting property observed [8] by K. Kikkawa and M. Yamasaki in 1984. It remains invariant under the replacement:

$$p \leftrightarrow w \ ; \ R \leftrightarrow \frac{\alpha'}{R}, \tag{4}$$

i.e., we exchange quantized momenta p by winding number w and at the same time R by α'/R. Physically this is quite a surprising symmetry since it indicates that the energies of string states are identical in a theory with compact radius R and in a theory with radius α'/R. It turns out that not only the spectrum but all physical properties of those two string configurations are identical. This is remarkable because this tells us that for a compactified string large R and small R are equivalent! This is the simplest example of what now is called T-duality [b]. If instead of one dimension of radius R one compactifies two dimensions, the $R \leftrightarrow \alpha'/R$ duality generalizes in an interesting way. The apropriate duality transformation is now [9] :

$$T \rightarrow \frac{1}{T} \quad ; \quad T \rightarrow T+1 \tag{5}$$

where now T is a complex ("modulus") field $T = b+iR^2$. Here R^2 is the overall compact volume and b is an axion-like field. The above two transformations generate the discrete infinite group $SL(2,Z)$. Again, this transformation has to be acompanied by the apropriate transformation of quantized momenta and winding numbers.

When applied to the list of ten-dimensional closed supersymmetric strings a number of equivalences were found. It turns out that Type IIA string compactified on a circle of radius R is equivalent (T-dual) to Type IIB compactified on a circle of radius α'/R

(and viceversa) [10] . Furthermore the heterotic $E_8 \times E_8$ compactified on a circle of radius R is T-dual to the $SO(32)$ heterotic compactified on a circle of radius α'/R (and viceversa) [11] . This is remarkable since it shows us that there are indeed only three types of supersymmetric strings which are disconnected: Type II, Type I and heterotic. Thus the number of candidates for a fundamental theory was substantially reduced. This nevertheless did not impress very much the string practitioners. Tacitally most stringers had in the back of their minds the idea that only the heterotic string had any chance of constituting the long sought for unified theory of all known interactions. Type II and Type I strings were bothersome theories that sooner or later would be proven to be inconsistent. In this way we would be left with a unique unified theory, the heterotic. This turned out not to be the case, as has been shown in the last four years.

---

[b]This symmetry was termed just duality. It was first called T-duality in ref.[12] to distinguish it from the newly proposed S-duality.

4 S-duality

In string theory the strength of interactions is not given by coupling constants but by the vacuum expectation value of a massless real scalar ϕ, the dilaton. It turns out that, although apparently totally different in origin, the dilaton behaves to some extent in a way quite similar to the compact radius R which appears in the field T described in the previous section. Let us consider the simplest compactification of the heterotic string down to four dimensions obtained by compactifying on a six torus. The low energy Lagrangian has $N = 4$ supersymmetry in $D = 4$. It turns out that the dilaton appears naturally as the imaginary part of a complex scalar field $S = \eta + i\phi$. Here η is an axion-like field. + Furthermore, at the classical level the field S appears in the Lagrangian in a way quite analogous to the fields T. Both have asociated symmetries of the type $SL(2,R)$. We already mentioned above that in the case of the T field a discrete subgroup of that symmetry, $SL(2,Z)$, was in fact exact, corresponding to T-duality. A natural conjecture would then be to assume that, also for the S field an $SL(2,Z)$ symmetry generated by the transformations:

$$S \rightarrow \frac{1}{S} \quad ; \quad S \rightarrow S + 1 \tag{6}$$

could be a symmetry of this class of string vacua and this was our proposal [12] in 1990. But now this symmetry is quite bizarre: notice that eq.(6) includes the transformation $\phi \leftrightarrow 1/\phi$. Since the dilaton ϕ gives a messure of the strength of the interactions, this is a symmetry which relates strongly coupled to weakly coupled string theory! Unlike the case of T-duality this symmetry is clearly non-perturbative in nature and hence difficult to check with purely perturbative methods.

The origin of the $SL(2,Z)$ symmetry in the case of T-duality had to do with the generalization of the simple symmetry $R \leftrightarrow \alpha'/R$ which appears when a closed string has one compact dimension of size R. Is there anything like this in S-duality? In fact it turns out that if one derives the Lagrangian of $D = 10$ supergravity starting from $D = 11$ supergravity and compactifies on a circle of radius R_{11}, the ten-dimensional dilaton $\phi = R_{11}$, and hence one would have hoped that perhaps a duality in R_{11} would have then implied a $\phi \leftrightarrow 1/\phi$ symmetry [12] . But in 1990 $D = 11$ supergravity had no room in the realm of string theory and we had to wait till 1995 to have a geometrical understanding of S-duality.

In the case of T-duality, the transformations of eq.(5) come along with the interchange of quantized momenta p and winding numbers w. What is the equivalent statement in S-duality? We proposed [12] that in the case of S-duality one has to exchange standard charged particles with solitonic monopole states which generically are present in this class of theories. In fact already in 1975 Claus Montonen and David Olive [13] had considered the posibility of an exchange symmetry between charged particles and monopoles in certain non-supersymmetric gauged field theories with scalars. They conjectured the possibility that the physics of charged particles at strong coupling g^2 could be equivalent to that of magnetic monopoles at small coupling. In the $N = 4$ supersymmetric version of these theories [14] one finds certain states, called BPS states whose masses are given by an expression of the form:

$$M^2 \propto q_e^2 g^2 + \frac{q_m^2}{g^2} \tag{7}$$

where $q_e(q_m)$ are the electric(magnetic) charges of the particles and g^2 is the gauge coupling constant[2]. Notice the explicit invariance under the exchanges $q_e \leftrightarrow q_m$, $g^2 \leftrightarrow 1/g^2$. It is obvious the analogy of this expression to that of the first two terms in the rhs. of eq.(3) . In the string case one would have the dilaton ϕ instead of g^2. Since in the heterotic string compactification that we are discussing there is also an effective $N = 4$ supersymmetric Lagrangian, we considered this as additional circumstantial evidence in favour of the S-duality symmetry [12] .

Being a non-perturbative symmetry, it was not obvious how to find evidence in favour of its reality. However A. Sen [15] studied in more detail the heterotic string compactified on the six torus and found that the spectrum of BPS particles was indeed invariant under S-duality. More importantly, starting with a purely electric state, S-duality predicts the existence of certain monopole configurations obtained by $SL(2, Z)$ transformations. Ashoke Sen found those solutions with the correct multiplicities [15]. His arguments applied even non-perturabatively because the mass-formulae for BPS states is necesarily exact due to $N = 4$ supersymmetry. These developements changed the initially skeptical atitude of the comunity concerning strong-weak coupling dualities and gave rise to the beginning of the so called second revolution of string theory [16] . It also inspired the seminal work of Seiberg and Witten [17] in which they constructed the exact effective Lagrangian of certain gauge field theories with $N = 2$ supersymmetry in four dimensions by using duality arguments.

5 Dualities and unification of all string theories

The S-duality symmetry of the heterotic string compactified on a six-torus which we discussed in the previous section was the first example of a general class of symmetries present in many string configurations. Starting in 1994 many such strong/weak coupling dualities were conjectured to hold. These conjectures have not been formally proven but a number of consistency checks have been made in a good number of cases making difficult to believe that they are purely accidental. It is dificult to briefly describe all the concepts about string theory that have changed in the last four years due to these dualities. We will limit ourselves here to review some of the results which are more directly related to the unification of all string theories into a single structure. In this context the pioneering work of M. Duff [18] , A. Strominger [19] , P. Townsend and C. Hull [20] , completed and systemathized by E. Witten in 1995 [21] were very important.

We already discussed how, due to T-duality, there are only three disconnected ten-dimensional supersymmetric string theories: Type I, Type II and heterotic. The new S-dualities further reduce the number of independent theories. It has been found that [16] :

1. Type I string theory and the heterotic $SO(32)$ are S-dual to each other [22] . This means that ten-dimensional Type I theory at small coupling ϕ is equivalent to heterotic $SO(32)$ at (large) $1/\phi$ coupling. This is in some way the least surprising of all dualities since both theories posses $SO(32)$ gauge bosons.

2. Type IIB theory is S-dual to itself [20] .

3. Type IIA theory gives rise to a surprise [21] . At weak coupling ϕ it is a ten-dimensional theory but increasing the coupling a new eleventh dimension reveals itself. Type IIA theory corresponds to a new theory in eleven dimensions in which one of the dimensions is compactified on a circle with radius $R_{11} = \phi$. The Kaluza-Klein states in this 11-th dimension have masses $\propto 1/\phi$ and that is why at weak coupling (small ϕ) this extra dimension is not seen. This is a new surprising phenomenon: as the strength of interactions increases new dimensions may appear. This 11-dimensional theory, whose low-energy Lagrangian turns out to coincide with that of 11-dimensional supergravity, is termed M-theory and a proper formulation for it is still lacking.

4. Something analogous to the Type IIA case happens for the $E_8 \times E_8$ heterotic [23] . As one increases the coupling ϕ a new eleventh dimension reveals itself with size $R_{11} = \phi$. However in this case the $E_8 \times E_8$ theory is obtained by compactifying the 11-dimensional M-theory on a finite segment of length R_{11}. At each of the two boundary points of the segment

one gets a gauge group E_8 and this explains the presence of the two factors.

With all the above connections and equivalences it is clear that all known string theories are connected to each other and to the misterious 11-th dimensional M-theory: Type IIA and $E_8 \times E_8$ theories are both connected to M-theory at strong coupling. On the other hand T-dualities connect those two theories with Type IIB and $SO(32)$ heterotic respectively. Finally, this last theory is connected by S-duality to Type I $SO(32)$ theory. Thus all five supersymmetric string theories are in one way or another connected to M-theory. All string theories are unified.

6 Dualities and p-branes

An important role in the duality developements is played by p-branes [16,24]. They are extended objects which are generalizations of the notion of string. We mentioned how the string coordinate $X^\mu(\tau,\sigma)$ depends on one time-like variable τ and a space-like variable σ which parametrizes the position along the string. As the string moves it sweeps a world-sheet parametrized by (τ,σ). One can as well consider other extended objects (p-branes) whose coordinates $X^\mu(\tau,\sigma_a)$, $a=1,..,p$ depend on more than one space-like variable σ_a. As they move they swep a $p+1$-dimensional "world-volume". Point-like particles are 0-branes, strings are 1-branes, membranes are 2-branes etc. In fact these p-branes appear as soliton-like solutions in the effective low-energy field theories from the different supersymmetric strings. This is very much analogous to monopoles, which also appear as soliton-like solutions of certain gauge theories with scalar fields. In particular, the massless sector of supersymmetric strings contain antisymmetric tensor fields $A_{\mu_1..\mu_n}$. It turns out that one finds p-brane solutions for $p=n-1$ ("electric") and for $p=D-n-3$ ("magnetic"), where D is the number of space-time dimensions. Thus, for example, in all $D=10$ closed strings there is a massless antisymmetric tensor field $B_{\mu\nu}$. Thus $n=2$ and we have a 1-brane (the fundamental string) and a 5-brane.

In the case of Type II strings in addition one has some extra massless antisymmetric fields with odd(even) number of indices for Type IIA(B). Thus there are p-branes with even(odd) p for TypeIIA(B) strings, with $p<10$. These are called Dirichlet branes [24], D-branes for short, and turn out to be particularly important because they allow for some specific perturbative computations which lead to checks of certain duality conjectures. One important property of these D-branes is that open Type I strings are forced to have their boundary ends lying necesarily on the portion of space ocupied by D-branes.

In the case of 9-branes their worldvolume ocupies all ten dimensions and hence the end-points of an open string may be anywhere. But for $p < 9$ open strings are forced to start and end on submanifolds of the full 10-dimensional space, the submanifold ocupied for the relevant D-brane worldvolume. Thus, for example, in the case of 3-branes whose world-volume can be made to coincide with the physical Minkowski space, open strings are forced to start and end on Minkowski space, they cannot start or end on the remaining $10 - 4 = 6$ dimensions (the famous "bulk").

7 Embeding the observed world into D-branes

D-branes have asociated gauge fields living in their worldvolume [24] . Due to this fact, open strings starting and ending on the same D-brane give rise to a massless $U(1)$ field. If we have N D-branes with overlapping worldvolume the gauge symmetry is enhanced to $U(N)$. The non-Abelian generators correspond to open strings going from one D-brane to another. A variety of Type II or Type I string vacua can be constructed containing different p-brane configurations considered as static classical objects. This new class of string vacua may have in general quite large gauge groups, groups with rank much higher than 28 which is the maximum for a $D = 4$ heterotic compactification. This already shows us that the possilities for embedding the physical $SU(3) \times SU(2) \times U(1)$ group into string theory are much wider than in the perturbative heterotic vacua in which the SM group was always a subgroup either of $E_8 \times E_8$ or $SO(32)$. Thus in principle one can now consider the SM group as coming from a configuration of e.g., four coinciding 3-branes. Although this is true in principle, getting chirality and 3 generations in this manner is not so obvious at the moment. New Type I $D = 4$, $N = 1$ string models have recently been built [25] with the new techniques which incorporate these new features although it is too soon to expect fully realistic models.

If indeed the SM gauge group corresponds to gauge fields living on the world-volume of 3-branes [c] some surprises concerning the structure of fundamental mass scales appear [26,27,28,29,30,31,32] . Let me first recall the relationship between Planck mass M_{Planck} and string scale $M_s = 1/\sqrt{\alpha'}$. In the perturbative heterotic string those two scales are unavoidably tied up by the equation $M_s^2 = \alpha_{GUT} M_{Planck}^2/8$, where α_{GUT} is the gauge coupling. This is in some way related to the fact that both gravitational and gauge interactions come from closed strings, there is a real unification between both interactions. Thus the string scale is necesarily closed to the Planck mass, $M_s \simeq 10^{17}$ GeV.

[c] Or (q+3)-branes with q dimensions wrapping on the compact dimensions.

An important point is that this relationship between M_s and M_{Planck} is (in first aproximation) independent of the compactification scale M_c which gives as the overall inverse size of the six extra compact dimensions.

This is not in general the case for Type I string vacua. In this case, if the gauge group comes from open strings starting and ending on a set of p-branes, one has [26,29,30,32] :

$$\frac{M_c^{(p-6)}}{M_s^{(p-7)}} = \frac{\alpha_p M_{Planck}}{\sqrt{2}} \tag{8}$$

where α_p is the gauge coupling of the corresponding gauge group. Consider for example the case in which we asume that our gauge group comes from a set of 3-branes ($p = 3$). Then one has

$$M_s^4 = \frac{\alpha_{GUT}}{\sqrt{2}} M_c^3 M_{Planck} \quad . \tag{9}$$

It is clear from this expression that one can lower the value of the string scale M_s as much as we wish by lowering accordingly the compactification scale M_c and still mantaining M_{Planck} fixed at its experimental value. Thus the standard statement that string physics happens at the Planck scale is not justified in the present situation. The value of M_s depends on what the size of the extra dimensions are.

One could argue that the compactification scale M_c cannot be below a few hundred GeV, since accelarators have not seen any Kaluza-Klein excitations at those energies. However this conclusion would have been incorrect, as recently emphasized in ref.[28] . Indeed, I already mentioned above that the gauge group coming from 3-branes lives only on Minkowski space since open strings can only start or end on the worldvolume of the 3-branes, which we have identified with Minkowski space. Open strings cannot propagate in the bulk. Other way of saying the same thing is that the gauge group has no Kaluza-Klein excitations. On the other hand the gravitational sector, which comes from closed strings, can propagate in the bulk and has both Kaluza-Klein and winding modes. However this Kaluza-Klein gravitational excitations are very weakly coupled and would not have been observed at accelerators even if M_c was as small as the weak scale.

What is then the scale of string theory M_s? It cannot be lower than 1 TeV or so since otherwise string excitations would have been observed at accelerators. Notice that although the gauge group has no KK excitations it has string excitations. Thus in principle any value of M_s above 1 TeV is possible. There are however three natural options for the value of M_s which look particularly appealing:

i) $M_s \simeq M_{GUT}$ [26] . Here M_{GUT} is the scale at which the extrapolated gauge couplings of the minimal supersymmetric standard model (MSSM) join. Numerically this is of order 10^{16} GeV. This is compatible with eq.(9) for values of M_c slightly below M_s. This possibility has the advantage that it incorporates the succesfull joining of coupling constants of the MSSM in a natural way. On the other hand the gauge hierarchy between M_W and M_{Planck} has to be blamed to some mechanism (like gaugino condensation) able to generate such large hierarchies.

ii) $M_s \simeq \sqrt{M_W M_{Planck}}$. This is the geometrical intermediate scale $\simeq 10^{11}$ GeV which coincides with the SUSY-breaking scale in models with a hidden sector and gravity mediated SUSY breaking. The interest of this choice has been recently emphasized in ref.[31] . In this case one has $M_c/M_s \simeq \alpha \simeq 0.01$ and the hierarchy between M_W and M_{Planck} may be understood without the necessity of any hierarchy-generating mechanism like gaugino condensation. In addition in this scheme it is easy to accomodate the axion solution to the strong CP problem. Indeed, cosmological and astrophysical constraints imply that the axion scale should be of order of the intermediate scale. On the other hand axion fields with this scale appear naturally in Type I string models with M_s of this order.

iii) $M_s \simeq 1$ TeV [27,28,29] . In this case the hierarchy problem is solved in an obvious way by lowering the fundamental scale of gravitation close to the weak scale. In this case eq.(9) gives (for an isotropical compactification) $M_c/M_s \simeq 10^{-5}$ and hence $M_c \simeq 10$ MeV. As we remarked above, such low values of M_c are perfectly compatible with accelerator data since gauge fields have no KK excitations. In fact one can even consider a non-isotropic case with four compact dimensions with $M_c \simeq 1$ TeV and the other two compact dimensions with $M_c \simeq 10^{-3}$ eV [28,29] . In this case string physics could be tested at future accelerators such as LHC. On the other hand the large hierarchy between M_c and M_s has now to be explained.

Each of these three possibilities has its own adventages and shortcomings, the first being the most conservative one. The possibility of finding traces of string theory at accelerator energies as may happen in the 1 TeV scenario is quite exciting [33] . On the other hand a number of theoretical issues like proton stability, gauge coupling unification, cosmology and generation of the M_c/M_s hierachy may be difficult to deal with in that scheme. All those questions seem more easy to deal with in the scheme with M_s equal to the intermediate scale. A lot of effort is at present being dedicated to the study of the different alternatives.

8 Epilogue

The duality revolution has changed not only our ideas about string theory but also about field theory. Many perturbative and non-perturbative properties of supersymmetric gauge theories have been reformulated in the last three years in the language of D-branes giving a geometrical interpretation to facts like the Higgs effect or Seiberg's duality. There are reasons to believe that the D-brane language may lead eventually to a more fundamental formulation of gauge theories themselves. A new duality has also been propossed by Maldacena [34] which relates certain gauge field theories to Type II string theories on certain backgrounds. That has lead even to attempts [35] to compute the glueball spectrum of QCD in terms of the dual Type IIB compactified theory. Although it is not clear whether the aproximations made in such computations are in a consistent regime, the mere fact of thinking that one can perhaps compute physical things like glueball masses show how string theory and duality are giving us unexpected tools for the study of field theories.

An important application of D-branes has been to the problem of information loss in blackholes. In particular, for certain classes of blackholes, it has been possible [36] to obtain a microscopic explanation of the Bekenstein-Hawking formula in terms of D-branes (see e.g. refs.[37] for an intuitive explanation of this and references). On the other hand we are still lacking a fundamental formulation of M-theory, the theory which unifies all supersymmetric string theories and 11-dimensional supergravity. There is a candidate for such a fundamental non-perturbative formulation of M-theory which goes under the name of M(atrix)-theory [38] (see refs.[37] for an intuitive idea of this). A lot of work remains to be done in order to understand what is the fundamental theory which incorporates the notion of dualities in a built-in manner. It is sure that many surprises are still to come both from the more theoretical as well as from the more phenomenological sides.

Acknowledgments

I thank the organizers of this meeting for their kind invitation to this wonderfull place. And I apologize to those members of the audience who got tangled up in the network of dimensions, dualities and compactifications they were exposed to during my talk.

References

1. For books on string theory consult:
   Superstring Theory, Vols. I and II, M.B. Green, J. H. Schwarz and E. Witten, Cambridge University Press, 1987;
   Lectures on String Theory, D. Lüst and S. Theisen, Springer (1989);
   String Theory, Vols I and II. J. Polchinski, Cambridge University Press, 1998.
2. M. Green and J. Schwarz, Phys. Lett. B109 (1982) 444 .
3. L. Alvarez-Gaumé and E. Witten, Nucl. Phys. B234 (1983) 269 .
4. M. Green and J. Schwarz, Phys. Lett. B149 (1984) 117
5. D. Gross, J. Harvey, E. Martinec and R. Rohm, Nucl. Phys. B256 (1985) 468 .
6. P. Candelas, G. Horowitz, A. Strominger and E. Witten, Nucl. Phys. B258 (1985) 46.
7. For reviews on string phenomenology see e.g.:
   F. Quevedo, hep-ph/9707434; hep-th/9603074 ;
   K. Dienes, hep-th/9602045 ;
   J.D. Lykken, hep-th/9607144 ;
   G. Aldazabal, hep-th/9507162 ;
   L.E. Ibáñez, hep-th/9505098;
   Z. Kakushadze and S.-H.H. Tye, hep-th/9512155 .
8. M. Kikawa and M. Yamasaki, Phys. Lett. B149 (1984) 357;
   N. Sakai and I. Senda, Prog.Theor.Phys.75(1986)692.
9. A. Shapere and F. Wilczek, Nucl. Phys. B320 (1989) 669 .
10. M. Dine, P. Huet and N. Seiberg, Nucl. Phys. B322 (1989) 301 ;
    J. Dai, R. Leigh and J. Polchinski, Mod.Phys.Lett. A4 (1989) 2073.
11. P. Ginsparg, Phys. Rev. D35 (1987) 648 .
12. A. Font, L.E. Ibáñez, D. Lüst and F. Quevedo, Phys. Lett. B249 (1990) 35.
13. C. Montonen and D. Olive, Phys. Lett. B72 (1977) 117 .
14. E. Witten and D. Olive, Phys. Lett. B78 (1978) 97;
    H. Osborn, Phys. Lett. B83 (1979) 321 .
15. A. Sen, Int.J.Mod.Phys. A9 (1994) 3707 [hep-th/9402002] .
16. For reviews see:
    J. Schwarz, Nucl. Phys. BP (19r) oc. Suppl. 55B(1997) 1, hep-th/9607201;hep-th/9711029 ;
    J. Schwarz and N. Seiberg, hep-th/9803179;
    J. Polchinski, hep-th/9607050 ;hep-th/9611050;hep-th/9812104
    A. Sen, hep-th/9609176;hep-th/9802051 ;hep-th/9810356;

M. Duff, hep-th/9608117;hep-th/9805177
P. Townsend, hep-th/9612121 ;
S. Förste and J. Louis, hep-th/9612198;
S. Mukhi, hep-ph/9710470 ;
C. Vafa, hep-th/9702201;
M. Li, hep-th/9811019 ;
E. Sezgin, hep-th/9809204 .

17. N. Seiberg and E. Witten, Nucl. Phys. B426 (1994) 19;Nucl. Phys. B431 (1994) 484 .
18. M. Duff and J. Lu, Nucl. Phys. B357 (1991) 534;
M. Duff, Nucl. Phys. B422 (1995) 47 .
19. A. Strominger, Nucl. Phys. B343 (1990) 167 .
20. P.K. Townsend, Phys. Lett. B350 (1995) 184;
C. Hull and P. Townsend, Nucl. Phys. B438 (1995) 109 .
21. E. Witten, Nucl. Phys. B443 (1995) 85 .
22. J. Polchinski and E. Witten, Nucl. Phys. B460 (1996) 525 .
23. P. Horava and E. Witten, Nucl. Phys. B460 (1996) 506;Nucl. Phys. B475 (1996) 94 .
24. J. Polchinski, Phys. Rev. Lett. 75 (1995) 4724 ; hep-th/9611050 .
25. M. Berkooz and R. G. Leigh, Nucl. Phys. B483 (1997) 187, hep-th/9605049;
C. Angelantonj, M. Bianchi, G. Pradisi, A. Sagnotti and Ya.S. Stanev, Phys. Lett. B385 (1996) 96, hep-th/9606169;
Z. Kakushadze, hep-th/9704059;hep-th/9804110 ;hep-th/9806044;
Z. Kakushadze and G. Shiu, hep-th/9706051;
G. Zwart, hep-th/970804;
G. Aldazabal, A. Font, L.E. Ibáñez and G. Violero, hep-th/9804026;
J. Lykken, E. Poppitz and S. Trivedi, hep-th/9806080;
L.E. Ibáñez, E. Rabadan and A.M. Uranga, hep-th/9808139 .
26. E. Witten, Nucl. Phys. B471 (1996) 135.
27. J.D. Lykken, Phys. Rev. D54 (1996) 3693, hep-th/9603133.
28. N. Arkani-Hamed, S. Dimopoulos and G. Dvali, hep-ph/9803315.
29. I. Antoniadis, N. Arkani-Hamed, S. Dimopoulos and G. Dvali, hep-ph/9804398 .
30. K. Dienes, E. Dudas and T. Gherghetta, hep-ph/9803466; hep-ph/9806292; hep-ph/9807522 ;
R. Sundrum, hep-ph/9805471; hep-ph/9807348 ;
G. Shiu and S.H. Tye, hep-th/9805157;
C. Bachas, hep-ph/9807415;
Z. Kakushadze and S.H. Tye, hep-th/9809147;

K. Benakli, hep-ph/9809582 .
31. C. Burgess, L.E. Ibáñez and F. Quevedo, hep-ph/9810535.
32. L.E. Ibáñez, C. Muñoz and S. Rigolin, hep-ph/9812397 .
33. G. Giudice, R. Rattazzi and J.D. Wells, hep-ph/9811291;
S. Nussinov and R. Shrock, hep-ph/9811323;
T. Han, J.D. Lykken and R.J. Zhang, hep-ph/9811350;
E.A. Mirabelli, M. Perelstein and M.E. Peskin, hep-ph/9811337;
N. Arkani-Hamed and S. Dimopoulos, hep-ph/9811353;
J. Hewett, hep-ph/9811356;
Z. Berezhiani and G. Dvali, hep-ph/9811378.
34. J. Maldacena, hep-th/9711200 .
35. C. Csaki, H. Ooguri, Y. Oz and J. Terning, hep-th/9806021 .
36. A. Strominger and C. Vafa, Phys. Lett. B379 (1996) 99.
37. J. Polchinski, hep-th/9812104;
E. Witten, hep-ph/9812208 ;
A. Sen, hep-th/9810356 .
38. T. Banks, W. Fischler, S. Shenker and L. Susskind, Phys. Rev. D55 (1997) 5112.

QCD

D. ESPRIU
Departament d'Estructura i Constituents de la Matèria and IFAE, Universitat de Barcelona, Diagonal 647, E-08028 Barcelona
E-mail: espriu@ecm.ub.es

This is the written version of the lecture on deep inelastic scattering and related topics in QCD, delivered in the course of the XXVI International Meeting on Fundamental Physics to an audience of young experimentalists. The aim is fundamentally pedagogical. I review the theoretical setting of the Altarelli-Parisi equations, discuss recent determinations of α_s from deep-inelastic scattering and then move to the kinematical region explored by HERA. In the way I mention some unsolved theoretical problems. I discuss low-x physics and to what extent $\log \frac{1}{x}$ resummations are called for.

1 Introduction

These notes are not an introduction to Quantum Chromodynamics (QCD), the theory of strong interactions. Many excellent textbooks exist where the interested reader can find clear expositions of the subject[1]. In fact we shall assume here that the reader is familiar with the basic technical tenets of perturbative QCD, such as Feynman diagrams, dimensional regularization, renormalization, beta function and so on. We pretend rather to give physical insight of the reasons behind a rather remarkable theoretical development which is nearly as old as QCD itself, namely deep inelastic scattering. Why is it so remarkable?

QCD is, in a way, a rather simple theory (specially when compared to the intricacies of the electroweak part of the Standard Model). It is just a simple extension of good old Quantum Electrodynamics. Instead of matter fields carrying electrical charge +1 (and anticharge -1), if we are talking about electrons, the matter fields of QCD (the quarks) carry a new quantum number: color. Color can take three different values (and their corresponding anti-values). Furthermore the intermediate bosons (the gluons), unlike photons which cannot change the charge of a particle, change the color of a quark. They may for instance turn a red quark into a blue quark. This simple fact implies that the gauge group of QCD is much larger that the $U(1)$ of QED. Since every quark comes in three copies, whose labels get exchanged, there is a $SU(3)$ invariance[a].

[a]The reason why the symmetry group is $SU(3)$ and not $U(3)$ —which would have nine

Simple as this theory may seem, it is not an easy matter in QCD to relate theory and experiment. It is well known that the fields and particles we know how to compute with (with the simplest tool at our disposal, perturbation theory) are not those that are observed by experimentalists in their detectors due to the phenomenon of confinement. Quarks and gluons are real, but they cannot be detected as free particles as they are known to be confined inside hadrons. In view of this is quite remarkable that there are theoretical techniques enabling us to put the theory to very stringent tests.

The phenomenon of confinement comes about because the coupling constant of QCD (which is relatively small at large values of the momentum transfer, the value quoted by the Particle Data Group[2] is $\alpha_s(M_Z) = 0.119 \pm 0.002$; notice the truly amazing precision, which may soon be reduced to a mere 1%) becomes strong as the energy decreases. The behaviour predicted by the renormalization-group is, at one loop,

$$\alpha_s(Q) = \frac{-\pi}{\frac{\beta_1}{2}\log(Q^2/\Lambda_{QCD}^2)}. \tag{1}$$

where Λ_{QCD} is a renormalization-group invariant, but scheme dependent, quantity and $\beta_1 = -11/2+N_f/3$. The preferred value is $\Lambda_{QCD} = 219^{+25}_{-23}$ MeV (5 flavours, $\overline{MS}$-scheme). The meaning of the scale-dependent, renormalized coupling constant is roughly the following: it is the "effective" coupling, relevant at the scale Q, namely, the one that (within the choosen scheme) minimizes further quantum corrections, in particular resumming all large logs. From (1) we see that precisely at the scale $Q^2 = \Lambda_{QCD}^2$, the effective coupling has a pole. Of course well before that scale is reached perturbation theory becomes completely unreliable, and the $1/r$ potential of the perturbative interaction change to a stronger behaviour, possibly to a linear $\sim r$ behaviour. When computing the production of any physical hadron (typically of mass $\sim$ few Λ_{QCD}), perturbation theory is completely useless.

One instance where perturbative QCD can be applied is to inclusive processes, provided that the characteristic momentum transfer is large enough. These will not be discussed in this lecture. The interested reader can look at the determination of α_s through R_{had} R_τ, for instance in.[3]

gluons— has to do with the mathematical fact that $\epsilon_{\alpha\beta\gamma}$ is not an invariant tensor of $U(3)$. This tensor is necessary to build combinations of three quarks which are antisymmetric, as required by Fermi statistics. For instance, $\Delta^{++} = |u^\uparrow u^\uparrow u^\uparrow\rangle$. Being the lightest hadron with this quark contents we expect to have the three quarks in the ground state, hence in a symmetric wave function. This is in contradiction with Fermi statistics. The contradiction can be solved if we admit the existence of a new quantum number α and $|u^\uparrow u^\uparrow u^\uparrow\rangle = \frac{1}{\sqrt{6}}\epsilon^{\alpha\beta\gamma}|u_\alpha^\uparrow u_\beta^\uparrow u_\gamma^\uparrow\rangle$.

A clear application of perturbative QCD is provided by deep inelastic scattering. The subject is now over twenty years old and, by now, perturbative QCD has been tested to a high degree. Furthermore, the commissioning of HERA has opened a new kinematical region where it may be possible to study the onset of non-perturbative effects in a controlled fashion. The exploration of this region is a fascinating subject interesting on its own right.

Due to the lack of time and space we have not included two sections that, in our view, should be in any general review of perturbative QCD and deep inelastic scattering. The first one concerns the so-called "spin of the proton" problem[4]. Another topic that is not covered at all is the study of the photon structure functions.[5] The list of references is very incomplete and those provided merely reflect personal tastes.

2 Logs in QCD

Beyond tree level most Feynman diagrams are ultraviolet divergent. Take for instance the one diagram contributing at one loop to the gluon propagator. Neglecting external momenta, the integral over the momenta of the internal particles is of the form

$$\int \frac{d^4k}{(2\pi)^4} \frac{k^\alpha k^\beta}{k^4} = \infty. \tag{2}$$

To make sense of the theory and get a finite result we must introduce a cut-off Λ and counterterms. A possible method is to perform a subtraction at some $q^2 = -\mu^2$. For instance, for the self-energy of the gluon propagator

$$\Pi(q^2) - \Pi(-\mu^2) \equiv \Pi_R(q^2) = \text{finite}. \tag{3}$$

Alternatively we can make sense of the integrals using dimensional regularization by continuing the dimensionality from 4 to $n = 4 + 2\epsilon$,

$$\int \frac{d^4k}{(2\pi)^4} \to \int \frac{d^nk}{(2\pi)^n}, \tag{4}$$

and subtract just the poles in $1/\epsilon$ (minimal subtraction, MS) or also the $\gamma_E - \log 4\pi$ that always accompanies the singularity in $1/\epsilon$ (improved minimal subtraction, $\overline{MS}$). For instance, for the quark contribution to the gluon self-energy, one has the following result after computing the integral in $n = 4 + 2\epsilon$ dimensions

$$\Pi(q^2) = -\frac{\alpha_s}{6\pi}\delta_{ab}\left(\frac{1}{\epsilon} + \gamma_E + \log\frac{m^2}{4\pi\mu^2} + \ldots\right). \tag{5}$$

Using the MS and $\overline{MS}$ schemes one gets

$$\Pi_{MS}(q^2) = -\frac{\alpha_s}{6\pi}\delta_{ab}(\gamma_E + \log\frac{m^2}{4\pi\mu^2} + \ldots) \qquad \Pi_{\overline{MS}}(q^2) = -\frac{\alpha_s}{6\pi}\delta_{ab}(\log\frac{m^2}{\mu^2} + \ldots) \tag{6}$$

The above expressions illustrate the appeareance of ultraviolet logs $\log q^2/\mu^2$ through the renormalization procedure. There are, however, other source of logs in QCD. They are of the form $\log q^2/\lambda^2$ where λ^2 can is some external momentum squared or a small (mass)^2 that we have given by hand to the (massless) gluon. While the former are associated to ultraviolet divergent integrals (integrals with a bad behaviour when the internal momentum is large), the latter are infrared logs and are related to Feynman diagrams with a bad behaviour when one or more external momenta vanish. Ultraviolet logs appear in any renormalizable field theory after renormalization. On the contrary, infrared logs appear whenever a theory has massless particles in the spectrum (such as photons or gluons). A given Feyman diagram can give rise to both type of singularities at the same time.

There actually two classes of infrared logs caused by massless particles. The so-called infrared divergences arise from the presence of a *soft massless* particle ($k^\mu \to 0$). For instance in the process $e^+e^- \to \mu^+\mu^-$ at the one loop level we have to compute the integral (figure 1)

Figure 1. Example of diagram leading to an infrared divergence.

$$\int \frac{d^4k}{(2\pi)^4}\frac{1}{k^2[(p_1+k)^2 - m^2][(p_2+k)^2 - m^2]}. \tag{7}$$

When $p_1^2 = p_2^2 = m^2$ the integral behaves for $k^\mu \to 0$ as

$$\int \frac{d^4k}{(2\pi)^4}\frac{1}{k^4}. \tag{8}$$

and diverges. This divergence is unphysical so it must be cancelled by something else. The Bloch-Nordsieck theorem[6] states that in inclusive enough cross-sections the infrared logs cancel. What do we mean by 'inclusive enough'? A detector will not be able to discern a 'true' muon from a muon accompanied by a soft enough photon (with $\vec{k} \to 0$). Therefore, we have to consider diagrams where a soft photon is radiated by the muon, square the modulus of the amplitude and integrate over the available phase space (which actually depends on the experimental cut). When this is done the result is infrared finite. The relevant diagrams are depicted in figure 2.

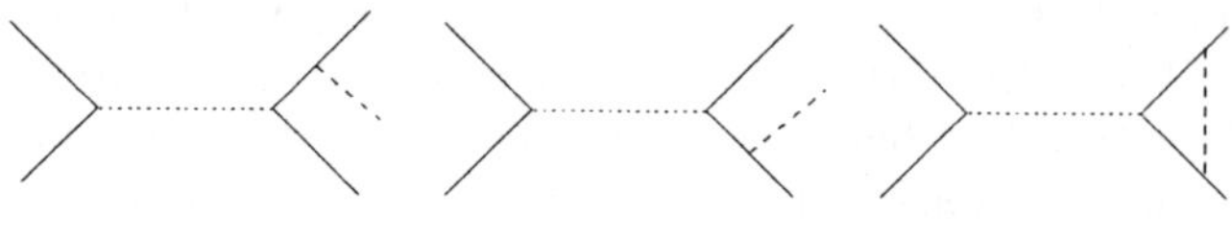

Figure 2. Real and virtual photons have to be included for IR safe results.

The other type of infrared logs are called mass singularities. They occur in theories with massless particles because two *parallel massless* particles have an invariant mass equal to zero

$$k^2 = (k_1 + k_2)^2 = ||(\omega_1 + \omega_2, 0, 0, \omega_1 + \omega_2)|| = 0. \tag{9}$$

The appeareance of such a mass singularity is illustrated in figure 3

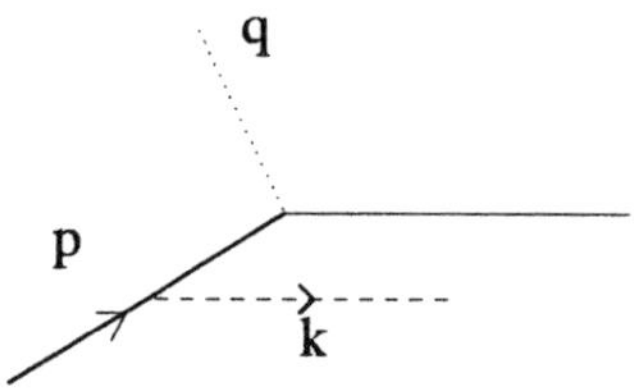

Figure 3. Diagram with a mass singularity.

$$\frac{1}{(p-k)^2} = \frac{1}{p^2 + k^2 - 2k^0p^0 + 2k^0p^0 \cos\theta}, \tag{10}$$

the denominator vanishes when we set all particles on shell ($p^2 = k^2 = 0$) and $\theta \to 0$ (i.e. $\vec{k}$ is parallel to $\vec{p}$). Even if one of the two particles is massive there is a singularity, provided the 3-momenta are parallel.

The Kinoshita-Lee-Nauenberg theorem[7] ensures that for inclusive enough cross section the mass singularities also cancel. Both for mass singularities and for infrared divergences there is a trade-off between λ^2, the infrared regulator of a massless particle, and the energy and angle resolution of the inclusive cross section ΔE, $\Delta\theta$.

In practice, it is better to regulate the infrared logs using dimensional regularization (introducing λ^2 leads to difficulties with gauge invariance). Real gluon emission diagrams are regulated by performing the phase space integration in n dimensions.

There is in fact a lot of physical insight hidden in the infrared logs. Physical arguments tell us that the probability of finding a 'bare' isolated muon should be *zero*. We know this because detectors are unable to tell apart a muon from a muon plus one soft photon or indeed from a muon plus any number of soft photons. Infrared divergences in QED can be summed up and then one sees that the probability of finding an isolated muon is indeed zero and not infinite as the one loop diagram led us to believe. Whenever a Feynman diagram is infrared divergent it means that we have forgotten something relevant.

Let us consider in QED the interaction of a charged fermion with an external source and let us expand in the number of *virtual* photons n (i.e. in the number of loops). The total amplitude will be expressed as

$$M(p,p') = \sum_{n=0}^{\infty} M_n(p,p') \tag{11}$$

then a calculation shows that

$$\begin{aligned} M_0 &= m_0, \\ M_1 &= m_0 \alpha B + m_1, \\ M_2 &= m_0 \frac{(\alpha B)^2}{2} + m_1 \alpha B + m_2, \\ &\dots \end{aligned} \tag{12}$$

The quantities m_n are IR-finite, while B is IR-divergent. The series in (11) can be summed up

$$M = \exp(\alpha B) \sum_{n=0}^{\infty} m_n, \qquad m_n \sim \alpha^n, \tag{13}$$

and B can be obtained just from the lowest order diagram. Introducing an IR cut-off λ, $B \sim -\log m^2/\lambda^2$, which indeed shows that when we remove the cut-off the probability of finding an isolated charged fermion is zero in QED. The addition of soft photons changes that result, multiplying the total amplitude by a factor $\sim (\Delta E/\lambda)^2$. There is a trade between the infrared regulator and $\Delta E, \Delta\theta$. The latter are, of course, detector-dependent quantities.

Although only partial results exist[8], it is believed that a similar exponentiation takes place in QCD. Due to the confinement subtleties it is unclear whether the suppression factor is compensated by radiation of soft gluons. Even if this compensation does actually take place that would not disprove confinement, only that confinement would have nothing to do with the structure of infrared singularities of the theory.

The previous discussion can be summarized in the following way: due to IR singularities one is forced to consider cross sections not of individual particles in the final state, but rather of bunches of particles, each 'hard' quark and gluon surrounded by a 'soft' cloud of gluons and, perhaps, quarks. These bunches are called 'jets'.

The Bloch-Nordsieck and Kinoshita-Lee-Nauenberg theorems guarantee the finiteness of the cross-sections. We have to define an energy and angle resolution. For instance, if p is the momentum of a primary quark we can impose that the energy of each soft particle in its jet satisfies $k_i^0 < \epsilon p_0$ and also that $\arg(\vec{p}, \vec{k}_i) < \delta$. We will get singularities of the form $\alpha_s \log\epsilon \log\delta$ when $\epsilon, \delta \to 0$. The specific details depend on the precise definition of the jet.

3 Free Parton Model

The counterpart of having an effective coupling constant which grows at low energies is that the theory becomes simple at high energies, making perturbative calculations possible (at least *some* of them). Indeed a brilliant confirmation of the existence of nearly free constituents inside the nucleon was provided more than twenty years ago by a series of experiments carried out at SLAC[9]. Then it became possible to scatter electrons off nucleons in fixed target experiments with a typical momentum transfer $\sim 1-10\ (\text{GeV})^2$, a kinematical range unexplored until that time. The kinematics of Deep Inelastic Scattering (DIS) processes is shown in fig. 4.

The virtual intermediate boson is far off its mass-shell and scatters off a quark or gluon in a time of $\mathcal{O}(1/\sqrt{-q^2})$. Typically quarks and gluons are themselves off-shell by an amount of $\mathcal{O}(\Lambda_{QCD})$. After the scattering the outgoing particles recombine into hadrons in a time of $\mathcal{O}(1/\Lambda_{QCD})$. Thus Deep Inelastic is a two-step process

- Short distance scattering occurs with a large momentum transfer. Well described by perturbation theory.
- Outgoing particles recombine. Not calculable in perturbation theory.

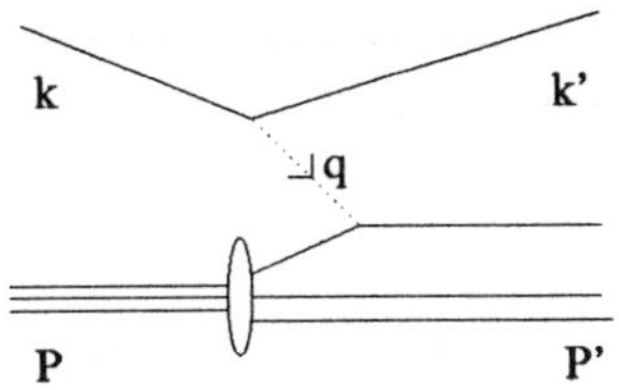

Figure 4. The kinematics of deep inelastic scattering.

However, the second part can be side-stepped all together for fully inclusive rates. Then perturbation theory is adecuate to describe many features of DIS.

If we place ourselves in the center of mass of the hadron and virtual intermediate boson both particles move very fast towards each other. Whatever components the hadron contains they will all have moments parallel to P^μ, up to transversal motion of $\mathcal{O}(\Lambda_{QCD})$. Let us write

$$p^\mu = xP^\mu. \tag{14}$$

The squared CM energy of the lepton and proton constituent will be

$$\hat{s} = (xP + k)^2 \simeq 2xPk \simeq xs. \tag{15}$$

We neglect masses (as well as the fact that constituents are off-shell by $\mathcal{O}(\Lambda_{QCD})$) The final momentum of the constituent is $xP + q$. Therefore

$$0 \simeq (xP + q)^2 \simeq 2xPq + q^2, \tag{16}$$

so $x = -q^2/2Pq$. If ν is the energy transfer in the LAB system, we can also write

$$x = \frac{-q^2}{2\nu m_N} \tag{17}$$

m_N being the nucleon mass. It is convenient to introduce

$$y = \frac{Pq}{Pk} = 1 - \frac{Pk'}{Pk} \tag{18}$$

In the lab frame $y = \nu/E$ and $0 \leq y \leq 1$. y is thus the relative energy loss of the colliding lepton.

Let us for the time being ignore altogether QCD interactions and let us assume that constituents of the nucleons (which we will call partons) are free.

DIS will then be described by an incoherent sum over elementary processes. The partonic differential cross sections in the LAB frame will be

- νq, $\overline{\nu} q$-scattering

$$\frac{d\hat{\sigma}_\nu}{dy} = (\frac{g^2}{4\pi})^2 \frac{\pi m E}{(q^2 - M_W^2)^2}[g_L^2 + g_R^2(1-y)^2], \tag{19}$$

$$\frac{d\hat{\sigma}_{\overline{\nu}}}{dy} = (\frac{g^2}{4\pi})^2 \frac{\pi m E}{(q^2 - M_W^2)^2}[g_R^2 + g_L^2(1-y)^2]. \tag{20}$$

- eq-scattering

$$\frac{d\hat{\sigma}_e}{dy} = Q^2 \frac{4\pi\alpha^2 m E}{q^4}[1 + (1-y)^2]. \tag{21}$$

The neutral current sector is dominated by γ exchange below $q^2 = M_Z^2$, so we have not bothered to include Z exchange. In (21) Q is the quark electric charge (in units of e) and m is the target mass. Since $p^\mu = xP^\mu$, we just take $m = xm_N$. Then, for instance,

$$\frac{d^2\hat{\sigma}_e}{dxdy} = Q^2 \frac{4\pi\alpha^2 x m_N E}{q^4}[1 + (1-y)^2]. \tag{22}$$

Let $u(x)dx, d(x)dx, ...$ be the number of $u, d, ...$ quarks with momentum fraction between x and $x + dx$ in a nucleon. Then $xu(x), xd(x), ...$ will be the fraction of the nucleon momentum carried by $u, d, ...$ quarks. We, of course, identify quarks with partons and, since we assume that they are free, proceed to sum incoherently over the different scattering possibilities. For instance in $ep \to eX$

$$\frac{d^2\sigma}{dxdy} = \frac{2\pi\alpha^2}{s}\frac{1+(1-y)^2}{xy^2}[\frac{4}{9}(u(x)+\overline{u}(x)) + \frac{1}{9}(d(x)+\overline{d}(x)) + \frac{1}{9}(s(x)+\overline{s}(x))]. \tag{23}$$

(We neglect here the possible contribution from the sea of heavy quarks in the nucleon.) Other DIS processes weigh differently quarks and antiquarks. For instance, in $\nu p \to \mu X$ if $-q^2 \ll M_W^2$ we have

$$\frac{d^2\sigma}{dxdy} = x\frac{G_F^2 s}{\pi}[c_c^2 d(x) + s_c^2 s(x) + \overline{u}(x)(1-y)^2], \tag{24}$$

with $c_c = \cos\theta_c$, $s_c = \sin\theta_c$, the cosinus and sinus of the Cabibbo angle, respectively.

The parton distribution functions (PDF) $q(x)$ are quantities which are not calculable within perturbative QCD, as we will see.

Probably the first thing that one learns is that gluons are very important. From the SLAC-MIT data[9]

$$Q = U + D + S = \int_0^1 dxx(u(x) + d(x) + s(x)) \simeq 0.44, \tag{25}$$

$$\bar{Q} = \bar{U} + \bar{D} + \bar{S} = \int_0^1 dxx(\bar{u}(x) + \bar{d}(x) + \bar{s}(x)) \simeq 0.07. \tag{26}$$

The total fraction of momentum carried by quarks (and antiquarks) is only about 50% The rest is carried by gluons (parametrized by a PDF $g(x)$), showing that although the naive quark model works very well is just a gross simplification as a model of hadrons, at least at large $-q^2$. In fact we know the asymptotic values of (25) and (26) based in the equipartition of energy in a free theory (since, asymptotically, QCD is free)

$$\int_0^1 dxxq(x) \to \frac{3N_f}{16 + 3N_f} \qquad \int_0^1 dxxg(x) \to \frac{16}{16 + 3N_f}. \tag{27}$$

From the above limiting values we see and at higher energies the total momentum carried by *constituent* or *valence* quarks diminishes and that an equally important role is played by particles from the Dirac sea of the nucleon.

Another example where the quark model fails to describe some basic features of hadrons is provided by the 'spin of the proton' problem[4]. μ-scattering on polarized targets shows that the fraction of the total spin of the proton that can naively be associated to constituent quarks is surprisingly small. We shall not dwelve on this matter further here.

Nevertheless, there are some obvious sum rules for the parton distribution functions which can ultimately be explained in terms of the quark model. For the proton

$$\int_0^1 dx(u(x) - \bar{u}(x)) = 2, \tag{28}$$

$$\int_0^1 dx(d(x) - \bar{d}(x)) = 1, \tag{29}$$

$$\int_0^1 dx(s(x) - \bar{s}(x)) = 0. \tag{30}$$

On QCD grounds we expect that this free parton model description of the hadrons becomes more and more accurate when $-q^2 \to \infty$, $\nu \to \infty$, while keeping x fixed. This limit is known as Bjorken scaling and in the strict $-q^2 = \infty$ limit everything depends just on x.

Let us now try to rederive the previous results in a more theoretical setting. Let us consider for instance νp scattering. Then

$$\frac{d^2\sigma}{d(-q^2)d\nu} = \frac{G_F^2 m_N}{\pi s^2} L^{\mu\nu} H_{\mu\nu}, \tag{31}$$

where

$$L^{\mu\nu} = \frac{1}{8}\mathrm{Tr}[\gamma^\mu(1-\gamma_5)\gamma^\alpha\gamma^\nu(1-\gamma_5)\gamma^\beta]k_\alpha k_\beta \tag{32}$$

is the trace over the leptonic external lines, and $H_{\mu\nu}$ is given by

$$\sum_X \langle P|J_\mu(0)|X(P')\rangle\langle X(P')|J_\nu(0)|P\rangle = \int d^4z e^{iqz}\langle P|J_\mu(z)J_\nu(0)|P\rangle, \tag{33}$$

which is just $\mathrm{Im}\Pi_{\mu\nu}(q)$, with

$$\Pi_{\mu\nu}(q) = \int d^4z e^{iqz}\langle P|TJ_\mu(z)J_\nu(0)|P\rangle. \tag{34}$$

We decompose $H_{\mu\nu}$ as

$$H_{\mu\nu} = -g_{\mu\nu}F_1 + \frac{P_\mu P_\nu}{\nu m_N}F_2 + \frac{i}{2\nu m_N}\epsilon_{\mu\nu\rho\sigma}P^\rho q^\sigma F_3 \tag{35}$$

(If we assume that we are working with non-polarized targets P and q are the only vectors at our disposal.) F_1, F_2 and F_3 are called the nucleon structure functions. Using the kinematical relations $x = -q^2/2\nu m_N$ and $y = 2m_N\nu/s$ we get

$$d(-q^2)d\nu = \nu s dx dy, \tag{36}$$

$$\frac{d^2\sigma}{dxdy} = \frac{G_F^2 s}{2\pi}[F_1 x y^2 + F_2(1-y) - F_3 x y(1-\frac{y}{2})]. \tag{37}$$

Let us now compare with the free parton model. We see that (restoring the νp index, to make apparent that the structure functions are process dependent)

$$F_1^{\nu p}(x) = c_c^2(\bar{u}(x) + d(x)) + s_c^2(s(x) + \bar{u}(x)), \tag{38}$$

$$F_2^{\nu p}(x) = 2xc_c^2(\bar{u}(x) + d(x)) + 2xs_c^2(s(x) + \bar{u}(x)), \tag{39}$$

$$F_3^{\nu p}(x) = 2c_c^2(\bar{u}(x) - d(x)) + 2s_c^2(-s(x) + \bar{u}(x)). \tag{40}$$

For other processes the actual expressions may vary but the structure functions are always linear combinations of the parton distribution functions, i.e. $F_2(x) = x\sum_f \delta_f q_f(x)$, etc.

Note that in the free parton model

$$F_L(x) = F_2(x) - \frac{F_1(x)}{2x} = 0. \tag{41}$$

This is the Callan-Gross relation, which actually is not an exact one; it gets modified when the q^2 dependence is included, i.e. we depart from the strict $-q^2 = \infty$ limit.

An exact sum rule, which is easily expressed in terms of the structure function $F_2(x)$ was given by Adler

$$\int_0^1 \frac{dx}{x}(F_2^{\bar{\nu}I} - F_2^{\nu I}) = 4\langle I_3\rangle \tag{42}$$

where I_3 is the third component of the target of isospin I. Other sum rules are not exact, except in the strict free parton model, but their violations are computable within perturbative QCD. Two instances are the Gross-Llewellyn-Smith sum rule

$$\frac{1}{2}\int_0^1 dx(F_3^{\nu p}+F_3^{\bar{\nu}p}) = \int_0^1 dx(u(x)-\bar{u}(x)+d(x)-\bar{d}(x))+\mathcal{O}(\alpha_s) = 3+\mathcal{O}(\alpha_s), \tag{43}$$

and the Gottfried sum rule

$$\int_0^1 \frac{dx}{x}(F_2^{\mu p} - F_2^{\mu n}) = \frac{1}{3} + \mathcal{O}(\alpha_s) + \int_0^1 dx(\bar{u}(x) - \bar{d}(x)). \tag{44}$$

Violations to the different sum rules are a theoretically clean way to extract α_s. Of course in practice things are difficult because the sum rules involve an integral over all values of x, which is always poorly known in some range of the integrand and extrapolations are needed.

If we assume that α_s is extracted from some other source, the Gottfried sum rule provides some interesting information on the sea contents of light antiquarks in the nucleon. The collaboration NMC[10] has determined that for the proton

$$\bar{U} - \bar{D} \equiv \int_0^1 dx(\bar{u}(x) - \bar{d}(x)) = -0.15 \pm 0.036. \tag{45}$$

In the proton quark sea there are many more d-type antiquarks than u-type. This is in fact a large isospin violation, much larger than expected to the mass difference of the quarks and, in fact, goes, at least naively, in the opposite direction.

The isospin violation is larger at low values of x ($x \leq 0.2$) and also at low values of Q^2, which hints that long-distance physics (quite remote from perturbative QCD) is called for. The enhancement of d-type antiquarks is confirmed by other experiments. For instance Na51[11] finds at $x = 0.18$ $\bar{u}/\bar{d} = 0.51 \pm 0.09$, while NuSeaC[12] finds at $Q^2 = 7.4$ GeV^2 that $\bar{U} - \bar{D} = -0.10 \pm 0.024$.

Could this be due to the exclusion principle, as illustrated in figure 5, which makes harder for u-type antiquarks to appear in the sea? It could be, but it is very difficult to come up with quantitative results. A partial understanding is provided by the chiral quark model, including pion exchange (see figure 5), but this type of physics is still very poorly understood.

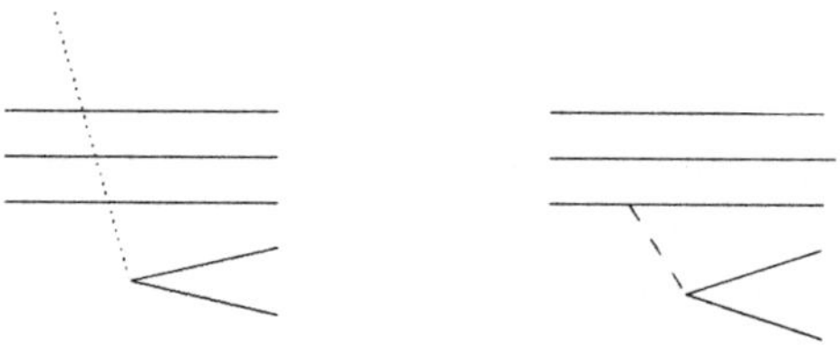

Figure 5. Valence and sea quarks. Contribution from the chiral quark model.

4 Scaling Violations

It is plain clear from the data that there is some $Q^2 = -q^2$ dependence in the structure functions. In other words, there are violations of Bjorken scaling and actually $F_f = F_f(x, Q^2)$. The free parton model is not completely correct (no big surprise, of course). Our job is to try to understand these violations in the framework of QCD.

Let us assume that we have isolated a parton with initial momentum $p = xP$. The probability of finding such a parton is given by $q(x)$. At the parton level the structure function F_2 is just $\hat{F}_2 = x\delta_f$ (δ_f is the appropriate charge. At the proton level, however, this partonic cross-section has to be multiplied by the probability of finding the parton with momentum fraction x, i.e. by $q(x)$. We write this in the form

$$F_2(x) = x\delta_f \int_0^1 d\xi q(\xi)\delta(x - \xi) = x\delta_f \int_0^1 \frac{d\xi}{\xi} q(\xi)\delta(1 - \frac{x}{\xi}). \qquad (46)$$

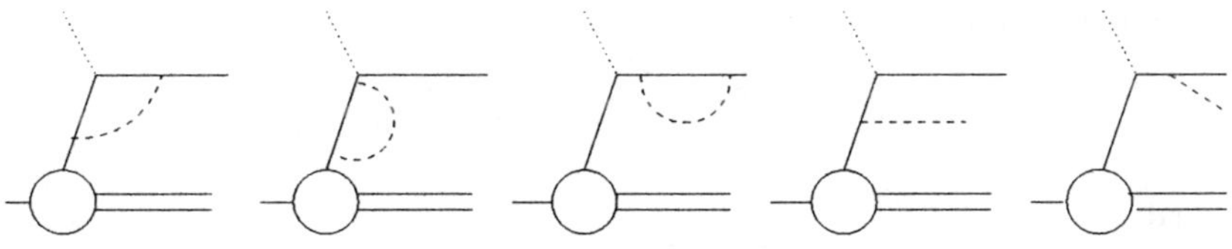

Figure 6. Contribution at $\mathcal{O}(\alpha_s)$ to the relevant DIS subprocess.

At $\mathcal{O}(\alpha_s)$ many diagrams contribute. They are given in figure 6. We are looking for scaling violations and, therefore, we must look for logs. In other words we must investigate ultraviolet, infrared and mass singularities of any kind. It turns out that ultraviolet singularities are proportional to the free result and simply renormalize δ_f. Infrared divergences cancel amongst all diagrams. Only the mass singularity present in diagram (d) when the momentum of the gluon is parallel to that of the gluon survives. Keeping only logarithmic terms, the caculation at $\mathcal{O}(\alpha_s)$ amounts to the replacement

$$\delta(1-\frac{x}{\xi}) \rightarrow \delta(1-\frac{x}{\xi}) + \frac{\alpha_s}{2\pi}P(\frac{x}{\xi})\log\frac{Q^2}{\lambda^2}, \tag{47}$$

where λ^2 is an infrared regulator and

$$P(z) = C_F[\frac{1+z^2}{(1-z)_+} + \frac{3}{2}\delta(1-z)]. \tag{48}$$

Only the logarithmic term is retained for this discussion. Now we understand the reason for writing things in apparently such a complicated way. First of all, scaling violations appear through the contribution of real soft particles. ξ is the original momentum fraction of the nucleon carried by the parton, which is reduced to $x \leq \xi$ after the emission of the soft gluon. The hard scattering takes place with the parton carrying fraction x. All along the discussion, the transverse motion of the partons inside the target is neglected as well as are all masses.

Then

$$F_2(x) = x\delta_f[q_0(x) + \int_x^1 \frac{d\xi}{\xi} q_0(\xi)\frac{\alpha_s}{2\pi}P(\frac{x}{\xi})\log\frac{Q^2}{\lambda^2}]. \tag{49}$$

In addition we have replaced $q(x)$ by $q_0(x)$, the bare PDF. If we now define the renormalized PDF by

$$q(x,\mu^2) = q_0(x) + \frac{\alpha_s}{2\pi}\int_x^1 \frac{d\xi}{\xi} q_0(x) P(\frac{x}{\xi}) \log\frac{Q^2}{\lambda^2}, \tag{50}$$

we can write

$$F_2(x,Q^2) = x\delta_f [q(x,\mu^2) + \int_x^1 \frac{d\xi}{\xi} q(x,\mu^2) \frac{\alpha_s}{2\pi} P(\frac{x}{\xi}) \log\frac{Q^2}{\mu^2}]. \tag{51}$$

No doubt the similarity with the usual renormalization process did not go unnoticed. Now the infrared regulator has been eliminated (hidden in the bare PDF) at the expense of introducing a renormalization-scale dependence. These are the sought after scaling violations.

5 Altarelli-Parisi Equations and Λ_{QCD}

At this point it is convenient to introduce the variable $t = \frac{1}{2}\log\mu^2/\Lambda^2_{QCD}$. It then follows from (50) that

$$\frac{\partial}{\partial t} q(x,t) = \frac{\alpha_s(t)}{\pi}\int_x^1 \frac{d\xi}{\xi} q(\xi,t) P(\frac{x}{\xi}), \tag{52}$$

which immediately translate into differential equations for the structure functions themselves. These are the Altarelli-Parisi equations[13]. They summarize the rate of change of the parton distribution functions with t.

We define the moments of the PDF's by

$$q(n,t) = \int_0^1 dx x^{n-1} q(x,t). \tag{53}$$

Introducing the anomalous dimension γ_n as

$$\gamma_n = \int_0^1 dx x^{n-1} P(x), \tag{54}$$

the convolution over the fractional momentum ξ transforms into a product

$$\frac{\partial}{\partial t} q(n,t) = \frac{\alpha_s(t)}{\pi} \gamma_n q(n,t). \tag{55}$$

This leads to following scaling behaviour for the moments of the structure functions

$$F_2(n,Q^2) = F_2(n,Q_0^2)\left(\frac{\alpha_s(Q_0)}{\alpha_s(Q)}\right)^{\frac{\gamma_n}{\beta_1}}, \tag{56}$$

which is our final expression. Experiments agree on the whole very nicely with the scaling violations predicted by QCD. Taking into account all the subtle points of Quantum Field Theory that have gone into the analysis, this provides a beautiful check of the theoretical framework.

We have been considering F_2, but the same procedure can be repeated for any structure function. The expression (56) amounts to resumming the leading logs obtained by iteration of soft collinear gluons. The diagram is the one shown in figure 7.

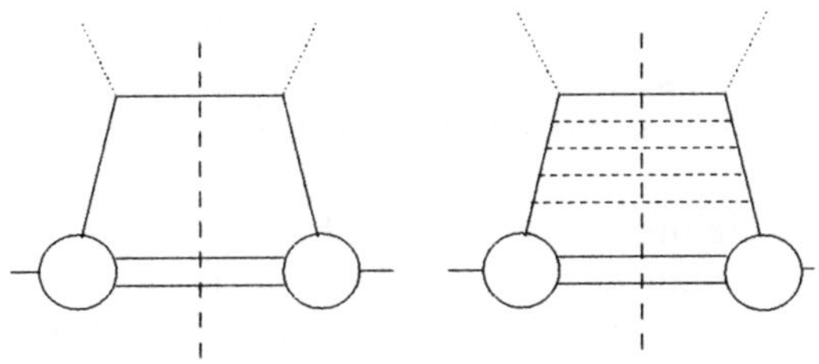

Figure 7. Handbag and ladder diagrams.

As a simplifying hypothesis we have neglected mixing. In fact, the evolution equation is a $(2N_f+1)\times(2N_f+1)$ matrix, involving quarks and gluons. In the flavour singlet case life is more complicated; there is mixing with gluon operators and therefore one must also consider gluon parton distribution functions as well

$$\frac{\partial q(x,t)}{\partial t} = \frac{\alpha_s(t)}{\pi}\int_x^1 \frac{dy}{y}[q(y,t)P_{qq}(\frac{x}{y}) + g(y,t)P_{gq}(\frac{x}{y})] \tag{57}$$

$$\frac{\partial g(x,t)}{\partial t} = \frac{\alpha_s(t)}{\pi}\int_x^1 \frac{dy}{y}[g(y,t)P_{gg}(\frac{x}{y}) + q(y,t)P_{qg}(\frac{x}{y})] \tag{58}$$

The detailed form of the Altarelli-Parisi kernels at leading and NLO order can be found in [14]. No complete calculation exists yet at the NNLO to my knowledge, just some partial results.

It is important to realize that the Altarelli-Parisi equations are not exact. They take into account the perturbative contribution only (and this up to a given order in perturbation theory). They also neglect transverse motion, which leads to corrections of $\mathcal{O}(\Lambda_{QCD}/Q^2)$ to the leading results. These are more easily dealt with in the perhaps more rigorous (but more cumbersome) treatment based in the Operator Product Expansion[15], which will not be discussed here. Target mass corrections should be equally taken into account. They are particularly important near thresholds (such as the charm and bottom thresholds). For instance the proper treatment of thresholds is highly relevant for HERA, since a big chunk of the data comes from a region close to these thresholds. And, of course, mass corrections are important for charm PDF from the sea, which have actually been recently measured.

The analysis of Deep Inelastic Scattering based on the Altarelli-Parisi equations (or, alternatively, on the Operator Product Expansion) has been one of the most clear tests of perturbative QCD and traditionally the best way of determining α_s, which, as we have seen, enters in the scaling violations. However, at present the value of $\alpha_s(M_Z)$ extracted from Z-physics is equally accurate if not more.

For some time a discrepancy was claimed between the value of α_s obtained from the analysis of scaling violations and the Z-pole value. For instance the value quoted by CCFR was $\alpha_s(M_Z) = 0.111 \pm 0.004$, well below the world average[2], and quite away, for instance, from measurements based on event shapes at the Z peak ($\alpha_s(M_Z) = 0.122 \pm 0.007$; for measurements of α_s at LEP, see e.g.[17]). The average α_s from DIS was given[18] just two years ago to be $\alpha_s(M_Z) = 0.113 \pm 0.005$. Theoretical speculations were fuelled and it was claimed that non-perturbative corrections could be larger than originally thought.

Deep inelastic scattering data have been reanalyzed recently and the quoted value for $\alpha_s(M_Z)$ from DIS from a global fit[19] to all data is $\alpha_s(M_Z) = 0.118 \pm 0.005$. The agreement with other determinations is now almost perfect. The discrepancy was apparently due, it is claimed, to the energy calibration of the detector (in the case of CCFR, at least; the new CCFR value is $\alpha_s(M_Z) = 0.119 \pm 0.005$[16]), a better understanding of higher twist corrections (not discussed here) and from the treatment of the contribution from heavy quarks from the sea, charm in particular. We will see later, when we discuss in somewhat more detail the form of the PDF, that one must actually make a number of hypothesis before being able to extract $\alpha_s(M_Z)$. While do not regard the issue as totally settled yet, because some of the older data sets are very poorly described by the now preferred value of $\alpha_s(M_Z)$, the most recent data coming from HERA[20] (H1 and ZEUS) give values which agree nicely with each other and in fact fall well in the high α_s range, almost on top of the new average value. These ongoing experiments will become statistically more and more significant in the near future in the determination of $\alpha_s(M_Z)$ as more and more data points pile up.

6 Parton Distribution Functions

We do not know in general how to compute the parton distribution functions, even for $-q^2 \to \infty$. Only their evolution can be reliably computed either through the Operator Product Expansion of the use of the Altarelli-Parisi equations and this for large enough values of $-q^2$. The scaling behaviour is governed by the anomalous dimensions. At leading order they are

$$\gamma_{qq}(j) = C_F[-\frac{1}{2} + \frac{1}{j(j+1)} - 2\sum_{k=2}^{j}\frac{1}{k}], \tag{59}$$

$$\gamma_{qg}(j) = T_R[\frac{2+j+j^2}{j(j+1)(j+2)}], \tag{60}$$

$$\gamma_{gq}(j) = C_F[\frac{2+j+j^2}{j(j^2-1)}], \tag{61}$$

$$\gamma_{gg}(j) = 2C_A[-\frac{1}{12} + \frac{1}{j-1} + \frac{1}{(j+1)(j+2)} - \sum_{k=2}^{j}\frac{1}{k}] - \frac{2N_f}{3}T_R, \tag{62}$$

where C_F, T_R and C_A are group-theoretical factors.

An interesting issue is the behaviour of the parton distribution functions at the endpoints $x = 0$ and $x = 1$. The large n behaviour of the moments probes the $x \to 1$ region. Since it is natural to expect that at the kinematical boundaries the parton distribution functions vanish, one can make the following ansatz for $x \to 1$

$$q(x, Q^2) \sim A(Q^2)(1-x)^{\nu(\alpha_s(Q^2))-1}. \tag{63}$$

Demanding that eq. (63) fulfills the q^2 evolution equation leads to

$$A(Q^2) = A_0\frac{[\alpha_s(Q^2)]^{-d_0}}{\Gamma(1+\nu(\alpha_s(Q^2)))} \qquad \nu(\alpha_s) = \nu_0 - \frac{16}{33-2N_f}\log\alpha_s(Q^2), \tag{64}$$

$$d_0 = \frac{16}{33-2N_f}(\frac{3}{4} - \gamma_E). \tag{65}$$

Likewise, for the gluons we have

$$g(x, Q^2) \sim A_0'\frac{[\alpha_s(Q^2)]^{-d_0}}{\Gamma(2+\nu(\alpha_s(Q^2)))}\frac{(1-x)^{\nu(\alpha_s(Q^2))}}{\log(1-x)}. \tag{66}$$

The constants A_0, A'_0 and ν_0 are not calculable on perturbative QCD and depend on the specific operator. d_0 is universal.

When $x \to 1$ the gluon distribution functions approach zero more rapidly than the quark ones. For large values of x the quark contents of nucleons is the relevant one. Second order corrections to this asymptotic behaviour can be derived in a similar way and are known. It turns out that the correction is arbitraryly large if one gets sufficiently close to $x = 1$. This is because the collinear gluon is, in addition, soft in that exceptional configuration thus giving rise to a $\log(1-x)$ singularity. Multiple emission is then kinematically favoured, since the log overcomes the α_s suppression.

For small values of x the opposite behaviour takes place, the gluon distribution function eventually becomes dominant. At LHC the cross-section will be greatly dominated by low-x physics and the important process there will be gluon-gluon scattering. At the current Tevatron run the quark contents of protons and antiprotons is still dominant. Let us see why gluons dominate completely at low x.

The key point is the appeareance of the singularity for $j = 1$. Indeed, as $j \to 1$

$$\gamma_{gg}(j) \sim \frac{2N}{j-1}. \tag{67}$$

Then

$$g(j,t) = g(j,t_0)\exp[\frac{N}{\pi\beta_1(j-1)}\log\frac{t}{t_0}]. \tag{68}$$

Then we proceed to evaluate $g(x,t)$ by performing an inverse Mellin transform

$$xg(x,t) = \frac{1}{2\pi i}\int_C dj x^{1-j} g(j,t). \tag{69}$$

The integration circuit is a line in the direction of the imaginary axis in the complex j plane, to the right of the $j = 1$ singularity. The saddle point method can now be used provided that $\log(1/x)$ is large and this is the reason why this procedure gives only the small x behaviour. Working things out we see that the gluon parton distribution function for low x behaves as

$$g(x) \sim \frac{1}{x}\exp\sqrt{C(Q^2)\log\frac{1}{x}}, \tag{70}$$

where $C(Q^2)$ is calculable. Unfortunately, this answer is not totally satisfactory because something must stop the growth in $g(x)$ for low x, or else one runs into unitarity problems sooner or later, and thus eq. (70) it is not credible all the way to $x = 0$. Technically speaking, there must be corrections that destabilize the saddle point solution. Physically, the uncontrolled growth of the gluon distribution is an infrared unstability. The density of soft gluons is too large. Shadowing and non-linear evolution equations are the buzzwords here[21].

The double scaling limit (high Q^2, low x) is well supported by the data. See for instance[22] for a recent analysis coming from measurements of F_2 at H1.

Except in these two limiting cases, PDF's have to be parametrized. The way one proceeds is by proposing a given parametrization at some reference value $Q_0 =^2$, then evolve to all desired values of Q^2 using the Altarelli-Parisi equations, then perform a global fit of the parameters describing the PDF and, at the same time, determine α_s.

7 Confinement

Λ_{QCD} sets a natural scale in the theory. Well above Λ_{QCD} perturbation theory makes sense. Of course perturbative QCD at large enough energies describes a world of quasi-free quarks, interacting with Coulomb-like forces. We know very well that hadronic physics is a very different world where quarks are confined into colorless hadrons. As soon as $q^2 \sim \Lambda_{QCD}^2$ perturbation theory is unreliable. It simply cannot explain confinement.

What confinement means is that there is a force between quarks that does not decrease with distance. There is indeed phenomenological evidence (which is supported by lattice analysis) that the interquark potential at large distances in QCD is of the form

$$V(r) \sim a\Lambda_{QCD}^2 r - \frac{b}{r} + \dots \tag{71}$$

The first term is a confining quark potential. The constant a has to be ~ 1 because Λ_{QCD}^2 is the only dimensional quantity at our disposal. The Coulombic part is called the Lüscher term and plays a crucial role in heavy quark spectroscopy[23].

At short distances the behaviour of the interquark potential is totally different. It is Coulomb-like

$$V(r) \sim -\frac{\alpha_s}{r}. \tag{72}$$

Different quarks probe the different regimes. Indeed, because $\alpha_s(m_t)$ is so small (say ~ 0.1), the corresponding Bohr radius is $r_0 \sim 10^{-2}$ fm, much smaller than Λ_{QCD}^{-1}. The coulombic part of the interquark potential largely dominates. (At such short distances the linearly rising potential is not at work, the leading confinement effects are $\sim r^3$, as discussed by Leutwyler some time ago[24], but they can be safely neglected at first approximation.)

Bottom and charm are in a somewhat intermediate position. $\alpha_s(m_b)$ is still relatively small. The Bohr radius is 10^{-1} fm, smaller but comparable to Λ_{QCD}^{-1}. Spectroscopy is basically perturbative, at least for the lowest levels, but some non-perturbative effects are visible. Charm is really no-man's land. Both perturbative and non-perturbative effects compete even for the ground state $n = 1$. For light quarks the Bohr radius is several fm and the confining potential is fully at work.

The existence of a confining potential leads to very large multiplicities and jets. One can imagine a quark-antiquark being formed at the primary vertex then moving apart. Part of their kinetic energy is deposited in the interquark potential as they move away. Very quickly a separation r_m is reached where the energy deposited is enough to form a new quark-antiquark pair,

$$\Lambda_{QCD}^2 r_m \simeq 2m_q, \tag{73}$$

at that moment the quark-antiquark 'string' breaks and the process is repeated until the average relative momentum is small enough and hadronization takes place.

There is a lot of physics in the string picture. We can think of color forces being confined in some sort of tube or string joining the two moving quarks. The chromodynamic energy is thus stored in a relatively small region of space-time. If this picture is correct we should expect hadronization to take place in this region in preference to any other. This is indeed the case; in three jet events (which originate from $\bar{q}qg$, with a hard gluon) there is a clear enhancement of soft gluon and hadron production in the regions between color lines (representing the gluon by a double color line, or $\bar{q}q$ state), and a relative depletion in other regions. This phenomenon is called color coherence[25].

8 Dual Models

We now backtrack in history, to the pre-QCD days, and recall that in the 60's the duality hypothesis was much in fashion. The hypothesis stated that in strong interactions a sum over intermediate states in the s-channel should reproduce the sum over resonances in the t-channel. Mathematically,

$$A(s,t) = \sum_J \frac{g_J^2 s^J}{t - M_J^2} = \sum_J \frac{g_J^2 t^J}{s - M_J^2}. \tag{74}$$

Of course for this to have even a chance of being true an infinite number of intermediate states is required. It should be stated right away that the evidence for this peculiar property was (and still is) rather weak.

However, in 1968, Veneziano[26] took the idea seriously and proposed the following amplitude

$$A(s,t) = \frac{\Gamma(-\alpha(s))\Gamma(-\alpha(t))}{\Gamma(-\alpha(s)-\alpha(t))}, \qquad \alpha(s) = \alpha(0) + \alpha' s. \tag{75}$$

This amplitude is manifestly dual. In 1969 and 1970 Y.Nambu and others unveiled the relation between the Veneziano amplitude and open string theory. It was later generalized to closed strings by Koba and Nielsen.

This of course is the way to make contact with the long-distance properties of QCD that we have discussed in the previous section. If in some kinematical regime QCD can be described by some type of string theory, an amplitude of the Veneziano type should describe strong interactions in a regime where perturbation theory is not valid.

Unfortunately life is not so easy. First of all, consistency of string theory requires $\alpha(0) = 1$ and then the Veneziano amplitude exhibits poles in the s-channel whenever $s = (n-1)/\alpha'$, $n = 0, 1, \ldots$. There is a tachyonic scalar/pseudoscalar particle and a massless vector particle. In addition, the amplitude does not exhibit the proper chiral behaviour (Adler zero), i.e. $A(0,0) = 0$, if we are to interpret the pseudoscalars as pions. It is a complete phenomenological failure.

Lovelace and Shapiro[27] finally proposed an amplitude with the correct behaviour or, at least, not manifestly incorrect. It is inspired from the theory of supersymmetric strings and the equivalent of the Veneziano amplitude now reads

$$A(s,t) = \frac{\Gamma(1-\alpha(s))\Gamma(1-\alpha(t))}{\Gamma(1-\alpha(s)-\alpha(t))}, \qquad \alpha(s) = \alpha(0) + \alpha' s, \tag{76}$$

where $\alpha(0)$ in principle also equals 1 here. Again this amplitude (which is tachyon-free — this is the main virtue of the supersymmetric string) is not physically acceptable because the amplitude does not have the correct Adler zero. However, it does reproduce the right chiral behaviour if one replaces by hand this value by 1/2 and is then appropriate to describe pion scattering. Unfortunately this replacement cannot be derived at present from any known string theory. The corresponding trajectory is called the Regge trajectory and corresponds to the exchange of open strings and, as discussed, has an intercept $\alpha_R = 1/2$. Physically this is interpreted as the exchange of quark-antiquark pairs In addition there is a Pomeron trajectory which has an intercept $\alpha_P = \alpha_R + 1/2$, and is due to the exchange of closed strings (interpreted as glueballs in QCD).

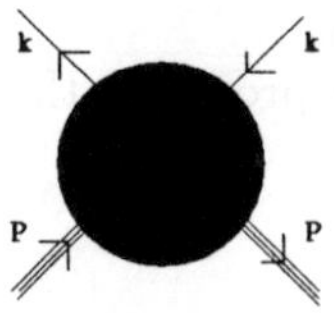

Figure 8. Nucleon-parton scattering process.

Let us now return to DIS and let us see what all this has to do with it. Let us consider the behaviour of the amplitude $A(s,t)$ for large s and fixed t

$$A(s,t) \sim s^{\alpha' t+\alpha_0}. \tag{77}$$

Let us now assume that the elastic nucleon-parton amplitude (figure 9) is described by such an amplitude. Kinematically, $s = (P-k)^2$. If we decompose

$$k = xP - \frac{k_T^2}{2x} n + k_T, \tag{78}$$

where n is a vector such that $n^2 = 0$, $n \cdot P = 1$ and $k_T \cdot n = k_T \cdot P = 0$, then, with the usual approximations,

$$s = -2k \cdot P = -\frac{k_T^2}{x} \tag{79}$$

We see then that, at a finite value of k_T, low x corresponds to the large s behaviour. On the other hand, for this subprocess $t = 0$, and we of course realize that the amplitude is directly related to the cross-section for parton + proton $\rightarrow$ anything. If the parton is a quark only the Reggeon will contribute. If there is mixing with partonic gluons we will have a contribution from both the Regge and Pomeron trajectories. After a short calculation we shall conclude that

$$F_2(x) \sim A_P x^0 + A_R x^{\frac{1}{2}}. \tag{80}$$

And, therefore that, at low values of x,

$$g(x) \sim x^{-1}, \qquad q(x) \sim x^{-\frac{1}{2}}. \tag{81}$$

These are the implications of Regge theory for the parton distribution functions. Notice that there is no Q^2 dependence anywhere, so the question poses itself as to which is the appropriate value of Q^2 to compare with. The answer is not very well defined, but it should correspond to the typical range of energies where Regge phenomenology is known to be valid in other contexts, i.e. a few GeV. In fact, a fit to the gluon PDF shows that at $Q^2 = 4$ GeV^2, $g(x) \sim x^{-1.17}$. Not bad.

9 Low x region

The region below 10^{-2} had not been explored experimentally until very recently; a first look at these low-x values has been provided by the commissioning of HERA. HERA is a machine ideally suited for an in-depth analysis of structure functions. It should be possible to arrive at very low values of x (down to $x \sim 10^{-5}$).

Most parametrizations have traditionally performed very poorly when extrapolated to the low x region. Typically they predict an increase as $x \to 0$ which is lower than what is actually seen. The behaviour $F_2(x) \sim x^{-\lambda}$, with $\lambda \sim 1/2$ as $x \to 0$, which is predicted from the BFKL evolution equation[28] seemed at some point (see e.g.[29]) to stand the comparison with HERA results best. However, this behaviour is still incompatible with unitarity and cannot hold all the way to $x = 0$ either. If fact we know now that the predictions from BFKL cannot be trusted. This has prompted a renewed interest in trying to extract the behaviour at low x from conventional Altarelli-Parisi evolution. The consensus now seems to be that even for the low values of x analyzed at HERA there is no real evidence of any results beyond ordinary perturbative QCD.

It is easy to understand why perturbative QCD must fail at some point. The expansion of the splitting function $P(z)$ in powers of α_s at the NLO actually resums all terms of the form $(\alpha_s \log Q^2)^n$ and $\alpha_s^n \log^{n-1} Q^2$. Looking at (10) we see that the propagator causing the mass singularity is $(p = \xi P)$

$$\frac{1}{2pk} = -\frac{2x}{\xi k_T^2}. \tag{82}$$

Apart from the parametric integrals, we have

$$\int \frac{d^2 k_T}{k_T^2}. \tag{83}$$

This is the origin of the $\log \lambda^2$ and, eventually, of the $\log Q^2$.

The leading $\log^n Q^2$ will thus be produced by one single region in integration

$$\int \frac{d^2 k_T^n}{(k_T^n)^2} \int \frac{d^2 k_T^{n-1}}{(k_T^{n-1})^2} \cdots \int \frac{d^2 k_T^1}{(k_T^1)^2}, \tag{84}$$

with $|Q| \gg |k_T^n| \gg |k_T^{n-1}| \gg \ldots |k_T^1|$.

At sufficiently large x logarithms of $1/x$ necessarily appear. We have actually seen them in the double scaling limit. They must, at some point, spoil the predictivity of the perturbative expansion. One must then identify the regions in integration capable of giving rise to terms of the form $(\alpha_s \log \frac{1}{x})^n$, and eventually to $\alpha_s^n \log^{n-1} \frac{1}{x}$ and so on.

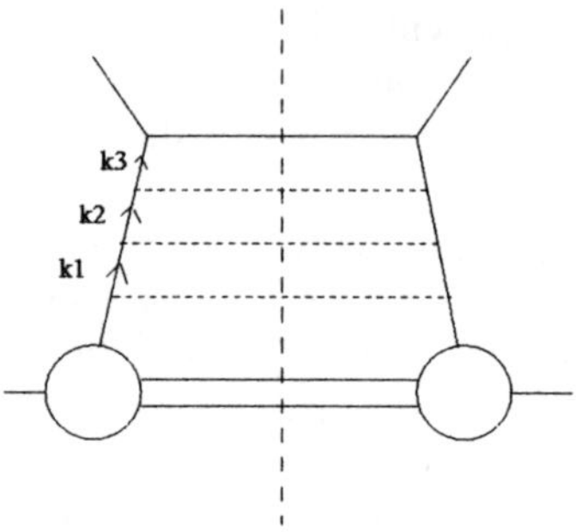

Figure 9. Ordering leading to the most singular $\log 1/x$ contribution.

Lipatov and coworkers[28] (see also[30] for alternative derivations) have identified such a contribution. It corresponds to the diagram depicted in figure 10, more specifically to the region

$$k_i = \alpha_i P + \beta_i n + K_{iT}, \tag{85}$$

$$\alpha_1 \gg \alpha_2 \gg \ldots \gg \alpha_{n-1}, \qquad k_{iT} \sim k_{jT}, \qquad \beta_1 \ll \beta_2 \ll \ldots \ll \beta_{n-1}. \tag{86}$$

This leads to splitting kernels similar to those of the Altarelli-Parisi equations

$$F_2(x, Q^2) = \int d^2k_T \int_x^1 \frac{d\xi}{\xi} C(\frac{x}{\xi}, Q^2, k_T) F_2(\xi, k_T), \tag{87}$$

where $F_2(\xi, k_T)$ obeys the differential equation

$$\xi \frac{\partial}{\partial \xi} F_2(\xi, k_T') = \int d^2k_T K(k_T', k_T) \frac{{k_T'}^2}{k_T^2} F_2(\xi, k_T). \tag{88}$$

The BFKL kernel is now known to leading and subleading order.[28,31] The leading asymptotic solution is

$$F(x, k_T) \sim x^{-4N \log \frac{2\alpha_s}{\pi}}. \tag{89}$$

Unfortunately the corrections implied by the next-to-leading calculations are gigantic[31]. There is no way of doing anything useful with BFKL scaling at present.

As previously discussed this does not seem to be a problem for HERA data since a careful analysis shows that —perhaps surprisingly— the data is well accounted for by ordinary perturbative QCD (the matter is however somewhat controversial to this date), but it will come the day where $\log 1/x$ corrections will be essential. The subject is thus still open.

Acknowledgements

It is a pleasure to thank S.Forte for discussions on some of the topics presented in this lecture. This work has been supported in part by CICYT grant AEN98-0431 and CIRIT project 1998SGR 00026. Being at the meeting has been, as always, a great pleasure. Thanks are due to B.Adeva and the rest of the organizing committee for the invitation extended to me. Thanks are also due to Mercedes Fatas, the efficient secretary of the XXVI International Meeting on Fundamental Physics for her patience in dealing with this long overdue contributor to the proceedings.

References

1. See for instance: F.J.Ynduráin, *The Theory of Quark and Gluon Interactions*, Springer Verlag; G.Sterman, *Quantum Field Theory*, Cambridge University Press, T. Muta, *Foundations of Quantum Chromodynamics*, World Scientific; E. de Rafael, in Proceedings of the GIFT Workshop on Quantum Chromodynamics, Jaca, 1979, Springer Verlag; G.Sterman, TASI lectures 1995, hep-ph/9606312.
2. C.Caso et al. (the Particle Data Group), Europ. Phys. J. C 3 (1998) 1.
3. E. Braaten, S.Narison and A.Pich, Nucl. Phys. B373 (1992) 581. See also.[14]
4. D.Adams et al. (the SMC collaboration), CERN-PPE-94-59; For theoretical discussions see for instance: G. Altarelli, in Proceedings of the International School of Subnuclear Physics, Erice 1989; A. Manohar, in Proceedings of the Polarized Collider Workshop, University Park, 1990; G. Altarelli et al., hep-ph/9803237; B.Lampe and W.Reya, hep-ph/9810270; J.Kodaira and K.Tanaka, hep-ph/9812449.
5. See for instance: E.Laenen et al, Phys. Rev.D49 (1994) 5753, and references therein. M.Gluck, and A.Vogt, Phys. Lett. B306 (1993) 391; M. Krawczyk, M. Staszel and A. Zembrzuski, hep-ph/9806291.

6. F.Bloch and A. Nordsieck, Phys. Rev. 52 (1937) 54.
7. T.Kinoshita, J. Math. Phys. 3 (1962) 650; T.D.Lee and M.Nauenberg, Phys. Rev 133B (1964) 1549.
8. T.Appelquist et al. Phys. Rev. Lett. 36 (1976) 768; Nucl. Phys. B120 (1977) 77. See also G.Sterman, *loc. cit.*.
9. G.Miller et al. (the SLAC-MIT collaboration), Phys. Rev. D5 (1972) 528.
10. P.Amaudruz et al. (the NMC collaboration), Phys. Rev. Lett. 66 (1991) 2712.
11. A.Baldit et al. (the Na51 collaboration), Phys. Lett. B332 (1994) 244.
12. E.A.Hawker at al. (the NuSea collaboration), Phys. Rev. Lett. 80 (1998) 3715; J.Peng et al (the NuSea collaboration), hep-ph/9804288.
13. G.Altarelli and G.Parisi, Nucl. Phys. B126 (1977) 298.
14. See for instance: R.K.Ellis, W.J.Stirling and B.R.Webber, in *QCD and Collider Physics*, Cambridge Univ. Press, 1996.
15. See for instance: F.J.Ynduráin, *loc. cit.*.
16. W.G.Seligman et al. (the CCFR collaboration), hep-ex/9701017.
17. T.Hebbeker, Physics Reports 217 (1992) 69; R.Barate et al. (the ALEPH collaboration), Phys. Rep. 294 (1998) 1.
18. A.Milsztajn and M.Virchaux, Phys. Lett. B274 (1992) 221.
19. W.J.Stirling, hep-ph/9709429
20. B. Foster, Int. J. Mod. Phys. A13 (1998) 1543.
21. For some recent review see e.g.: A Capella et al., hep-ph/9707466.
22. H1 collaboration, S.Aid et al., Phys. Lett. B354 (1995) 494.
23. See e.g.: S.Titard and F.J.Ynduráin, Phys. Rev D49 (1994) 6007.
24. H.Leutwyler, Phys. Lett. 98B (1981) 447.
25. See e.g.: R.K.Ellis and W.J.Stirling, Lectures given at the CERN School of Physics, FERMILAB-Conf-90/164-T.
26. G.Veneziano, Nuovo Cim. 57A (1968) 190.
27. C.Lovelace, Phys. Lett. 28B (1968) 265; J.Shapiro, Phys. Rev. 179 (1969) 1345.
28. E.Kuraev, L.Lipatov and V.Fadin, Sov. Phys. JETP 45 (1977) 199; E.Levin, in Proceedings of the Blois Conference on Elastic and Diffractive Scattering, Providence, 1993. For recent reviews see: V. Del Duca, hep-ph/9503226; L.N.Lipatov, Phys. Rep. 286 (1997) 131; S.Catani, hep-ph/9608310, hep-ph/9712442; M.Ciafaloni, hep-th/9510025.
29. A.Martin, W.J.Stirling and R.Roberts, Phys. Rev. D50 (1994) 6734.
30. R.D.Ball and S.Forte, hep-ph/9703417. S.Forte, hep-ph/9812382.
31. V.S.Fadin and L.N.Lipatov, hep-ph/9802290; R.D.Ball and S.Forte, hep-ph/9805315.

Strong WW interactions at the LHC

ANTONIO DOBADO
Departamento de Física Teórica
Universidad Complutense de Madrid
28040 Madrid, Spain

J. R. PELÁEZ[1]
Stanford Linear Accelerator Center
Stanford University, Stanford, California 94309

We present a brief pedagogical introduction to the Effective Electroweak Chiral Lagrangians, which provide a model independent description of the WW interactions in the strong regime. When it is complemented with some unitarization or a dispersive approach, this formalism allows the study of the general strong scenario expected at the LHC, including resonances.

1 Introduction

As is well known, the Standard Model (SM), which is a $SU(3)_C \times SU(2)_L \times U(1)_Y$ quantum gauge theory, is able to describe all our knowledge about the strong and electroweak interactions, even at the high level of precision reached at LEP (see for instance [1]). The SM can be divided in three sectors: The first one is the matter content (quarks and leptons), whose elementary particles interact between themselves by mediating bosons that belong to the second sector. These fields are the eight gluons associated to the $SU(3)_C$ group of the strong interactions as well as the W^+, W^- and Z bosons together with the photon, which are associated to the $SU(2)_L \times U(1)_Y$ group of the electroweak interactions. Finally, there is the so-called Symmetry Breaking Sector (SBS). It is responsible for the spontaneous symmetry breaking of the electroweak gauge group $SU(2)_L \times U(1)$ down to the electromagnetic group $U(1)_{em}$. Through the Higgs mechanism it provides masses for the W^+, W^- and Z bosons while leaving the photon massless. In addition, the SBS is also connected with the matter sector through Yukawa couplings which give rise to the quark and lepton masses, quark mixing (Cabbibo-Kowayashi-Maskawa matrix) and eventually to CP violation because of the complex phase in this matrix.

Now we arrive to the first important remark of this lecture. In contrast with the matter and the gauge sectors, *the SBS is very poorly known from the*

[a]E-mail:pelaez@slac.stanford.edu. On leave of absence from the Departamento de Física Teórica. Universidad Complutense. 28040 Madrid. Spain.

experimental point of view. In fact, several different theoretical scenarios have been widely discussed in the literature. Generically they can be grouped in three kinds, the Minimal Standard Model (MSM), the Minimal Supersymmetric Standard Model (MSSM)[2] and QCD-like theories [3]. Let us briefly review the main features and problems of these scenarios.

1.1 Minimal Standard Model

It contains the minimum ingredients to explain the present data. However, it does not shed much light on possible new physics effects and it does not address several problems, among others:

- The Higgs potential is introduced *ad hoc.* It is not a gauge interaction as the rest of the the known forces in Nature such as the strong, the electroweak or even gravity. The origin and nature of this Higgs field remains a mystery.

- Keeping the mass of this scalar field at scales close to the electroweak symmetry breaking (100 GeV to 1 TeV) requires a very fine tuning, since radiative corrections tend to make its mass of the order of the next new physics scale. This is known as the naturalness problem.

- The electron and top masses fall five orders of magnitude apart. The problem of why the masses present such a hierarchy is not addressed.

- There are hints in the literature suggesting that the simple realization of the Higgs sector in the MSM could indeed be a trivial (non-interacting) quantum field theory.

Nowadays the existence of the Higgs is taken for granted by many people as it was the case for the top quark. However the Higgs is not the top. By this we mean that the SM model would have not been consistent without the top quark (it becomes an anomalous gauge theory), whereas the Higgs boson is not a theoretical need. It is possible to postulate different versions of the SM differing in the SBS which are theoretically consistent. Indeed, in most of the physical systems that present an spontaneous symmetry breaking (like chiral symmetry breaking in QCD or, in solid state physics, the Cooper pair formation, magnetization, etc..,), there is nothing analogous to a fundamental Higgs field.

Nevertheless, for its simplicity, this model is very useful to describe the data, without additional assumptions.

1.2 Minimal Supersymmetric Standard Model

In this model, an additional symmetry relating fermions and bosons is introduced. As a consequence the Higgs potential is related to the gauge couplings and the scalar particles appear in a natural way. The advantage of this new symmetry is that for each fermion loop there is a corresponding boson loop with similar couplings and masses but opposite sign, thus avoiding the naturalness problem. However:

- Nature is *not* supersymmetric. "Soft" breaking terms must be added by hand in order to break spontaneously the $SU(2)_L \times U(1)_Y$ gauge symmetry, without spoiling too much the cancellations needed to solve the naturalness problem. Those terms break supersymmetry explicitly.

- The values of the parameters in those soft breaking terms (more than a hundred) are unknown, and they severely limit the predictive power of these models. The origin of those soft breaking terms are the origin of further speculation.

- Probably the most robust (soft breaking parameter independent) prediction is that a Higgs should appear below around 120 GeV. Thus this particle could have been produced at LEP. So far nothing has been found, but there is still a small room for discovery. However, if nothing of this kind is found at the next generation of colliders the low-energy supersymmetric scenarios would be in serious trouble.

1.3 QCD-like scenario

These models mimic the spontaneous chiral symmetry breaking of QCD and are generically known as Technicolor (TC). The Higgs does not exist as a fundamental field although some other composite fields with different quantum numbers play a similar role. However

- There is no completely consistent and universally accepted Technicolor model.

- Predictions are very vague due to the strong nature of the interactions.

- The simplest versions, like a direct rescaling of QCD, are ruled out by the LEP data or by the appearance of undesired flavor changing neutral currents.

Therefore we arrive to the second main remark of this lecture: *it possible that the SBS of the SM has nothing to do with our current theoretical expectations.*

At this point one could ask which are the main experimental constraints on the SBS or, in other words, what we really know about this sector. The main pieces of our knowledge are the following [4]:

1. First of all there must be a physical system coupled to the SM displaying a spontaneously symmetry breaking pattern from a *global* G group to a subgroup H. This symmetry breaking triggers the Higgs mechanism that breaks the electroweak *gauge* group $SU(2)_L \times U(1)_Y$ down to the electromagnetic group $U(1)_{em}$. Thus we have $SU(2)_L \times U(1)_Y \in G$ and $U(1)_{em} \in H$.

2. Since we need three would-be Goldstone bosons in order to give masses to the W^+, W^- and the Z gauge bosons, we have $dim\ G \ - \ dim\ H \ = \ 3$.

3. Experimentally we know that the ρ parameter (which measures the relative strength of the charged and neutral weak currents) is very close to one - apart from some radiative corrections proportional to the hypercharge coupling g' squared. Probably the most natural explanation for this fact is to assume that the unbroken H group of the SBS contains the so called custodial group $SU(2)_{L+R}$ (as it happens in the MSM). Any other assumption leads to some fine tuning.

   With these conditions on G and H it is very easy to show that the only possible solution is $G = SU(2)_L \times SU(2)_R$ and $H = SU(2)_{L+R}$.

4. Finally, from the muon mean life it is possible to obtain the dimensional parameter $v \simeq 250\,\text{GeV}$ which sets the scale of the SBS dynamics in the SM.

At this point it is reasonable to think whether it is possible to build a model independent description of the SBS. As we will see this can be done by using the Effective Electroweak Chiral Lagrangian (EChL), which is based on a similar formalism used in low-energy hadron physics, namely, Chiral Perturbation Theory (ChPT) [5]. As we will see, this approach is especially useful when the SBS is strongly interacting.

2 The Electroweak Chiral Lagrangian

The EChL provides a phenomenological description of the Goldstone boson dynamics associated to the symmetry breaking of $SU(2)_L \times SU(2)_R$ down to

$SU(2)_{L+R}$. As far as we are not introducing any other field, it has to be realized nonlinearly. That will limit the applicability of the approach up to the energies where the other relevant degrees of freedom show up. In the case of strong dynamics, we expect these other modes to appear at energies much higher than $v \simeq 250\,\text{GeV}$ and the formalism will be very useful. In contrast, for theories with, for instance, a light Higgs (as in the MSSM), there is no applicability region for this formalism, but in that case we will have additional information to disentangle the SBS physics when measuring these light modes.

Therefore, we will be assuming a strong SBS. For simplicity, let us then switch off momentarily the gauge fields, whose interactions with the SBS are comparatively weak. In such case, no other degrees of freedom are present at low energies except the Goldstone bosons $\omega^a(x)$, which will be gathered in the $SU(2)_{L+R}$ matrix

$$U(x) = \exp\left(\frac{i\omega^a(x)\sigma^a}{v}\right), \tag{1}$$

where the σ^a are the Pauli matrices.

A low energy expansion of the amplitudes is nothing but a derivative expansion of the Lagrangian. Then, the simplest G-invariant Lagrangian relevant at low energies (with two derivatives) can be written as

$$\mathcal{L}_2 = \frac{v^2}{4}\text{tr}\,\partial_\mu U \partial^\mu U^\dagger. \tag{2}$$

From this Lagrangian it is possible to obtain the exact behavior of the elastic low-energy scattering amplitude for the Goldstone bosons. Indeed, using the SU(2) and crossing symmetries, any amplitude can be obtained from that of $\omega^+\omega^- \to \omega^0\omega^0$, which is given by

$$A(s,t,u) = \frac{s}{v^2} + O\left(\frac{s^2}{v^4}\right). \tag{3}$$

Of course, the Goldstone bosons are not directly observable, since through the Higgs mechanism they are "eaten" by the $W^\pm$ and Z longitudinal components, that we will denote, generically, by W_L. Indeed, the so-called Equivalence Theorem [6] (ET) relates the Goldstone bosons amplitudes with the corresponding longitudinal components of the electroweak gauge bosons for the MSM, as follows

$$A(W_L^a W_L^b \to W_L^c W_L^d) \simeq A(\omega^a\omega^b \to \omega^c\omega^d) + O\left(\frac{M_W}{\sqrt{s}}\right). \tag{4}$$

This result is a consequence of the Slavnov-Taylor identities coming from the $SU(2)_L \times U(1)_y$ gauge symmetry. The $O(M_W/\sqrt{(s)})$ corrections can be understood by noting that the Goldstone bosons are massless in contrast with the gauge bosons, whose mass is $O(100\,\text{GeV})$. Note that the ET is a high energy limit, whereas the EChL is a low energy limit. Indeed, for the EChL the formulation of the Equivalence Theorem is not so simple [7] but we will not discuss the details here since, later we will unitarize the amplitudes of the effective lagrangian and in such case the above formilation is valid (the interested reader can find a complete account of this issue in [5] and [8]). At this point the following comments are in order:

- First we see that the low energy dynamics of the Goldstone bosons is dictated by symmetry and the scale v only. In this sense, it is *universal*, i.e. independent of the details of the SBS. The amplitudes obtained from eq.(2) are called the Low Energy Theorems.
- The amplitudes above grow with the energy. Thus, if we assume that no other particles modify this behavior at low energies, they give rise to strong interactions for the Goldstone bosons as well as for the longitudinal components of the electroweak gauge bosons, according to the Equivalence Theorem.
- However, the growth of this amplitudes is in conflict with unitarity around $O(1\text{TeV})$ energies.

In conclusion, provided there are no other light modes in the SBS, we expect strongly interacting $W_L's$. From unitarity constraints we also expect new physics at the TeV scale, possibly in the form of resonances.

3 Beyond the Low Energy Theorems

In order to switch on the gauge fields in the low energy EChL we change the derivatives in eq.(2) into the appropriate covariant derivatives, D_μ, containing the electroweak gauge fields. That is

$$\mathcal{L}_2 = \frac{v^2}{4}\text{tr}D_\mu U(D^\mu U)^\dagger. \tag{5}$$

In addition we can introduce the next to leading order (four derivative) terms [9] in the EChL

$$\mathcal{L}_4 = L_1\left(\text{tr}D_\mu U D^\mu U^\dagger\right)^2 + L_2\left(\text{tr}D_\mu U D^\nu U^\dagger\right)^2 + ... \tag{6}$$

where we have only displayed those terms that give rise to $O(s^2/v^4)$ contributions to the Goldstone boson elastic scattering amplitude. These $O(s^2/v^4)$ terms depend on several L_i constants, which parameterize our ignorance on the SBS. For special values we recover some particular models. For example, the MSM with a 1 TeV Higgs corresponds to $L_1 = 0.007$ and $L_2 = -0.0022$ whereas the simplest TC model with three technicolors has $L_1 = -0.001$ and $L_2 = 0.001$. In addition, under renormalization these parameters can absorb the divergences appearing in the one-loop contributions to the amplitudes coming from $\mathcal{L}_2$, which are $O(s^2/v^4)$. From precision test of the SM it is possible to set bounds on some of these parameters. However, these bounds are too weak for L_1 and L_2, which are expected to lie in the 10^{-3} to 10^{-2} range. That precision could only be reached after a few years of LHC running at full luminosity.

4 Unitarization and dispersion Relations

Customarily, the longitudinal gauge boson amplitudes are given in a basis of states of definite angular momentum, J, and the "weak isospin", I, associated to the SU(2) group. These "partial waves", t_{IJ} are also obtained as an expansion of the form

$$t_{IJ}(s) = t^{(2)}_{IJ}(s) + t^{(4)}_{IJ}(s) + O(s^3), \tag{7}$$

where the superscript refers to the corresponding energy (momentum) power. As we have already seen, they grow with the energy and violate unitarity around 1 TeV. In this basis, the elastic unitarity constraint can be easily written for physical values of s; it reads

$$\mathrm{Im}\, t_{IJ}(s) = \mid t_{IJ}(s) \mid^2 \quad \Rightarrow \quad \mathrm{Im}\frac{1}{t_{IJ}(s)} = -1, \tag{8}$$

which is basically the Optical Theorem. Although the results obtained from the Chiral Lagrangian break unitarity, they are nevertheless unitary in the perturbative sense

$$\mathrm{Im} t^{(4)}_{IJ}(s) = \mid t^{(2)}_{IJ}(s) \mid^2 \quad \Rightarrow \quad \frac{\mathrm{Im}\, t^{(4)}_{IJ}(s)}{\mid t^{(2)}_{IJ}(s) \mid^2} = -1. \tag{9}$$

It is however possible to obtain unitary amplitudes from the effective Lagrangian. Note that from eq.(8) we know exactly the imaginary part of the inverse of the amplitude. As a consequence, any unitary amplitude will satisfy

$$\frac{1}{t_{IJ}(s)} = \mathrm{Re}\frac{1}{t_{IJ}(s)} - i \quad \Rightarrow \quad t_{IJ}(s) = \frac{1}{\mathrm{Re}\, t^{-1}_{IJ}(s) - i}. \tag{10}$$

That is, *we only have to approximate the real part of the inverse of the amplitude* t_{IJ}^{-1}, by means of eq.(7). Formally: $\mathrm{Re}\, t_{IJ}^{-1} = (t_{IJ}^{(2)})^{-1}[1 - \mathrm{Re} t_{IJ}^{(4)}/t_{IJ}^{(2)} + \ldots]$. Finally, using eq.(7) we can write

$$t_{IJ}(s) = \frac{t_{IJ}^{(2)}}{1 - t_{IJ}^{(4)}/t_{IJ}^{(2)}} \tag{11}$$

which is known as the Inverse Amplitude Method (IAM). It can be derived alternatively, by writing a two subtracted dispersion relation for the inverse amplitude. Using some extra hypothesis and approximations it is possible to solve the dispersion relation for $t_{IJ}(s)$ to find the same result.

This partial wave is strictly unitary and has the proper analytical structure with the appropriate cuts. In addition it is able to reproduce poles which can be interpreted as resonances generated dynamically. Note also that, by expanding this amplitude in power of s, *we recover the chiral low-energy expansion.*

4.1 The Inverse Amplitude Method at work

The IAM method has been successfully applied in a completely different physical context: the low-energy hadron dynamics[10]. As it is well know, QCD is the proper theory to describe strong interactions, but it cannot be applied directly at low energies due to the breaking of standard perturbation theory. However in the limit where the three lightest quarks are massless, the QCD Lagrangian possesses a global symmetry (chiral symmetry) which rotates right quarks or left quarks between themselves. The symmetry group is $SU(3)_L \times SU(3)_R$ and for different reasons it is known that it is spontaneously broken to the $SU(3)_{L+R}$ group. The corresponding Goldstone bosons are identified with the pseudoscalar mesons $\pi^0, \pi^\pm, K^0, \bar{K}^0$ and η and their relative low physical masses compared with the typical hadronic scale of 1 GeV, can be considered as a perturbation effect due to the very small, but non-zero, quark masses. Note that the symmetry pattern is very close to that of the SBS (it would be the same if we just considered two quarks).

As we did before we can gather the ω^a mesons fields in an $SU(3)$ matrix as $U(x) = \exp(i\omega^a\lambda^a/F)$, where λ^a are the Gell-Man matrices and F is basically the pion decay constant. Once more we can describe the low energy hadron dynamics in terms of a chiral Lagrangian. This approach is known as Chiral Perturbation Theory (ChPT)[13]. At the lowest order this Lagrangian is given by:

$$\mathcal{L}_2 = \frac{F^2}{4} \mathrm{tr}\, \partial_\mu U \partial^\mu U^\dagger. \tag{12}$$

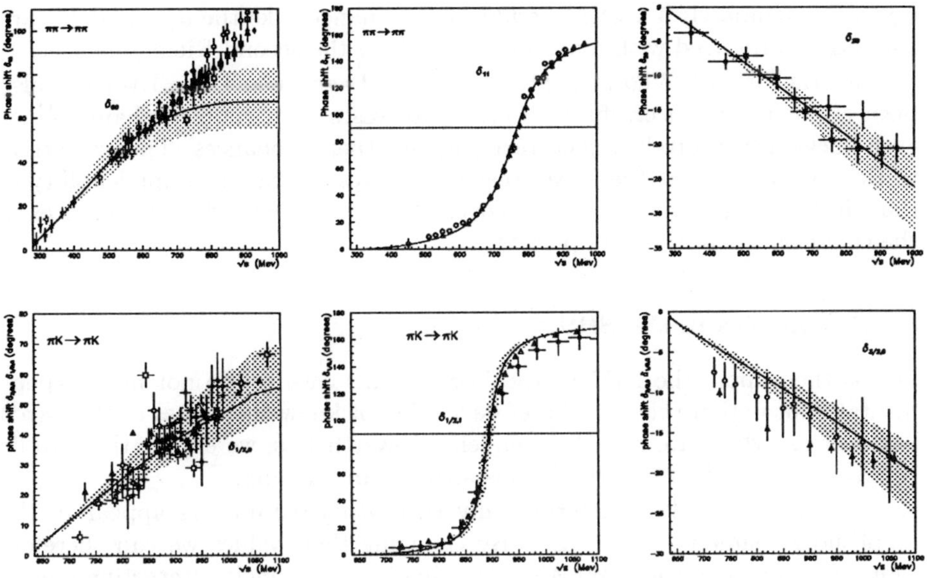

Figure 1: IAM fit to the phase shifts for $\pi\pi \to \pi\pi$ and $\pi K \to \pi K$. For data references see[5].

which reproduces the well know current algebra results in a very simple way. At the next order (four derivatives) one has additional terms whose precise form is not relevant here, although some of them have the same structure of those in eq.(6). As a matter of fact, the formalism that we have presented for the SBS is inspired in the massless limit of SU(2) ChPT, although rescaled from $F \simeq 93\,\text{MeV}$ up to $v \simeq 250\,\text{GeV}$. The main difference is the existence of real data on meson physics, from which it is possible to determine the values of the $\mathcal{L}_4$ ChPT Lagrangian parameters, whereas they are undetermined for the SBS.

The amplitudes can now be obtained as a truncated series in powers of the momentum p^2 over $4\pi F \simeq 1\,\text{GeV}$. This formalism is only suitable at low-energies up to about 500 MeV. We should not extrapolate them naively to higher energies since they would severely violate unitarity and they would not reproduce resonances.

However, we can use the IAM to extend the applicability of the effective Lagrangian approach. In Fig.1 we show an example of the results obtained with the IAM when applied to $\pi\pi \to \pi\pi$ (analogous to the Goldstone boson $\omega\omega \to \omega\omega$) and $\pi K \to \pi K$ scattering. Note that now the data is reproduced up to approximately 1 GeV. In addition, resonances like the σ, ρ and K^* are correctly reproduced with an associated pole in the second Riemann sheet.

Starting from the corresponding effective Lagrangians, the IAM has also been applied very successfully to other processes with coupled channels[11] or even nucleons[12], reproducing correctly many other resonances. Thus we arrive to the conclusion that the IAM greatly improves the range of applicability of the effective Lagrangians and, moreover, it is able to reproduce resonances in the channels where they are present.

5 Resonances in the SBS

Let us then apply the IAM to the SBS. In this case we do not have experimental information yet, and therefore we do not know the values of the $O(p^4)$ parameters, which are model dependent. Nevertheless, we expect them to lie between 10^{-2} and 10^{-3} if the SBS is strongly interacting.

For elastic $W_L W_L$ scattering only two $O(p^4)$ parameters appear in the amplitudes, namely, L_1 and L_2. By changing their values we can therefore reproduce the behavior of $W_L W_L$ scattering in any strongly interacting model. In Fig.3 we show the phase shifts δ_{IJ} that we expect in three different models. The first one corresponds to the SM with 1 TeV Higgs and, consistently, we see a resonance in the scalar isoscalar channel (a "Higgs"). The second set of values mimics a simple TC model with three technicolor and thus presents features very similar to ChPT (compare with the $\pi\pi$ curves in Fig.1), mainly, a vector resonance (a "techni-ρ). The last model is chosen to show two behaviors that deserve further comments. First, it could happen that a resonance becomes so broad that it may be hard to identify as a resonance, in such case we say there is a "saturation" of unitarity. (The situation with the σ particle in QCD is of this kind). Second, we have to remember that in the effective Lagrangian approach we only have a finite theory order by order in energy, but it is not renormalizable in the strict sense. Thus, the set of possible consistent fundamental theories is "smaller" than that of effective theories. By that we mean that there could be a choice of parameters which are *not* the low-energy limit of any fundamental theory. That may seem obvious if we take an absurd value for some parameters, like $L_1 = 10^6$. But it could also occur for values that look "reasonable". In such case, however, we would find inconsistencies in the effective theory. That happens indeed for the last model in Fig.3, which yields a pole in the *first* Riemann sheet of the $I = 2$ channel, which should not be present in a renormalizable quantum field theory. That can be used as a criterium to exclude a set of parameters. There are several arguments that support this interpretation[14].

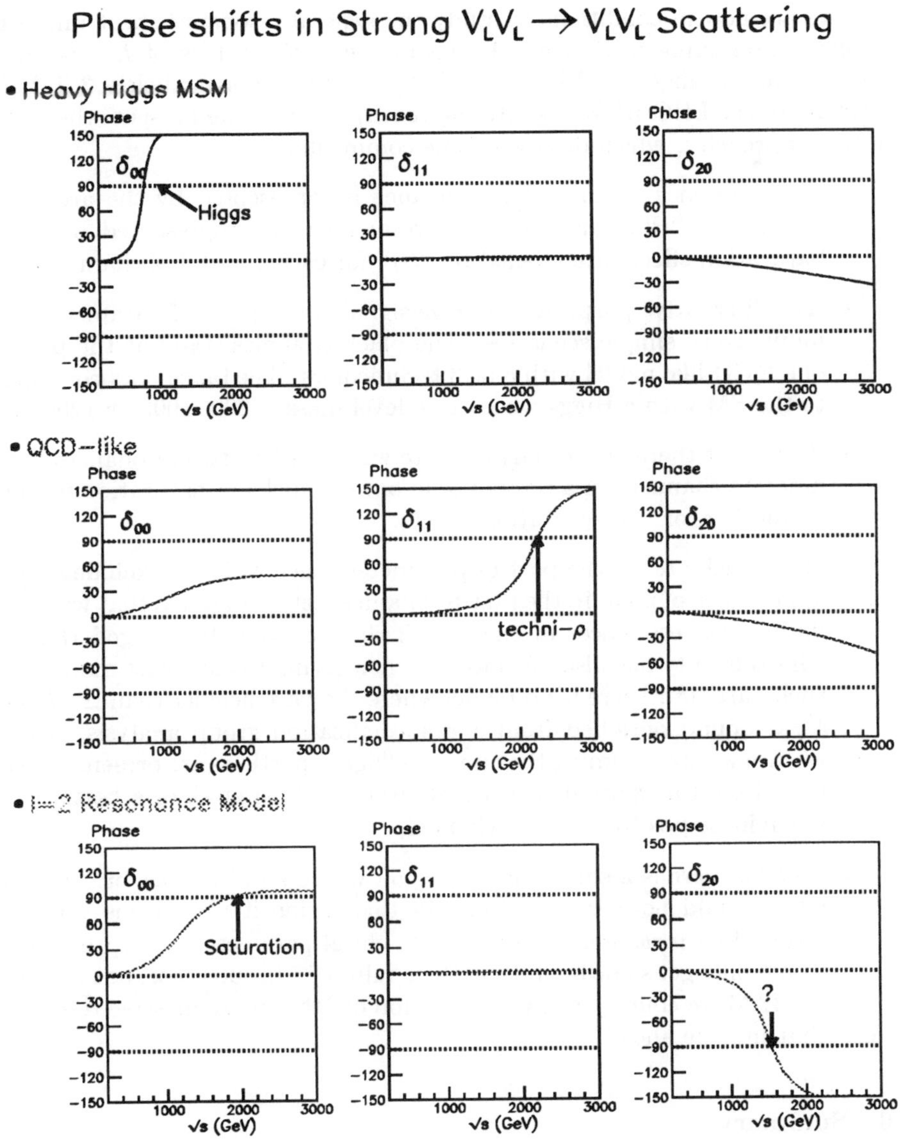

Figure 2: Phase shifts expected for different choices of electroweak chiral parameters.

We are now in conditions to study the general resonance spectrum of the strongly interacting SBS. We only have to vary the values of L_1 and L_2 in their expected ranges, and identify what resonances appear below 3 TeV (an estimate of the LHC $W_L W_L$ scattering reach). In Fig.5 we presents the results of this approach[14], which deserves some comments:

- The presence of an scalar resonance is represented by the areas that contain an "H", whereas vector resonances are represented by a "ρ". Saturation effects are labeled by "S_I" for each $I = 0, 1, 2$ channel.
- For illustrative purposes, we have signaled the pairs of parameters that mimic some simple scenarios. The black triangles stand at the position of a QCD-like model with 5 or 3 technicolors. The black dots correspond to the SM with a Higgs whose tree level mass is 800, 1000 or 1200 GeV.
- Note that there are scenarios where we could find two resonances in two different channels, or a resonance in one channel an a saturation behavior in another, or two saturation effects.
- The black area is the part of parameter space which is excluded by the appearance of poles in the first Riemann sheet, it suggests that we cannot find *heavy* resonances in the $I = 2$ channel (doubly charged Higgses). This difficulty has also been found when trying to construct models with such particles: there is no model where they are heavier than $\simeq 375$ GeV [15], a bound obtained from a renormalization group analysis. In this lecture we are assuming that no such "light" particles are present. In such case, from the figure it seems that either nothing at all or a "saturation" behavior is possible in that channel.
- Finally, there is a small shaded region where no resonance or saturation effect would be clearly visible. In this region it also seems very hard to obtain a measurement of just the chiral parameters and probably we would only get some bounds on their values[16]. In such case, not even with the IAM we could get any information of other more massive resonances that may lie ahead.

6 Summary

The main conclusions from the discussion below are the following:

- There is not any fundamental reason for the Higgs (Standard or Supersymmetric) to exist. We should therefore keep an open mind to alternative scenarios.

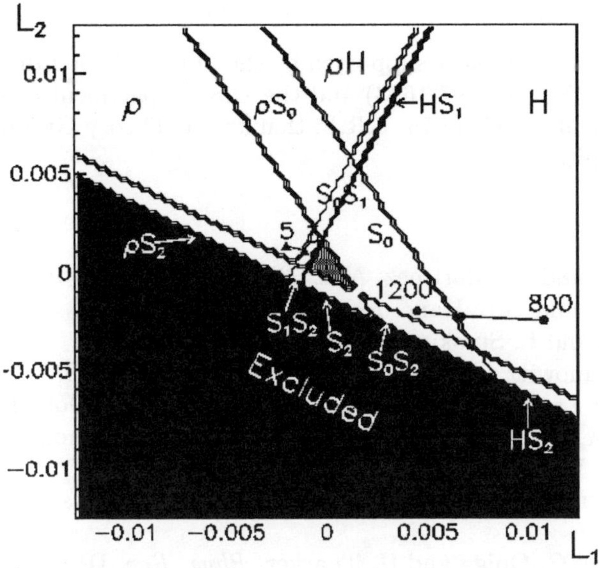

Figure 3: Resonance spectrum of the strong SBS in the L_1, L_2 plane. The black area is excluded. On the white areas, we have represented broad resonances or saturation effects in the I channel by S_I; Higgs-like narrow resonances by H and ρ-like narrow resonances by ρ. In the grey area there is no saturation of unitarity, nor resonances, below 3 TeV. The black dots represent the MSM with $M_H = 800, 1000, 1200$GeV and the triangles a QCD-like model with 3 or 5 technicolors.

- However unitarity requires new physics to appear before below 1 TeV.

- This new physics could be new particles in the best of the worlds or even a completely new and unexpected physics.

- In the worst case we will have an enhancement of the WW production. It will be difficult to observe at the LHC but not impossible. A lot of work should be done in this direction and chiral Lagrangians, supplemented with the inverse amplitude method, can provide a model independent approach to new phenomena like the strong $W_L W_L$ scattering and the resonances that may appear over the LHC energy range.

- In any case we have to wait for the LHC with an open mind. Nature will tell us.

Acknowledgments

This work has been partially supported by the Ministerio de Educación y Ciencia (Spain)(CICYT AEN97-1693) and the U.S. Department of Energy under contract DE-AC03-76SF00515. J.R.P. thanks the Theory Group at SLAC for their kind hospitality.

References

1. C. Maña and M. Martínez, *Nuc. Phys. Proc. Suppl.* B31 (1993)163.
2. H. Haber and G.L. Kane, *Phys. Rep.* 117 (1985) 75.
3. E. Farhi and L. Susskind, *Phys. Rep.* 74 (1981) 277.
4. M.S. Chanowitz, *Ann. Rev. Nucl. Part. Sci.* 38 (1988) 323.
5. *Effective Lagrangians for the Standard Model* A. Dobado, A. Gómez-Nicola, A.L. Maroto. and J.R. Peláez. Texts and Monographs in Physics. Springer Verlag (1997).
6. J.M. Cornwall, D.N. Levin and G. Tiktopoulos, *Phys. Rev.* D10 (1974) 1145
   B.W. Lee, C. Quigg and H. Thacker, *Phys. Rev.* D16 (1977) 1519
   M.S. Chanowitz and M.K. Gaillard, *Nucl. Phys.* B261 (1985) 379
7. H. J. He, Y. P. Kuang, X. Y. Li , *Phys. Lett.* B329 (1994) 278. A. Dobado and J.R. Peláez,*Phys.Lett.* B329 (1994) 469; *Nucl. Phys.* B425 (1994) 110.
8. A. Dobado, J.R. Peláez and M.T. Urdiales, *Phys.Rev.* D56 (1997) 7133.
9. A. Dobado and M.J. Herrero, *Phys.Lett.* B228 (1989) 495.
10. A. Dobado and J.R. Peláez, *Phys. Rev.* D56 (1997) 3057.
11. J.A. Oller, E. Oset and J.R. Peláez, *Phys. Rev. Lett.* 80 (1998) 3452; hep-ph/9804209, SLAC-PUB-7787; F. Guerrero, J. A. Oller, hep-ph/9805334.
12. E. Oset and A. Ramos, nucl-th/9711022 .
13. S. Weinberg, *Physica* A96 (1979) 327
    J. Gasser and Leutwyler, *Ann. Phys.* 158 (1984) 142.
14. J.R. Peláez, *Phys. Rev.* D55 (1997) 4193.
15. H. Georgi and M. Machaceck, *Nucl. Phys.* 262 (1985) 463. R.S. Chivukula, M.J. Dugan and M.Golden, *Phys. Lett.* B366 (1994) 62.
16. A. Dobado, M.J. Herrero, J.R. Peláez, E. Ruiz Morales and M.T. Urdiales, *Phys.Lett.* B352 (1995) 400.

Neutrino Oscillation Industry

Jose A. Hernando
Santa Cruz Institute for Particle Physics (SCIPP),
University of California at Santa Cruz, Santa Cruz, CA-95064,
E-mail: hernando@scipp.ucsc.edu

After an introduction to the theory of neutrino oscillation, the experimental evidences in its favor are summarized. The latest results from the SuperK, CHORUS and NOMAD experiments are presented. The discovery of the zenith angle dependence of the atmospheric neutrino flux made by the SuperK experiment is the strongest evidence so far in favor of neutrino oscillations. Finally, an outlook on the future of the field is given.

On the 5th of June, during the 98 neutrino conference, held at Takayama, Japan, the SuperK collaboration made public evidence for atmospheric neutrino oscillations [1]. It was during the same week that we were at the beautiful A Toxa island in Spain for the XXVI International Meeting on Fundamental Physics. This announcement caused great excitement in the physics community and a rebirth of interest in neutrino oscillations. The existence of oscillations implies that neutrinos have mass which in turn would be the first evidence for physics beyond the Standard Model and would require a revision of our concept of the universe.

The SuperK results do not give answers to all the unresolved problems in neutrino physics. The history of neutrino oscillations is long and not yet finished [2]. It started 25 years ago, when the Homestake experiment, in South Dakota, first reported a solar neutrino flux deficit [3]. Nowadays there are three "indications" of a possible existence of neutrino oscillations: a) the deficit in the solar neutrino flux observed by solar neutrino experiments, b) the zenith angle dependence of the atmospheric neutrino flux measured by the SuperK experiment, c) the claim of a direct observation of the oscillation $\nu_\mu \rightsquigarrow \nu_\tau$ by the LSND collaboration [4]. Any two of the above mentioned results are compatible with each other within the framework of the Standard Model, but the SM can not accommodate all three results at the same time. A new generation of experiments will investigate the solar neutrino problem in the next few years. In addition, there are proposals at FNAL and at CERN for experiments to check the LSND result in the near future. Beyond this, a number of experiment proposals to investigate further neutrino oscillations are currently being planned and discussed.

This note is organized as follows: the first section gives the theoretical framework of neutrino oscillations; in the second section, the experimental results on the neutrino oscillation problem are summarized; the third section is dedicated to the latest results from the SuperK, NOMAD and CHORUS experiments. The note concludes with an outlook on the future of the field.

1 Neutrino Oscillations

The neutrino might be regarded as one of the most "intriguing" components of matter. It can be defined as *almost nothing*. According to the Standard Model the neutrino has no mass, no charge and no color and has only weak isospin, i.e., it only interacts *weakly*. Measurements have demonstrated that, indeed, the neutrino mass, if at all different from zero, must be very small ($m_{\nu_e} < 5$ eV). On the other hand, there is no fundamental principle that forces the neutrino to be massless. It is straight forward to incorporate massive neutrinos in the SM Furthermore, the "see-saw" mechanism can be invoked as a way of explaining the smallness of the neutrino mass compared to other fermions.

Massive neutrinos would have dramatic implications on astrophysics and particle physics. Only 10% of the critical mass of the universe is visible; the remaining 90% is believed to be dark matter. There is evidence in favor of this assumption, such as the velocity of the galactic planes that can only be explained by the existence of "invisible" mass in the halo of the galaxies. Present models favor scenarios in which the dark matter is made of 30% "hot" matter (matter that was relativistic at the moment it decoupled from the universe expansion, such as the neutrinos) and 70% of "cold" matter. An excellent candidate for hot dark matter would be a neutrino of 7 eV mass.

In particle physics, if neutrinos have mass, oscillations between families may occur. The weak and the mass eigenstates do not need to be the same, in which case, they are related via a unitary matrix. Considering the simple case of two families, the ν_μ and ν_τ neutrinos can be expressed by a combination of two massive neutrinos ν_1, ν_2:

$$\nu_\mu = \cos\theta\ \nu_1 + \sin\theta\ \nu_2 \tag{1}$$

$$\nu_\tau = -\sin\theta\ \nu_1 + \cos\theta\ \nu_2 \tag{2}$$

An initially pure ν_μ beam is, in fact, composed of a fraction of both ν_1 and ν_2 neutrinos. As the beam propagate, its composition changes if the m_{ν_1} is different from m_{ν_2}, since the neutrinos evolve in time according to their energies ($e^{-iE\,t}$, where $E = \sqrt{m^2 + p^2}$). One can think of this as the ν_2, given that it is the heaviest neutrino, traveling slower than the ν_1. Therefore, there is a non-zero probability of finding ν_τ's at a distance L from the source. Simple algebra leads to the following formula:

$$P_{\nu_\mu \nu_\tau} = \sin^2 2\theta \sin^2(1.27 \frac{\Delta m_{12}^2 L}{E}), \tag{3}$$

where, L is the distance from the detector to the source (in km), E is the neutrino energy (in GeV) and $\Delta m_{12}^2 = |m_2^2 - m_1^2|$ is the difference of the squared neutrino masses (in eV^2).

There are four important parameters in the oscillation probability. Two parameters describe the underlying physics: Δm_{12}^2 and $\sin^2 2\theta$. These parameters define the two dimensional plane in which the results of neutrino oscillation experiments are usually depicted. The other two parameters are the neutrino energy E and the distance between the source and the detector L.

Natural neutrino sources are the Sun (and all the stars), and neutrinos stemming from the interactions of cosmic rays with particles in the atmosphere. According to formula 3, solar neutrino experiment are only sensitive to very small differences of squared masses ($\Delta m_{eX}^2 \sim 10^{-11} - 10^{-5}$ eV^2), since the ν_e's produced in the Sun, with average energies of 0.6 MeV, travel $\sim 1.5 \times 10^9$ Km, on their way from the Sun to Earth. Atmospheric neutrino experiments are sensitive to $\Delta m^2 \sim 10^{-3}$ eV^2, since the atmospheric neutrinos are ν_e's and ν_μ's with average energies of ~ 1 GeV and travel at least 15 km before detection. Man-made neutrino sources are nuclear reactors and particle accelerators. Reactors produce $\bar{\nu_e}$ of $E < 10$ MeV. Usually the detectors are positioned close to the source, because otherwise the neutrino flux is too small, and they explore $\Delta m_{eX}^2 \sim 1$ eV^2. The most energetic man-made neutrinos are produced at particle accelerators, where a charged particle beam hits a target and a neutrino beam arises from the decay of π, K and μ's. The CERN SPS neutrino beam, for example, is composed of ν_μ's of ~ 30 GeV. Different distances between source and detector allow us to explore different regions in Δm^2. The so-called Short Base Line Experiments (SBLE) position the detector typically at 1 km to explore "large" $\Delta m^2 \sim 10$ eV^2. Long Base Line Experiments (LBLE) with distances between source and detector of typically 1000 km, study the region $\Delta m^2 \sim 10^{-2}$ eV^2.

2 The Neutrino Oscillation Problem.

The LSND and SuperK experiments have reported evidence of neutrino oscillations. LSND, an accelerator experiment in Los Alamos, New Mexico, claims to have observed the oscillation $\bar{\nu_e} \rightsquigarrow \bar{\nu_\mu}$ with $\Delta m^2 \sim 0.8$ eV^2. SuperK, by measuring the atmospheric neutrino flux, has found evidence for a $\nu_\mu \rightsquigarrow \nu_X$ oscillation (where X cannot be e) with $\Delta m^2 \simeq 2.3 \times 10^{-3}$ eV^2. In addition,

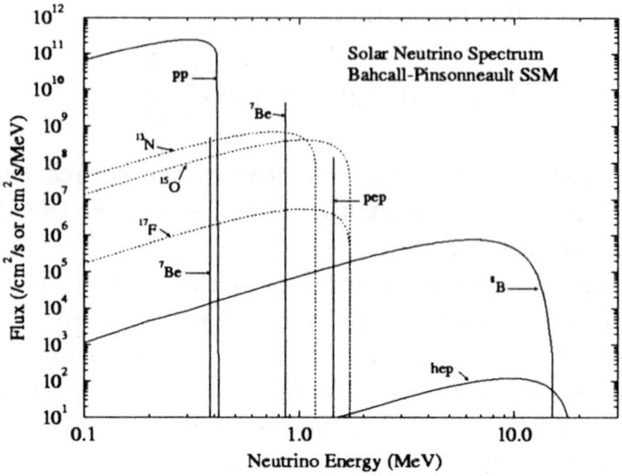

Figure 1: Solar neutrino energy spectrum as calculated from the SSM.

the observed deficit in the solar neutrino flux measured by several experiments, can be explained by a $\nu_e \rightsquigarrow \nu_X$ oscillation with $\Delta m^2 < 10^{-5}$ eV^2. All three of these results do not fit into the Standard Model, where only *two* differences of squared masses can be defined with three neutrino families. This is known as the neutrino oscillation problem. What follows is a more detailed discussion of the different experiments and their results.

2.1 Solar Neutrino Experiments

The Sun is an intense source of neutrinos. It produces 6.4×10^{10} $\nu_e/\text{cm}^2\text{s}$ with average energies of 0.6 MeV. As has been mentioned before, the Δm^2 exploration range of solar neutrino experiments is $10^{-5} - 10^{-11}$ eV^2. In order to compensate for the low interaction rate of these neutrinos, their detectors are large and heavy. The only possible way of detecting oscillations is via disappearance. This implies a good knowledge of the solar neutrino flux expected for massless neutrinos. This flux is calculated in the framework of the so-called Standard Solar Model, SSM [5]. Fig. 1 shows the solar neutrino energy spectrum which has the following components: *pp* neutrinos (85%), produced during the *pp* cycle, *Be* neutrinos (15%), from the *Be* cycle, and *B* neutrinos which amount to only 0.02% but are the most energetic and thus the easiest to detect.

The SAGE and GALLEX [6] experiments (*Ga* experiments) have a detection threshold of 233 KeV and detect mainly *pp* neutrinos. The Homestake experi-

Experiment	Data/SSM ratio
GALLEX	$0.55 \pm 0.04 \pm 0.03$
Homestake	$0.28 \pm 0.02 \pm 0.14$
Kamiokande	$0.42 \pm 0.06 \pm 0.07$
SuperK	$0.368 \pm 0.08 \pm 0.013$

Table 1: Ratio of measured solar neutrino flux and predicted value by the Solar Standard Model, for the different experiments.

ment (*Cl* experiments) with a large threshold of 815 keV, is mainly sensitive to the *Be* neutrinos. In addition, Kamiokande [7] and SuperK (Čherenkov experiments) measure at the tail of the energy spectrum, i.e, mainly *B* neutrinos. Table 1 shows a comparison of the flux detected by these experiments and the predicted flux according to the SSM. The detected flux, in all these cases, is smaller than expected for massless neutrinos. This deficit is found to depend on the energy. This is known as the solar neutrino problem. The fit to the SSM spectrum suggests that the *Be* neutrinos either are strongly suppressed or not produced at all [9]. This is a surprising conclusion, since the *Be* cycle starts the chain reaction in the Sun that later produces *B* neutrinos. Consequently, if the *B* neutrinos are detected on Earth, the *Be* neutrinos should be detected as well.

The deficit of the observed solar neutrino flux can be explained by $\nu_e \rightsquigarrow \nu_X$ oscillations. To understand its dependence on energy the Mikehyev-Smirnov-Wolfenstein (MSW) effect can be employed [8]. This effect takes into account that the ν_e neutrinos may interact with the electrons of the surrounding atoms which changes their travel velocity and thus their oscillation parameters. The MSW predicts the existence of a resonance condition that enhances (or diminishes) the probability of the oscillations. If the *Be* neutrino energy is such that the resonance condition is fulfilled, then all the *Be* neutrinos will have oscillated before reaching earth.

The solar neutrino problem can be interpreted as a $\nu_e \rightsquigarrow \nu_X$ oscillation with three possible solutions [9]: a) the vacuum solution (which does not use the MSW effect) with a large mixing angle ($\sin^2 2\theta \simeq 1$) and $\Delta m^2 \simeq 10^{-11}$ eV^2; b) a MSW solution with a small mixing angle $\sin^2 2\theta \simeq 8 \times 10^{-3}$ and $\Delta m^2 \simeq 5 \times 10^{-6}$ and c) a second MSW solution with large mixing angle $\sin^2 2\theta \simeq 0.63$ and $\Delta m^2 \simeq 1.6 \times 10^{-5}$ eV^2. Fig. 2 shows the regions allowed by the results of the different experiments. The MSW solutions correspond in the figure to the intersections of the different allowed areas.

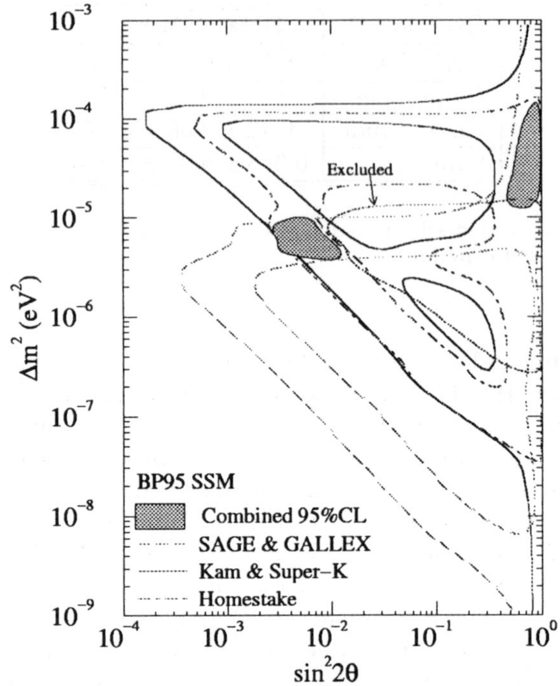

Figure 2: Regions of Δm^2-$\sin^2 2\theta$ plane allowed at 90% C.L. by the solar neutrino results using the SSM preditions that incorporate the MSW effect.

2.2 Atmospheric Neutrinos

An interaction of cosmic rays (mainly protons) with the particles of the atmosphere generates a hadronic cascade that, while evolving, produces ν_μ and ν_e neutrinos. The uncertainty in the neutrino flux that reaches the ground is quite large ,~ 25 %, due to poor understanding of the cosmic ray flux and shower fluctuations. Instead of using the flux, the flux ratio ν_μ/ν_e is employed. In this way the common errors cancel and the uncertainty in the ratio amounts to 5%. The ratio can be computed in a rather naïve way by assuming that every $\pi^- \to \mu^- + \nu_\mu$ decay is followed by $\mu^- \to e^- + \nu_\mu + \nu_e$; so there are two ν_μ's for every ν_e. As the energy increases, the ratio increases because some muons arrive at the ground prior to decaying. A common way to present the results is to compare the data ratio $\nu_\mu/\nu_e|_{data}$ with the one expected from MC simulations for massless neutrinos ($\nu_\mu/\nu_e|_{MC}$). This is called *the double ratio.*

SuperK has superseded the results obtained by previous experiments with smaller errors. SuperK has measured a double ratio of 63% which implies a deficit in the ν_μ rate (the ν_e rate measured at the ground agrees well with the simulation), which can be explained by a $\nu_\mu \rightsquigarrow \nu_X$ oscillation. Furthermore, the angular distribution of these events shows a dependence on the neutrino incident angle or zenith angle. There are less muons coming from "underneath" than from "above". Neutrinos from "above" have crossed $\sim$ 15 km of atmosphere, while those from "underneath" come from the other side of the Earth and have traveled $\sim$ 13000 km. This result indicates that the length of the oscillation wave is comparable to the radius of the Earth. The zenith angle distribution favors the parameters: $\Delta m^2 \simeq 2 \times 10^{-3}$ eV^2 and a large mixing angle.

2.3 The LSND result.

LSND operated at the Los Alamos Meson Physics Facility (LAMPF), New Mexico. The Los Alamos beam is created from a 800 MeV proton beam which is stopped in water. The neutrinos, mainly $\bar{\nu_\mu}$, arise from the π^+ and μ^+ decays that follow the stopping of the proton beam.

The LSND detector is located 30 m upstream the beam stop. Its main component is a tank of 167 tons of liquid scintillator. The walls are covered by 1220 8" PTMs that collect both Čherenkov and scintillator light. An anti-coincidence veto system surrounds the detector from all sides except from the bottom.

LSND searches for the appearance of $\bar{\nu_e}$ via the charged current interaction: $\bar{\nu_e} + p \rightarrow e^+ + n$. The e^+ emits Čherenkov light which is detected by the PTMs. From their signal the vertex position and the energy and direction of the electron can be measured. The n is re-absorbed via $n + p \rightarrow d + \gamma$. The γ of 2.2 MeV energy is detected by the PTMs. The search criteria requires correlation in time and location of the γ with the e^+ in order to reject the background. With an electron energy above 36 MeV, 22 events have been observed. This number should be compared with the number of expected background events, which is 4.5 ± 0.6. The probability that this excess is caused by a statistical fluctuation is $P < 4.1 \times 10^{-8}$.

The LSND collaboration explains this result as a consequence of the neutrino oscillation $\bar{\nu_\mu} \rightsquigarrow \bar{\nu_e}$. The oscillation probability has been computed to be: $P_{\nu_e\nu_\mu} = 3.3 \pm 1.1(stat.) \pm 0.5(syst.)$. The area allowed by this result is shown in Fig. 3. Most of the area has been already excluded by previous experiments (Budgey [10] and E776 [11]) and recently by KARMEN2 [12]. These measurements reduce the allowed area to a small strip centered around the values: $\Delta m^2 \simeq 0.8$ eV^2 and $\sin^2 2\theta \simeq 5 \times 10^{-2}$.

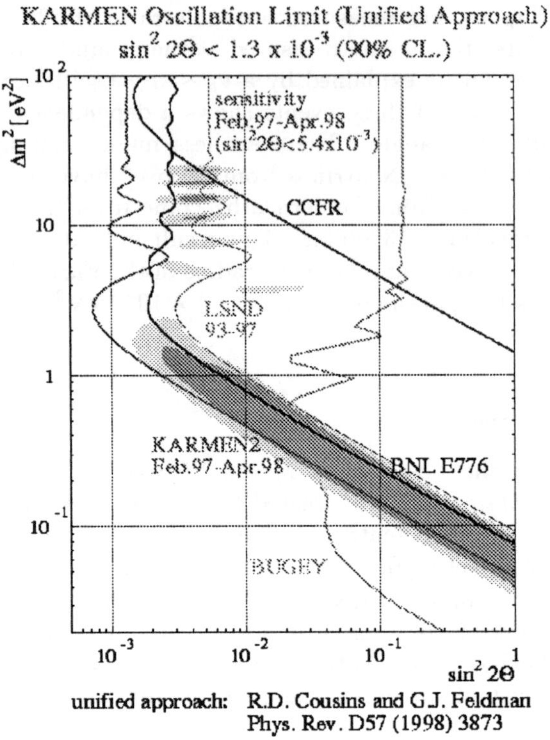

Figure 3: Region of Δm^2-$\sin^2 2\theta$ plane allowed by the LSND (shaded area) at 90 and 95 C.L., together with the exclusion regions form the previous experiment and KARMEN2.

3 SuperK, CHORUS and NOMAD

The SuperK, CHORUS and NOMAD experiments search in different areas of the Δm^2-$\sin^2 2\theta$ plane. SuperK studies atmospheric neutrinos while CHORUS and NOMAD search for a $\nu_\mu \rightsquigarrow \nu_\tau$ oscillation in the relevant cosmological range. While SuperK can be regarded as an example of a LBLE, CHORUS and NOMAD are SBLEs. Compared with SuperK, they employ lighter detectors with fine granularity.

3.1 SuperK

SuperK is the upgraded version of Kamiokande. Kamiokande (89-95 Kamioka, Japan) was a successful experiment. Designed to observe proton decay, as predicted by some Grand Unification Theories, it became the first experiment to detect neutrinos originating from a supernovae. Furthermore, it measured the atmospheric neutrino flux and found out its zenith angle dependence. However, a proton decay was never seen. SuperK is more than one order of magnitude bigger in size than Kamiokande; it has better energy (14% vs 19%), angular (26° vs 27°), and position (87 cm vs 100 cm) resolution. Its dimensions are impressive. SuperK uses a giant tank of 50000 tons of pure water, located 1000 m under ground level, in the Kamioka mine in Japan. The inner volume of the detector is a cylindrical tank 33 m in diameter and 36 m in height. It contains 32000 tons of water. Its walls are covered by 11146 20" PTMs that collect Čherenkov light. The outer volume, filled with 18000 tons of water, acts as an anti-coincidence veto system.

In two years of data taking, since April 1996, the quality and number of events recorded by SuperK has exceeded those of Kamiokande. The results of SuperK have confirmed and improved those of Kamiokande. In particular the confirmation of the zenith angle dependence of the atmospheric neutrinos flux is the strongest evidence of the existence of neutrino oscillations so far. SuperK has also observed the solar neutrino flux deficit. In addition, SuperK has recently presented preliminary results on the solar neutrino flux with sufficient precision to check some of the MSW effect predictions.

Atmospheric neutrinos

SuperK detects on average 8.53 atmospheric neutrinos per day. They are classified as e-like or μ-like events according to their Čherenkov light. The electron light for example is fuzzier and more spread out than the light produced by muons. The misidentification rate has been estimated to be < 2 %. From these events, the ratio $\nu_\mu/\nu_e|_{data}$ is calculated and compared to the MC expectations ($\nu_\mu/\nu_e|_{MC}$). The data sample is divided into a subGeV sample (i.e. the total event energy is lower than 1.33 GeV) and a multiGeV sample (i.e. the energy is greater than 1.33 GeV). In 533 days of data taking, 1231 events have been classified as e-like and 1158 as μ-like for the subGeV sample; in the multiGeV sample, the numbers are: 290 μ-like, 230 e-like. The double ratios obtained are: $\frac{\mu/e|_{Data}}{\mu/e|_{MC}} = 0.63 \pm 0.026(stat.) \pm 0.05(syst.)$ for the subGeV and $\frac{\mu/e|_{Data}}{\mu/e|_{MC}} = 0.65 \pm 0.05(stat.) \pm 0.08(syst.)$ for the multiGeV. These values are in agreement with those obtained by Kamiokande.

Furthermore, the angular distribution of these events is found to depend on the zenith angle (or the neutrino incident angle) (see Fig. 4). It has been observed that there are less up-going muons (coming from "underneath") than down-going muons (coming from "above"). For example, The fit of the data

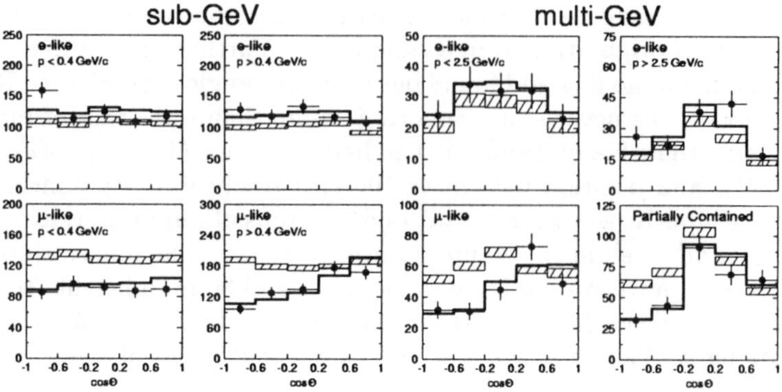

Figure 4: Zenith angle distributions of μ-like, e-like events for subGeV and multiGeV data sets. Upgoing particles have $cos\Theta < 0$. The shaded region shows the Monte Carlo expectation for no oscillations with statistical errors. The bold line is the best-fit expectation for $\nu_\mu \rightsquigarrow \nu_\tau$ oscillations.

to the distribution predicted by the simulations, gives a 5×10^{-4} C.L. in the multiGeV sample. On the other hand, the data can be nicely fit with a distribution obtained assuming $\nu_\mu \rightsquigarrow \nu_X$ oscillations. The best-fit favors the parameters: $\Delta m^2 \simeq 2.2 \times 10^{-3}$ eV^2 and a large mixing angle. This value of Δm^2 is one order of magnitude smaller than the one obtained by Kamiokande; although both results are consistent within errors, this shift has serious implications for future LBLEs (see discussion in next section). Early this year, the Chooz collaboration, a LBL reactor experiment that started data taking in 1997 in France, excluded the possibility of a $\nu_\mu \rightsquigarrow \nu_e$ oscillation on the area allowed by the SuperK result. Therefore, the results from SuperK can only be explained by an oscillation: $\nu_\mu \rightsquigarrow \nu_X$, where X stands for a τ or a sterile neutrino (sterile neutrinos are neutrinos that even do not interact weakly).

Solar Neutrinos

SuperK detects solar neutrinos with energies greater than 6.5 MeV, so it is mainly sensitive to B neutrinos. The neutrinos interact via elastic scattering with electrons. This process has the advantage that the ejected electron carries most of the energy and direction of the incident neutrino. The collection of Čherenkov light emitted by the electron, makes it possible to measure its energy and direction, and to compute the interaction vertex. The detector has been calibrated with a LINAC electron beam.

In 504 days of operation, SuperK has recorded 87051 solar neutrino events.

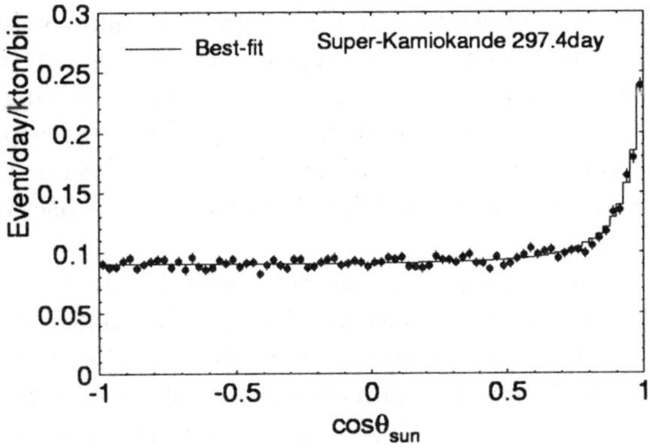

Figure 5: The cosine of the angle between the electron direction and a radius vector from the Sun. One obtaines a clear peak from solar neutrinos. The solid line shows the best fit to the data.

Fig. 5 shows the angular distribution of the electron direction with respect to the position of the Sun. The peak that stands out over a flat background corresponds to solar neutrinos. The measured flux has been compared to the SSM prediction [13], the ratio data/SSM is $0.368 \pm 0.08(stat.) \pm 0.013(sys.)$. In conclusion, SuperK has also confirmed the solar neutrino flux deficit observed by previous experiments. SuperK has improved the Kamiokande results with its larger statistics and smaller errors.

The MSW effect predicts some distortions in the flux and energy spectrum of the solar neutrinos. First, the ν_e flux should be enhanced during the night observations, since some ν_e's could be regenerated traversing the Earth. For the moment, SuperK has not seen any solar neutrino flux excess during any period of night. This result reduces the allowed area of the MSW-large angle solution. Second, the neutrino energy spectrum should be distorted with less neutrinos of energies close to the Be neutrino energy. SuperK has recently made public preliminary results where no evidence of this distortion in the ν_e spectrum has been seen [15]. However, the precision of this result is not good enough to yet exclude the existence of the MSW effect.

3.2 CHORUS and NOMAD

CHORUS and NOMAD search for $\nu_\mu \rightsquigarrow \nu_\tau$ oscillations in the CERN SPS neutrino beam at CERN. They are located, one behind the other, 700 m away from the neutrino source. The beam consist mainly of ν_μ's with a broad energy spectrum with average energy of 27 GeV and maximum values of 120 GeV. The two experiments search for the appareance of a ν_τ in the beam, which will interact with material of the detector via a charged current interaction: $\nu_\tau + N \to \tau + X$ (were X is a hadronic jet). These experiments detect τ's using complementary techniques. CHORUS tries to observe the tracks of a τ decay in a bulk of emulsion, while NOMAD attempts to discover τ's only with kinematical criteria.

CHORUS started taking data in 1994 and finished in 1997; NOMAD started one year later in 1995 and has finished in 1998. When the experiments were proposed, in the beginning of the decade, the fashionable theory of the universe suggested that the main component of dark matter would be neutrinos with masses up to ~ 30 eV. Assuming that the ν_τ is much heavier than the other two neutrinos, the most promising way of observing oscillation, would be: $\nu_\mu \leadsto \nu_\tau$. CHORUS and NOMAD were designed to search for oscillations at $\Delta m^2 \sim 10$ eV^2 up to very small mixing angle values ($\sin^2 2\theta > 4 \times 10^{-4}$). The discovery by COBE of anisotropies in the early universe, a few years ago, limited the possible contribution of neutrinos to dark matter to about 30%. As a consequence, the neutrino mass can not exceed 7 eV. The sensitivity of CHORUS and NOMAD at the new favored range $\Delta m^2 \sim 1$ eV^2, became rather marginal ($\sin^2 2\theta > 10^{-1}$).

With more than half of the data analyzed, CHORUS and NOMAD have not found indications of $\nu_\mu \leadsto \nu_\tau$ oscillation. They have reported an oscillation probability limit of $P_{\nu_\mu \nu_\tau} < 6 \times 10^{-4}$ at 90% C.L.; or in other words, for large $\Delta m^2 > 10$ eV^2 they have excluded a mixing angle of $\sin^2 2\theta > 1.2 \times 10^{-3}$. If no evidence is found in the rest of the data, the probability limit will be: $P_{\nu_\mu \nu_\tau} < 2 \times 10^{-4}$.

CHORUS

The main CHORUS [16] component is a 80 kg emulsion bulk, that has been replaced twice in a period of four years. The emulsion is followed by a tracking device composed of emulsion sheets (replaced every two months) and a fiber tracking system. The detector also includes electromagnetic and hadronic calorimeters and a muon spectrometer. The tracking device is located inside a 0.1 Tesla hexagonal magnet. A preselection is performed on the data using kinematical criteria (à la NOMAD) which reduces the number of events to be scanned by more than one order of magnitude. A new scanning technology makes it possible to follow the tracks back into the emulsion and record the track trajectories in a computer file that can be treated later as a collection of hits of a real-time detector. The one micron resolution of the emulsion allows

CHORUS to observe the τ decay vertex, since the τ life distance at SPS energies is $\sim$ 1 mm. In the case that the τ decay to a μ or a π the τ signal is a kink in the trajectory of these particles. If the *tau* decays to three tracks, a secondary vertex can be found.

CHORUS has scanned half of the data with one muon and 1/5 of the events with no muons. No candidate events have been found. In total 68 k events with one muon have been scanned and 7.2 k with no muons. The probability limit quoted by CHORUS is $P_{\nu_\mu \nu_\tau} < 6 \times 10^{-4}$ at 90 % C.L [17]. The CHORUS collaboration has estimated that the probability limit could be reduced to $P_{\nu_\mu \nu_\tau} < 4 \times 10^{-4}$ if no candidates are found in the rest of the data.

NOMAD

NOMAD stands for Neutrino Oscillation MAgnetic Detector[18]. The core of the detector consists of 44 drift chambers that fulfill the double purpose of being the target and the tracking device. The neutrino interactions take place in the wall material of the drift chambers via charged or neutral currents. This material is very light, only 2% of X_0 per chamber, in order to avoid γ conversions and secondary interactions that deteriorate the kinematical measurements. The detector has a total mass of 2.7 tons. The drift chamber dimensions are 2.6×2.6 m^2 by 10 cm thick. They are located inside a 0.4 Tesla magnetic field. The momentum resolution is 2% for 10 GeV muons. The detector is complemented by a TRD detector to identify electrons, an electromagnetic calorimeter (with resolution $\sigma_e/E = 3.2\%/\sqrt{E}$), a hadronic calorimeter and muon chambers ($>$ 97% efficient for muons with momentum greater than 5 GeV).

The NOMAD analysis is based on the kinematical differences between signal and background events. The neutrinos hit the detector horizontally, so the initial transverse momentum is zero, and thus the measured momenta of the event should be balanced in the transverse plane. NOMAD performs searches in every τ decay channel. For each channel the signal is different as are the backgrounds.

Let us consider as an example, the electron channel. The signal will be a $\nu_\tau + N \rightarrow \tau + X$, followed by the decay $\tau \rightarrow e + \bar{\nu}_e + \nu_\tau$. The main background for this channel are charged current interactions, $\nu_e + N \rightarrow e + X$, induced by the ν_e's present in the beam. This background contribution is one order of magnitude smaller than the one similar for the ν_μ's. Therefore, this channel has a low background contamination. In addition, NOMAD was optimized to identify electrons via the TRD detector. The main kinematical variables that separate the signal from the background are missing transverse momentum and the angle between the electron and the hadronic jet in the transverse plane. In the case of signal, there is some missing transverse momentum due to the two neutrinos that escape from the τ decay. For the ν_eCC background the only missing transverse momentum comes from undetected neutral particles and detector mismeasurements. In addition, the angle between the electron and the hadronic jet in the transverse plane is 180° for the ν_eCC, here the e direction is back to back to the hadronic direction due to momentum conservation. This is not the case for the signal, where the τ momentum is shared by the electron and two neutrinos, so the electron is no longer back to back to the hadronic jet.

A likelihood ratio function, defined with the variables described above, is used to separate signal and background. Exhaustive studies are carried out with large MC samples and with data that can fake the background, for example, a "data simulator" sample is obtained by replacing the μ from a ν_μCC by a τ particle. With these studies, the expected number of background events are estimated. NOMAD does not perform a zero background search in order to get a viable selection efficiency (which is on the order of 4%). NOMAD has analyzed 2/3 of its data [19]. In the four years of operation NOMAD collected 1.3×10^6 ν_μCC. No evidence of oscillation has been found so far. This result establishes a probability limit: $P_{\nu_\mu \nu_\tau} < 6 \times 10^{-4}$ at 90% C.L.

4 The future of neutrino oscillations

The solar neutrino problem will be addressed by two experiments in the next few years. The SNO experiment [20], in Ontario, which start operating in 1999, has the capability of measuring the total solar neutrino flux. SNO will detect the rate of neutral current events induced by solar neutrinos which interact with the d in the detector (a one kton tank of heavy water). Since the NC rate is independent of the neutrino flavor, SNO will check the total solar neutrino flux predicted by SSM. Another experiment, BOREXINO [21], at Gran Sasso in Italy, is under construction. It will measure the Be neutrino flux with large statistics ($\sim$ 50 events/day) to confirm if, indeed, the Be neutrinos are suppressed.

The LSND results will be checked in the next years. There are two proposed experiments to investigate the parameter region favored by the LSND result, one at FNAL, called BooNE [22], and another one at CERN [23]. The CERN experiment will re-use the PS accelerator that can produce ν_μ's with an average energies of 1.5 GeV. This experiment will observe a $\nu_\mu \rightsquigarrow \nu_e$ oscillation, by comparing the ν_μ/ν_e flux ratio measured at near and far detectors, which will be simple tracking/calorimeter devices.

LBLEs have marginal sensitivity to check the region allowed by the SuperK result on the zenith angle dependence of the atmospheric neutrinos. To observe a modulation on the ν_μ beam spectrum, whose energies are typically $\sim$ 10 GeV, compatible with an oscillation parameter of $\Delta m^2 \sim 10^{-3}$ eV^2, the detector should be located at $L \sim E/\Delta m^2 \sim 10^4$ km!. Some projects are being discussed in Japan (sending a neutrino beam from KEK to SuperK, 250 Km), in Europe [24] (from CERN to Gran Sasso, Italy, 732 Km) and in the USA (from FNAL to Soudan Mine, Minnesota, 730 Km). Only the last one, MINOS [25], will have enough sensitivity to explore the area of the oscillation parameters favored by SuperK. MINOS consists of near and far detectors. The far detector can be a massive (10 kton) sampling tracking/calorimeter. The oscillation can be observed by comparing the ν_μ flux or the NC/CC ratio or the energy spectrum, measured at both detectors. In addition, there is a new proposal for a LBL reactor Experiment, KamLAND [26], that could re-use the Kamioka cave by installing in it a 1 ktons scintillating liquid detector. This experiment could detect neutrinos of a few MeV which are produced at the nuclear plants located 150-200 Km around the Kamioka mine. KamLand could explore the region of large mixing angle and Δm^2 between $10^{-5} - 10^{-2}$ eV^2, i.e, it would cover the areas favored by the MSW-large angle solution of the solar neutrino problem and by the SuperK result.

Finally, a future SBLE that would explore the $\Delta m^2 \sim 1$ eV^2 region is unlikely after CHORUS and NOMAD reported negative results in their searches, and the SuperK evidence favored $\Delta m^2 \sim 10^{-3}$ eV^2 values. Despite that, the NOMAD and CHORUS collaborators have joined efforts, and have planned a future SBLE at CERN called TOSCA. The TOSCA experiment [27] could consist of six detector modules. Each module would have a 80 kg emulsion bulk and a tracker device with two silicon planes. In four years of operation at the CERN SPS neutrino beam, TOSCA would establish an oscillation probability limit one order of magnitude smaller than the one obtained by CHORUS and NOMAD, if no indication of oscillations are found.

5 Conclusions

Nowadays there are evidences for the existence of neutrino oscillations. The zenith angle dependence in the atmospheric neutrino flux, measured by the SuperK experiment, is strong evidence in favor of neutrino oscillations, and could be the first indication of physics beyond the SM. In the next few years, a new generation of experiments will check the LSND evidence and study in depth the solar neutrino problem. In addition, there are proposals for experiments which seek to confirm the SuperK results. Hopefully, these experiments will help to give an answer to the question of wheather or not the neutrino oscillations, indeed, do occur.

Acknowledgments

I thank J.J Gomez-Cadenas and M. Grothe for comments and their suffering correcting the english of the note. Also I thank G. Paliaga and T. Pulliam for final reading.

References

1. Super-Kamiokande Collaboration, Phys. Rev. Lett. **81** (1998) 1562-1567.
2. web site: www.hep.anl.gov/NDK/Hypertext/nuindustry.html.
3. R. Davis, Prog. Part. Nucl. Phys. **32** (1994) 13;
B.T. Cleveland et al., Nucl. Phys. **B** (Proc. Suppl.) 38 (1995) 47.
4. C. Athanassopoulos et at., Phys. Rev. Lett. **77**, 3082 (1996).
5. J.N. Bahcall and M.H. Pinsonneault, Rev. Mod. Phys. **64**, 885 (1992); J.N. Bahcall, *Neutrino Astrophysics*, (Cambridge University Press, Cambridge, 1989).
6. P. Anselmann et al., Phys. Lett. **B285** (1992) 376;
P. Anselmann et al., Phys. Lett. **B342** (1995) 440.
7. H.S. Hirata et al., Phys. Lett. **B280** (1992) 146.
8. L. Wolfenstein, Phys. Rev. **D17**, 2369 (1978); **D20**, 2634 (1979);
S.P. Mikheyev and A. Yu. Smirnov, Yad. Fiz. **42**, 1441 (1985) [Sov. J. Nucl. Phys. **42**, 913 (1986)]; Nuevo Cimento **9C**, 17, (1986).
9. N. Hata and P. Langacker, Phys. Rev. **D50**, 632 (1994).
10. L. Borodovsky et al., Phys. Rev. Lett. **68**, 274 (1992).
11. B. Armbruster et al., Nucl. Phys. B (Proc. Suppl.) **38**, 235 (1995).
12. B. Zeitnitz et al., Prog. Nucl. Phys. **40**, 169 (1998).
13. Super-Kamiokande collaboration, Phys. Rev. Lett. **81**, 1158 (1998).
14. Super-Kamiokande collaboration, hep-ex/9812009, submitted to Phys. Rev. Lett.
15. Super-Kamiokande collaboration, hep-ex/9812011.
16. CHORUS collaboration, N. Armenise et al., CERN-SPSC/90-42 (1990).
17. CHORUS collaboration, E. Eskut et al., Phys. Lett. B **424**, 202, (1998).
18. NOMAD collaboration, J. Altegoer et al., CERN-SPSC/91-21 (1991); Nucl. Instr. and Meth. A **430**, 96, (1998).
19. NOMAD collaboration, Phys. Lett. B **431**, 219, (1998).
20. web site: www.sno.phy.queensu.ca.
21. web site: almime.mi.infn.it.
22. web site: www.neutrino.lanl.gov/BooNE/.

23. SPSC memorandum M-627 CERN-SPSC/99-1.
24. ICARUS collaboration, G. Acquistapace et al., CERN-98-02.
25. MINOS collaboration, E. Ables et al., FNAL proposal (1995). web site: www.hep.anl.gov/NDK/HyperText/numu.
26. web site: www.awa.tohoku.ac.jp/html/KamLAND/index.html.
27. TOSCA collaboration, Letter of Intent SPSC-I213. web site: www.cern.ch/TOSCA/.

PHYSICS POTENTIAL OF A 500 GeV E^+E^- LINEAR COLLIDER

R. MIQUEL

Departament d'Estructura i Constituents de la Matèria, Facultat de Física, Universitat de Barcelona, Diagonal 647, E-08028 Barcelona (Spain)

Prospects for physics results at a generic Next Linear Collider are summarized. Top physics is reviewed with some detail. W and Z physics and searches for Higgs particles and Supersymmetry particles are also discussed.

1 Introduction

In the last ten years, a linear e^+e^- collider with centre-of-mass energies in the range 350–500 GeV and luminosities between 10^{33} and 10^{34} cm^{-2}s^{-1} has been proposed as a very interesting future high energy physics facility [1–4] with an impressive physics programme. From the accelerator point of view, the two challenges are luminosity and accelerating gradient. A strong final focus system providing the necessary luminosity seems possible. Concerning the gradient, several techniques are being investigated in several labs: S-band and X-band radiofrequency, superconducting cavities and a two-beam scheme being developed at CERN in which a low energy beam stores the energy that is then tranferred to the high energy beam.

In the period 1996–1998 several test facilities have been completed at CERN, DESY, KEK and SLAC. First designs with first firm estimates of cost and schedule for such a machine are expected by the year 2000.

The physics capabilities of an e^+e^- linear collider in the 500 GeV center-of-mass energy range can be summarized in the following way:

- The large amount of WW pairs (about 100K/year) and ZZ pairs (5K/year) produced will allow to put stringent limits in anomalous triple gauge-boson vertices. Sizable production of WWZ events (about 1000/year) will allow the first tests of the quartic couplings between Ws and Z's or photons.
- Very detailed studies of top-antitop production and decay will be made possible by the production of over 20K $\bar{t}t$ pairs per year. Studies at threshold will determine the top mass with unmatched precision and will measure parameters of the top-antitop QCD potential, almost free from non-perturbative uncertainties due to the large top quark width, which acts as an effective infrared cut-off.

- Capabilities for searching for Higgs particles and SUSY particles are excellent. In particular the Next Linear collider will either find the lightest SUSY Higgs, h, or rule out most Supersymmetry scenarios, easily. And if SUSY is right, the NLC can performe very detailed studies of its parameters.

Each of these topics will be now discussed in turn in some detail in the following.

2 W and Z Physics

Both the main goal and the methods in W and Z physics at the NLC will be very similar to those of LEP2. The main goal is to test the agreement of the couplings between three vector bosons and the Standard Model prediction. And the method relies on the measurement of the five-fold angular distribution of the Ws and their decay products.

The higher energy implies that the constraints coming form unitarity on these coupling are much more stringent and, therefore, that any deviation from the Standard Model values would have more dramatic consequences. Therefore, the sensitivity to anomalous couplings is greatly enhanced with respect to LEP2 energies. Since statistics is also greatly enlarged, the net result is the possibility of measuring much smaller deviations. Furthermore, all NLC designs consider the possibility of having longitudinally polarized electron beams, which also greatly helps in this analysis.

Figure 1 shows a comparison of the capabilities of several machines, including LEP2, LHC and NLC, in measuring $\Delta\kappa_\gamma$ and λ_γ separately. It can be seen that both LHC and NLC at 500 GeV can start probing the regions where those parameters are below 0.01, which is the region were is more natural to expect some deviations. There is a slight advantadge in sensitivity for NLC, which is more clear when going to the 1.5 TeV version. This is even more so when going to the determination of two parameters at the same time.

Feasibility studies of measurements of the quartic couplings have been presented in ref. [6]. And the prospects for disentagling the strong interactions of Ws and Zs at high energies, in no-Higgs models, have been thoroughly studied in ref. [7].

3 Top Physics

Top quarks will be copiously pair-produced at the NLC, allowing for detailed studies of their properties. The top quark being heavier than the intermediate

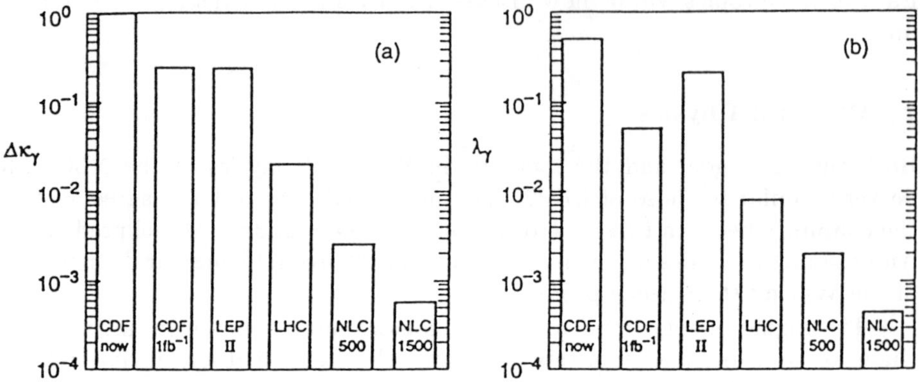

Figure 1: Comparison of the sensitivity of different machines to anomalous couplings. From ref.[5]

vector bosons, and maybe even heavier than the Higgs boson, it is not unreasonable to think that its properties might well be different from the ones of the lighter quarks and that it can shed some light over the mechanism of mass generation and provide valuable information for the understanding of flavour symmetries.

The dominant top production channel goes through the e^+e^- annihilation to a virtual photon or Z, which subsequently will decay to a $t\bar{t}$ pair. The $t\bar{t}$ production cross section is about 650 fb at $\sqrt{s} = 500$ GeV. At the foreseen luminosities of $10^{33} - 10^{34}$ cm^{-2} s^{-1} (or 10–100 fb^{-1} per year) the event sample is sufficient for detailed studies.

In the Minimal Standard Model (MSM) the top decays esentially 100% of the times into a bottom quark and a W. Depending on the decays of the two W's present in the event, the $t\bar{t}$ event will consist of six jets (44%); four jets, a high-momentum lepton and a neutrino (44%); or two jets, two high-momentum leptons and two neutrinos (11%). In all cases there will be two b jets.

The total top quark width is given approximately by $\Gamma_t \sim 0.18\,\mathrm{GeV}(m_t/M_W)^3$, which is about 2 GeV for $m_t = 180$ GeV. This large decay rate implies that the top quark decays before the toponium bound state has had time to form. More generally, Γ_t acts as an effective cut-off energy for non-perturbative QCD effects. This will make easier the studies, in particular, near the $t\bar{t}$ production threshold.

The following topics in top quark physics will be discussed here: i) the measurements of the top quark mass in a centre-of-mass energy scan around threshold; ii) the determination of top production and decay form-factors; and iii) the search for non-MSM top decays.

3.1 Threshold Scan

An energy scan around the $t\bar{t}$ can be used to do a precise measurement of the top quark mass. Three observables are available:

- The total production cross section, $\sigma_{t\bar{t}}(s)$, is particularly sensitive to m_t. However, it is also quite sensitive to the strength of the $t\bar{t}$ binding potential.

- The top momentum distribution, $d\sigma/dp(s)$, depends also on the top mass and the potential.

- The forward-backward asymmetry, $A_{FB}(s)$, due to the interference of S- and P-wave amplitudes, provides a third handle on the top mass.

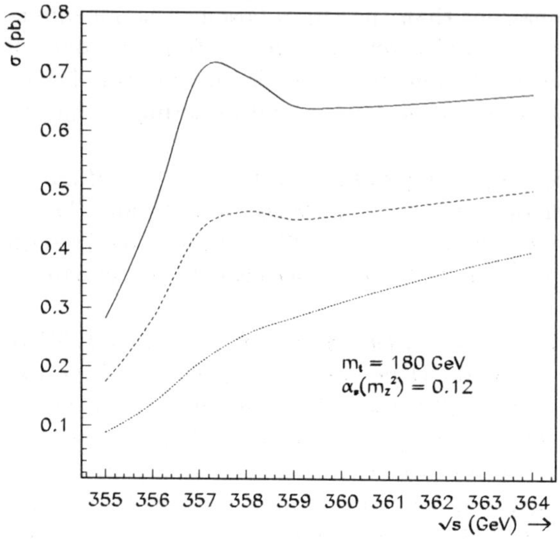

Figure 2: $t\bar{t}$ production cross section as a function of the centre-of-mass energy. Solid line: Born cross section; dashed line: with initial state radiation (ISR); dotted line: with ISR and beam effects.

The shape of the threshold is distorted by initial state radiation effects as well as by energy smearing due to the accelerator (fig. 1). The last effect is particularly important and leads to a substantially less steep threshold and a considerable reduction in the cross section. Both the top momentum distribution and the forward-backward asymmetry are also affected, although at a smaller level.

The cross section measurement uses events with six or five jets, one of which can consist of just a lepton. By asking that the event be spherical and using the two W mass constraints and the two top quark mass constraints, a sample of $t\bar{t}$ events can be selected with 33% overall efficiency and signal to background ratio in excess of 5.

For the measurements of $d\sigma/dp$ and A_{FB} only events in which one W decays leptonically are used. This way the charge of one of the tops is readily obtained from the lepton charge while the other top is used to measure the momentum. Backgrounds have been found to be negligibly small and the overall efficiency close to 15% [8].

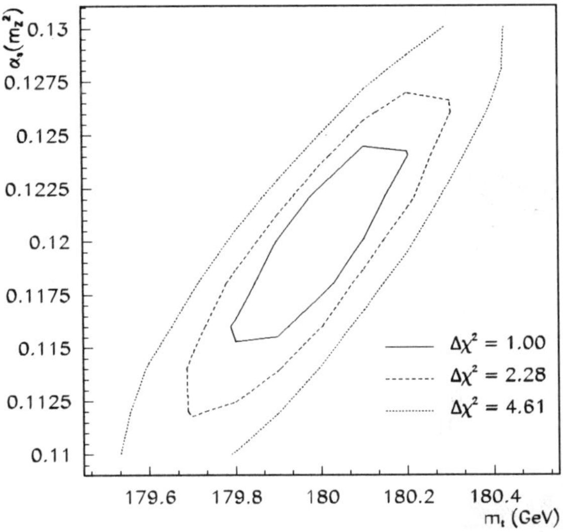

Figure 3: Results of the χ^2 fit using the total cross section, momentum distribution and forward-backward charge asymmetry.

The three observables are used in a χ^2 fit assuming a nine-point scan around $\sqrt{s} = 360$ GeV with an integrated luminosity of 5 fb^{-1} in each point. The fit has two free parameters: the top quark mass and α_s, which appears in the QCD $t\bar{t}$ potential. The resulting two-dimensional χ^2 contours can be seen in fig. 2. The overall precision expected in the top quark mass is around 200 MeV. Most of the sensitivity comes from the total cross section measurement, with a non-negligible improvement coming from $d\sigma/dp$.

3.2 Top Form Factors

The fact that the top decays before it hadronizes allows for clear tests of its decay structure and of its production properties.

Anomalous magnetic moment

An anomalous magnetic moment of the top would increase the probability for the helicity-flip process $e^+e^- \to t_L\bar{t}_L, t_R\bar{t}_R$. The angular distribution for this

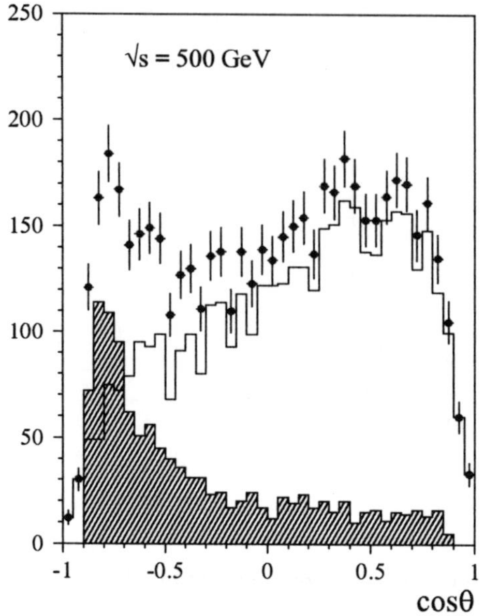

Figure 4: Top polar angle distribution. The solid line is the signal distribution, the shaded histogram represents the background and the dots the sum of both.

process is proportional to $\sin^2\theta$, θ being the top production angle:

$$\left.\frac{d\sigma}{d\cos\theta}\right|_{LL,RR} \propto \sin^2\theta \left[\frac{m_t}{E_b}F^V_{MSM} + \frac{E_b}{m_t}2F^V_{anom}\right]^2. \tag{1}$$

The relative weight of the anomalous term with respect to the MSM one increases quadratically with the beam energy.

Figure 3 shows the top production angular distribution from a simulation analysis using 4-jet, lepton, neutrino events at $\sqrt{s}$ = 500 GeV [9]. For an integrated luminosity of 50 fb^{-1}, a maximum likelihood fit to the shape of the distribution allows to measure the non-MSM component with an accuracy of about 4%, which is at the level of what is expected in the MSM for the QCD-induced anomalous magnetic moment.

It is clear that high electron beam polarization, which has not been considered in the previous analysis, could be used to improve this measurement.

Decay form factor

It is well known that the existence of a V+A component in the top decay form factor would modify the energy spectrum of the leptons coming from top decay. It has also been pointed out that other variables, such as the energy and angular distribution of the neutrino are even more sensitive to a possible V+A component [10]. The neutrino energy and direction can be estimated from the missing momentum in the event.

Monte Carlo studies at $\sqrt{s} = 362$ GeV [9] show that indeed an overall fit to the energy and angle of charged leptons and neutrinos increases the sensistivity to a V+A component by a factor of two if compared to the analysis using only the charged lepton energy. For an integrated luminosity of 50 fb^{-1}, the V+A component can be probed with a precision better than 2%.

CP violation in the production vertex

Any sign of electric (or weak) dipole moment in the top production form factor would indicate the presence of CP violation linked to the top system. The converse is also true. Using 4-jet, lepton, neutrino events, CP-violating asymmetries can be constructed with the momenta of the incoming positron and the produced lepton and jets [11]:

$$\langle A_1 \rangle = \langle \hat{p}_+ \cdot (\hat{q}_+ \times \hat{q}_X) \rangle - \langle \hat{p}_+ \cdot (\hat{q}_- \times \hat{q}_X) \rangle \,, \tag{2}$$

where $\hat{p}_+$ is the direction of the positron; $\hat{q}_{+(-)}$ is the direction of the positive (negative) lepton from top decay; and $\hat{q}_X$ is the direction of the top that has decayed hadronically. If $\langle A_1 \rangle$ is not zero, there is an non-vanishing electric or weak dipole moment. It has been shown [12] that, using 10 fb^{-1} of data, a limit in the top quark electric/weak dipole moment of order $5 \cdot 10^{-18}$ e·cm could be obtained.

3.3 Non Minimal Standard Model Decays

The high mass of the top quark means that it could well decay into particles which are not yet known. In particular two options can be mentioned:

- $t \to bH^+$. Since the top is heavy, its coupling to a charged Higgs is very strong.
- $t \to \tilde{t}\tilde{\chi}_1^0$. In some supersymmetric models the scalar top is the lightest squark, so that it could be even lighter than the top quark. The branching ratio could then be in the few per cent range.

In the first process a search is made for decays of the kind $H^+ \to \tau^+\bar{\nu}_\tau$. Compared to the standard decay of W into leptons, the signal would consist in an excess of tau leptons among the four jet, lepton, neutrino events. The significance of the excess is maximum for large values of $\tan\beta$.

An analysis based on identifying tau leptons as jets with just one track which is not identified as an electron or a muon [13] reached the conclusion that a charged Higgs with a mass below $m_t - 15$ GeV could be seen over most of the $\tan\beta$ range with 50 fb^{-1}.

As for the second process, the stop decays $\tilde{t} \to c\tilde{\chi}_1^0$ and $\tilde{t} \to b\tilde{\chi}_1^+$ with the subsequent decay $\tilde{\chi}_1^+ \to \tilde{\chi}_1^0 f_1 \bar{f}_2$ have been studied in ref. [13]. Since there are two neutralinos in the event for all stop decay channels, and the lightest neutralino is supposed to be stable and invisible, there is a clear missing energy signature to look for. For the two case examples worked out in ref. [13], the conclusion is that with 50 fb^{-1} a branching ratio as small as 1% should be enough to see the decay of a top quark into stop and neutralino.

4 Higgs searches

Production of the Standard Model Higgs particle, H, proceeds at NLC energies, through the three diagrams drawn in fig. 5. The WW fusion diagram dominates for light Higgs, while the Higgstrahlung, or Bjorken, diagram, dominates for heavy Higgses. At a center-of-mass energy of 500 GeV, for Higgs masses between 100 and 350 GeV, the number of events per year drop from about 4000 to about 400.

The search analysis proceeds in a very similar way as the corresponding LEP2 analysis. The conclusion is that the Standard Model Higgs can be easily found in the whole range

$$90 \leq M_H \leq 350 \text{ GeV},$$

and its mass can be measured to better than 1 GeV.

It can well be, however, that the Standard Model Higgs (or, in general, a Higgs particle) will have been found before NLC starts operation; at LHC, for instance. However, this discovery, based on a statistical excess of events (in the $\gamma - \gamma$ channel, for example), will not have allowed for detailed studies of the particle discovered. NLC can test its spin, its parity and its coupling to the Z, to draw convincing evidence that it is indeed a scalar particle, with non-zero vacuum expectation value, which gives mass to W and Z. As an example, fig. 6 shows a simulation study where one can measure the spin of the Higgs by looking at the distribution of the production angle in the process $e^+e^- \to HZ$ with Higgs and Z decaying to quark-antiquark.

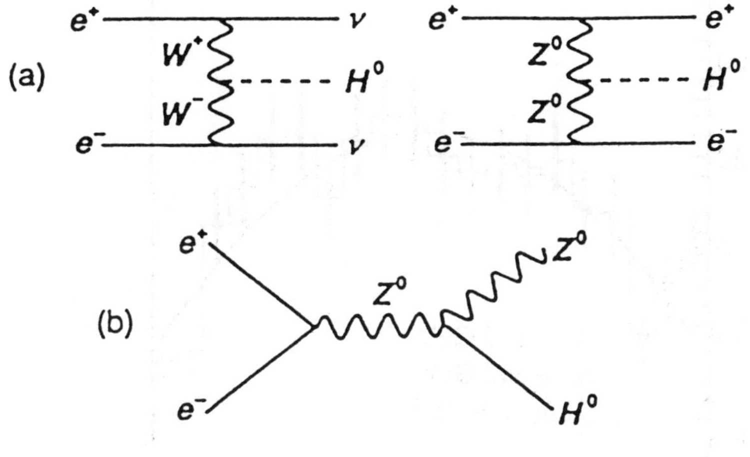

Figure 5: Feynmann diagrams contributing in lowest order to MSM Higgs production at NLC.

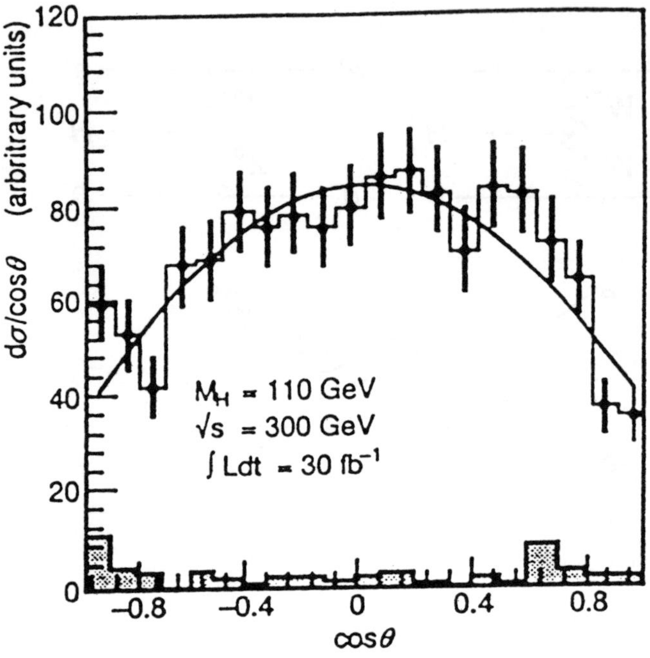

Figure 6: Simulated measurement of the distribution of Higgs production angle, in events in which both H and Z decay hadronically. From ref. [14].

The lightest SUSY Higgs, h, can be searched also in the same way as at LEP2. With only a two-week run at a center-of-mass energy above 300 GeV, the whole range of h masses, from 80 GeV (which will be covered by LEP2) to 150 GeV, the maximum allowed in most SUSY models, can be thoroughly explored. If h is not found at NLC, the most minimal SUSY models will be proved wrong.

If, on the contrary, a Higgs particle is found in this mass range, it will be possible, at least in some cases, to decide whether it is a SUSY Higgs or the Standard Model Higgs. The easiest way to see it is a SUSY Higgs is to find another Higgs particle, for example the CP-odd A particle. A can be easily found up to a mass close to half the center-of-mass energy. If M_A is larger than that, measurements of the branching ratios of the Higgs found can help disentagling h from H_{SM}, although only until M_A reaches about 400 GeV. For heavier A, the differences in branching ratios are two small and is, in general, not possible to know whether the Higgs found with mass below 150 GeV (otherwise it cannot be the lightest SUSY Higgs) is a SUSY Higgs or the Standard Model Higgs.

5 SUSY searches

Searches for the standard weak SUSY particles (charginos, sleptons, stop, etc.) can be performed in a similar way as at LEP2 up to the respective kinematical limits. The reach is, in a way, complementary to that of LHC, which can find more easily strongly interacting SUSY particles: squarks and gluinos.

However, if SUSY is found, only the NLC can perform truly detailed studies of the characteristics of the new particles: their spin, their quantum numbers, their coupling to ordinary particles. If all these properties agree with the definite predictions of Supersymmetry, then SUSY will have been found. As an example, by measuring the cross section and forward-backward charge-asymmetry for the process $e^+e^-_L \to \chi^+\chi^-$, one has access to the coupling between lepton, sneutrino and the wino component of χ^+, and one can see whether it is equal, as it should, to the coupling between lepton, neutrino and W, that is, $g = e\sin\theta_W$.

6 Summary

The Next Linear Collider has a rich physics program, continuing with the LEP2 program with a number of additional topics. The sensitivity to TGVs will be much improved, to few parts per mille. There are comprehensive top quark studies to be done, including precise determination of its mass and couplings.

Last but not least, the Next Linear Collider is certain to either find the lightest SUSY Higgs or rule out most SUSY models. If Supersymmetry indeed describes nature, then lots of new particles can be thoroughly studied and stringent tests of Supersymmetry can de made.

Acknowledgements

I would like to thank Bernardo Adeva and all the members the organizing committee for the excellent organization of the meeting, and, very specially, Mercedes Fatás for running the conference so efficiently and, yet, charmingly.

References

1. R. Orava, P. Eerola, M. Nordberg (editors), *Proceedings of the Workshop on Physics and Experiments with Linear e^+e^- Colliders at 500 GeV* (Saariselkä, Finland, September 1991), World Scientific, Singapore, 1992.
2. F. A. Harris, S. L. Olsen, S. Pakvasa, X. Tata (editors), *Proceedings of the Workshop on Physics and Experiments with Linear e^+e^- Colliders at 500 GeV* (Waikoloa, Hawaii, April 1993), World Scientific, Singapore, 1993.
3. A. Miyamoto, Y. Fujii, T. Matsui, S. Iwata (editors), *Proceedings of the Workshop on Physics and Experiments with Linear e^+e^- Colliders at 500 GeV* (Morioka, Japan, September 1995), World Scientific, Singapore, 1996.
4. P. M. Zerwas (editor), *Proceedings of the Workshops on e^+e^- Physics at 500 GeV: The Physics Potential* (Hamburg, 1991, 1993, 1995), reports DESY 93-123A, 93-123B, 93-123C, 93-123D.
5. D. L. Burke, in ref. [3], pp. 347–366.
6. A. Miyamoto, in ref. [3], pp. 654–663.
7. T. L. Barklow, in ref. [3], pp.105–128.
8. P. Igo-Kemenes, M. Martinez, R. Miquel and S. Orteu, in ref. [4], report DESY 93-123C, pp. 319–329.
9. M. Schmitt, in ref. [4], report DESY 93-123D, pp. 67–74.
10. M. Jezabek, J. H. Kuhn, Phys. Lett. B329 (1994) 317.
11. W. Bernreuther, O. Nachtmann, Phys. Lett. B268 (1991) 124.
12. W. Bernreuther, P. Overmann, "CP Violation in Top Quark Pair Production", in ref. [4], report DESY 93-123C, pp. 287–295.
13. A. Venturi, "Experimental Aspects of Rare Top Quark Decays at Future e^+e^- Linear Colliders", in ref. [4], report DESY 93-123C pp. 269–281.
14. K. Kawagoe, in ref. [2], pp. 660–665.

LIST OF PARTICIPANTS

Adeva, Bernardo
Universidad de Santiago de Compostela
adeva@fpegaes1.usc.es

Aguilar-Benítez, Manuel
CIEMAT
aguilar@uae1.ciemat.es

Bañuls, M. Carmen
IFIC/U.Valencia/CSIC
banuls@titan.ific.uv.es

Barreiro, Fernando
Universidad Autónoma de Madrid
barreiro@hepdc1.ft.uam.es

Bernabéu, José
Universidad de Valencia
bernabeu@evalvx.ific.uv.es

Blanch, Oscar
IFAE/UAB
blanch@ifae.es

Cabral, Luis Gustavo
Universidad de Valencia
cabral@titan.ific.uv.es

Carmona, Emiliano
IFIC/U.Valencia/CSIC
carmona@evalz1.ific.uv.es

Castro, José A.
Universidad de Santiago de Compostela
castro@fpsunae2.usc.es

Colomer, Marcos
CSIC/IFIC - U. Valencia
mcolomer@evalz1.ific.uv.es

Conde, Patricia
Universidad de Santiago de Compostela
Patricia@fpsunae2.usc.es

Costa, María José
IFIC/U.Valencia/CSIC
mjcosta@evalz2.ific.uv.es

Cuevas, Javier
Universidad de Oviedo
Javier.Cuevas@cern.ch

de la Calle, Ignacio
Universidad Complutense de Madrid
ignacio@gae.ucm.es

Deshpande, Abhay
Yale University
abhay.deshpande@yale.edu

Dobado, Antonio
Universidad Complutense de Madrid
dobado@eucmax.sim.ucm.es

Dosil, Mireia
IFAE/UAB
dosil@ifae.es

Espriu, Domenec
Universidad de Barcelona
Domenec.Espriu@cern.ch

Falagan, Miguel Angel
CIEMAT
falagan@ae.ciemat.es

Fernández, Marcos
CIEMAT
mfg@ae.ciemat.es

Fernández de Troconiz, Jorge
Universidad Autónoma de Madrid
troconiz@hepdc1.ft.uam.es

Ferrer, Toni
IFIC/U.Valencia/CSIC
ferrer@evalo1.ific.uv.es

Fonseca, Victoria
Universidad Complutense de Madrid
fonseca@eucmdx.gae.ucm.es

Fuster, Joan
IFIC/U.Valencia/CSIC
fuster@evalo1.ific.uv.es

Galas, Abraham
Universidad de Santiago de Compostela
abraham@fpsunae2.usc.es

Gallas, Venancio
Universidad de Santiago de Compostela
Gallas@fpsunae2.usc.es

García Irastorza, Igor
Universidad de Zaragoza
igarcia@public.ibercaja.es

García López, Gastón
Universidad Autónoma deMadrid
garcia@hepdc1.ft.uam.es

Gil Botella, Inés
IFIC/U.Valencia/CSIC
gilb@evalz1.ific.uv.es

Gómez Dumm, Daniel
Universidad de Valencia
daniel.gomez@uv.es

Gómez-Ceballos y Retuerto, Guillelmo
Instituto de Física de Cantabria
retuerto@esanw5.ifca.unican.es

González, José Carlos
Max-Planck-Institut für Physik
gonzales@HEGRA1.MPPMU.MPG.DE

González de la Hoz, Santiago
IFIC/U.Valencia/CSIC
sgonzale@evalz1.ific.uv.es

Guerrero, Francisco
Universidad de Valencia
fguerrer@titan.ific.uv.es

Hamacher, Klaus
Bergische Univ.-Gh. Wuppertal
Klaus.Hamacher@cern.ch

Hernández, José María
Universidad Autónoma de Madrid
jose.hernandez@desy.de

Hernando, José Angel
University of California - Santa Cruz
hernando@scipp.ucsc.edu

Ibáñez, Luis
Universidad Autónoma de Madrid
ibanez@madriz1.ft.uam.es

Ibarra, Aitor
Universidad Complutense de Madrid
ibarra@gae.ucm.es

López, Angeles
Universidad de Santiago de Compostela
lopez@fpsunae3.usc.es

Martínez Resco, José María
Universidad del País Vasco
wtbmarej@lg.ehu.es

Merino, Carlos
Universidad de Santiago de Compostela
merino@fpaxp1.usc.es

Miquel, Ramón
Universidad de Barcelona
miquel@ecm.ub.es

Mir, Lluisa M.
IFAE / UAB
mir@ifae.es

Navas, Sergio
IFIC/U.Valencia/CSIC
navas@evalz1.ific.uv.es

Padilla, Cristóbal
DESY
Cristobal.Padilla@desy.de

Peñaranda, Siannah
Universidad Autónoma de Madrid
siannah@delta.ft.uam.es

Peralta, Daniel
Universidad de Barcelona
peralta@ecm.ub.es

Pló, Máximo
Universidad de Santiago de Compostela
Max@gae.usc.es

Redondo, Ignacio
Universidad Autónoma de Madrid
redondo@hepdc1.ft.uam.es

Reyes, Juan
Universidad de Valencia
reyes@jupiter.ific.uv.es

Rodier, Stephane
Universidad Autónoma de Madrid
stephan@hepdc1.ft.uam.es

Rodriguez, José Manuel
Universidad de Santiago de Compostela
Jose@fpsunae2.usc.es

Ros, Eduardo
IFIC/U.Valencia/CSIC
Eduardo.Ros@cern.ch

Ruiz, Hugo
IFAE/UAB
ruiz@ifae.es

Sánchez Guillén, Joaquin
Universidad de Santiago de Compostela
joaquin@gaes.usc.es

Scopel, Stefano
Universidad de Zaragoza
scopel@posta.unizar.es

Sousa, Dolores
Universidad de Santiago de Compostela
dolores@fpaxp1.usc.es

Tobar, María José
Universidad de Santiago de Compostela
Cote@fpsunae2.usc.es

Zas, Enrique
Universidad de Santiago de Compostela
zas@fpegaes1.usc.es